Developmental Patterning of the Vertebrate Limb

NATO ASI Series

Advanced Science Institutes Series

A series presenting the results of activities sponsored by the NATO Science Committee, which aims at the dissemination of advanced scientific and technological knowledge, with a view to strengthening links between scientific communities.

The series is published by an international board of publishers in conjunction with the NATO Scientific Affairs Division

A	Life Sciences	Plenum Publishing Corporation
B	Physics	New York and London
C	Mathematical and Physical Sciences	Kluwer Academic Publishers
D	Behavioral and Social Sciences	Dordrecht, Boston, and London
E	Applied Sciences	
F	Computer and Systems Sciences	Springer-Verlag
G	Ecological Sciences	Berlin, Heidelberg, New York, London,
H	Cell Biology	Paris, Tokyo, Hong Kong, and Barcelona
I	Global Environmental Change	

Recent Volumes in this Series

Series A: Life Sciences

Developmental Patterning of the Vertebrate Limb

Edited by

J. Richard Hinchliffe

University College of Wales
Aberystwyth, Wales, United Kingdom

Juan M. Hurle

University of Cantabria
Santander, Spain

and

Dennis Summerbell

National Institute for Medical Research
London, United Kingdom

Plenum Press
New York and London
Published in cooperation with NATO Scientific Affairs Division

Proceedings of a NATO Advanced Research Workshop on
Developmental Patterning of the Vertebrate Limb,
held September 23–26, 1990,
in Santander, Spain

Library of Congress Cataloging-in-Publication Data

NATO Advanced Research Workshop on Developmental Patterning of the
   Vertebrate Limb (1990 : Santander, Spain)
    Developmental patterning of the vertebrate limb / edited by J.
   Richard Hinchliffe, Juan M. Hurle, and Dennis Summerbell.
        p.   cm. -- (NATO ASI series. Series A, Life sciences ; v.
   205)
    "Published in cooperation with NATO Scientific Affairs Division."
    "Proceedings of a NATO Advanced Research Workshop on Developmental
   Patterning of the Vertebrate Limb, held September 23-26, 1990, in
   Santander, Spain"--CIP t.p. verso.
    Includes bibliographical references and index.
    ISBN 0-306-43927-1
    1. Extremities (Anatomy)--Congresses. 2. Vertebrates-
   -Development--Congresses. 3. Extremities (Anatomy)--Evolution-
   -Congresses. 4. Morphogenesis--Congresses. 5. Pattern formation
   (Biology)--Congresses. I. Hinchliffe, J. R. II. Hurle, Juan M.
   III. Summerbell, Dennis. IV. North Atlantic Treaty Organization.
   Scientific Affairs Division. V. Title. VI. Series.
   QL950.7.N38 1990
   596'.03'32--dc20   91-22049
                                                               CIP

ISBN 0-306-43927-1

© 1991 Plenum Press, New York
A Division of Plenum Publishing Corporation
233 Spring Street, New York, N.Y. 10013

Printed in the United States of America

PREFACE

Following pioneering work by Harrison on amphibian limbs in the 1920s and by Saunders (1948) on the apical ridge in chick limbs, limb development became a classical model system for investigating such fundamental developmental issues as tissue interactions and induction, and the control of pattern formation. Earlier international conferences, at Grenoble 1972, Glasgow 1976, and Storrs, Connecticut 1982, reflected the interests and technology of their time. Grenoble was concerned with ectoderm-mesenchyme interaction, but by the time of the Glasgow meeting, the zone of polarizing activity (ZPA) and its role in control of patterning was the dominant theme. Storrs produced the first intimations that the ZPA could be mimicked by retinoic acid (RA), but the diversity of extracellular matrix molecules, particularly in skeletogenesis, was the main focus of attention.

By 1990, the paradigms had again shifted. Originally, the planners of the ARW saw retinoic acid (as a possible morphogen controlling skeletal patterning), the variety of extracellular matrix components and their roles, and the developmental basis of limb evolution as the leading contemporary topics. However, as planning proceeded, it was clear that the new results emerging from the use of homeobox gene probes (first developed to investigate the genetic control of patterning of Drosophila embryos) to analyse the localised expression of "patterning genes" in limb buds would also be an important theme. At the other end of the biological scale, in the field of evolutionary theory, the developmental basis of limb evolution had become a live issue for discussion, reflecting the dissatisfaction of a number of evolutionary biologists with the omission of development as a factor in the neo-Darwinian evolutionary synthesis. Thus one section of the conference aimed to examine the role of mutant genes in limb patterning, "developmental constraints" on the structural variation of the limb, and the developmental implications of the very recent palaeontological findings that the Devonian tetrapod ancestors had 7 or 8 digits, unlike the 5 digits until recently believed to characterise all tetrapods.

This monograph contains the papers and poster abstracts of the NATO Advanced Research Workshop on "Developmental Patterning of the Vertebrate Limb" held at the International University (Universidad Internacional Menendez Pelayo) on the Magdalena site, Santander, Spain, from 23-26 September 1990. The ARW was organised in collaboration with the University of Cantabria, Santander, which kindly provided additional funding, as did the British Council and the Spanish Ministry of Science and Education (DGICYT). Though this was a NATO Conference, with the relaxation of tension following the ending of the cold war in Europe, it was a pleasure to be able to include four participants from Czechoslovakia and from Russia. We thank all the participants for their contributions and enthusiasm which made for the success of the conference. The chairmen were Dennis Summerbell for the first day (molecular basis), John F. Fallon for the second day (extracellular matrix) and Pere Alberch for the final day (evolution and development). They also kindly agreed to review each day's proceedings. The burden of local organisation fell mainly on the Department of Anatomy and Cell Biology of the Faculty of Medicine, Santander. Despite the language barrier, the staff of the International University gave participants a most friendly reception and coped marvellously with last minute changes of plan by the organisers. Our Conference Organiser, Ms. Pilar Garcia, worked tirelessly to organise successfully some highly complex travel arrangements, made more difficult by road blocks in Cantabria on two days of the conference. Ms. Ann Walker typed all the early conference drafts and many of the subsequent manuscripts. We thank all these for their efforts. The extraordinarily beautiful conference site of the International University should have the last word; spectacular views over the Atlantic and the Bay of Santander with its distant mountains emerging from early morning mist, the park-land setting and the warm September sunshine on the nearby beaches all helped to create a successful conference atmosphere.

The next international limb development conference will be organised by a committee including Dr. John F. Fallon (Madison) and will take place in California in June 1992.

Richard Hinchliffe
Juan Hurle
Dennis Summerbell

CONTENTS

1. GENERAL OVERVIEW

2. THE MOLECULAR BASIS OF PATTERNING

A. Introduction

B. Pattern of Gene Expression

Summary

---

* indicates speaker in multi-authored paper

SOME PROBLEMS IN LIMB DEVELOPMENT

Lewis Wolpert

Department of Anatomy and Developmental Biology
University College and Middlesex School of Medicine
LONDON, UK

The problem is the relationship between genes and fingers, or more generally, how genes control the pattern and form of the limb. What is the generative programme for making a limb? How does one go from genes to morphology? The problem is difficult in itself, but made more so because of the limited possibilities for genetic analysis in vertebrates. While there are a number of mutations affecting limb development in mice they are, unlike the chick, not easily amenable to experimental analysis. The hope must thus lie in an underlying unity of developmental mechanisms, and so genes isolated from organisms as seemingly remote from vertebrates as Drosophila, can be used to identify genes in limb development. Already the pattern of expression of homeodomain containing genes supports this rather optimistic, but exciting, approach (Dolle et al, 1989).

In posing problems about limb development, I shall make three basic assumptions, whose own validity is, of course, problematic. (i) The mechanisms involved in limb development are essentially similar to those used elsewhere in the embryo. No special developmental mechanisms were invented during evolution to bring about limb development. (ii) Limb development is controlled by a single positional field. The assumption is that the pattern of cartilage, muscle, tendon and ectodermal cell differentiation is controlled by the same patterning mechanism. So whatever mechanisms control the pattern of cartilage also control the pattern of muscle. This is consistent with the observation that polarizing region grafts which result in mirror image duplication of the cartilage elements also result in a similar mirror image pattern of muscles and tendons, and in pigment cell patterns (Richardson et al, 1990). (iii) Patterning of the elements of the limb - cartilage, muscle, tendons, ectoderm - occurs long before overt differentiation takes place. Thus the pattern of cartilage elements would be specified long before the early signs of cartilage differentiation, such as condensation, take place. A contrary view has been proposed by Oster et al, (1985) who suggest that cartilage patterning is a physico-mechanical process that takes place at the time of condensation. We have recently shown that this view is inconsistent with the observation that in double anterior limbs, prepared long before cartilage condensation takes place, two humeri may develop (Wolpert and Hornbruch, 1990).

POSITIONAL INFORMATION

There is quite good evidence that during development of the limb bud the cells acquire positional information in the progress zone beneath the

apical ectodermal ridge. Grafts of the polarizing region from the posterior
margin of the bud to more anterior positions show that the pattern of digits
is determined by their position with respect to the graft and host
polarizing regions (Tickle et al, 1975; Wolpert and Hornbruch, 1981). The
polarizing zone is clearly a region with special properties and there is a
well defined pattern of activity in early embryos well before limb buds
develop and activity becomes confined to the regions posterior to the buds
(Hornbruch and Wolpert, in preparation). Retinoic acid, when applied
locally, can mimic the action of the polarizing region (Tickle and Eichele,
1985; Smith et al, 1989). Moreover retinoic acid is present in the limb bud
and is more concentrated in the posterior half. Although the possibility
that retinoic acid is a morphogen, providing the positional signal along the
anter-posterior axis, is very attractive, it should be treated with
caution. While retinoic acid can clearly alter positional values its action
on the limb may be indirect; it could, for example, convert adjacent cells
into a polarizing region. To establish a substance as a morphogen it will
be necessary to fulfil a number of quite rigorous criteria (Slack and
Isaacs, 1989; Wolpert, 1989). For example, it is necessary to show that the
cells in the limb are in fact responding to a putative morphogen.

A long standing problem with any mechanism of patterning based on
graded distribution of a morphogen acting as a positional signal is how such
a continuous signal is converted by the cells into discontinuities with
discrete regions. If the cells can respond to different thresholds then
such a conversion is possible. Recent evidence suggests that cells can
detect and differentially respond to changes in concentration of activin
(Green and Smith, 1990) and there is increasing evidence for responses to
small changes in concentration in insect development (Driever and Nusslein-
Volhard, 1985; Struhl et al, 1989).

The proposed mechanism for the specification of position along the
proximo-distal axis is based on time spent in the progress zone. Direct
evidence for this mechanism, in spite of its consistency with some
experiments (Wolpert et al, 1979) remains rather weak. On the other hand it
links very well with the possibility that homeobox containing genes are
involved in recording positional value. The attractive link lies in the
observation that the sequence of activation of the Hox 5 complex in the
mouse limb is similar to that order of the genes along the chromosome and
that the activation of the genes is based on a timing mechanism (Dolle et
al, 1989).

PREPATTERN

A prepattern refers to an underlying pattern that reflects the observed
pattern. The distinction between generating a pattern by positional
information and a prepattern can be illustrated with reference to the
digits. To specify the digits by positional information there could be a
monotonic concentration gradient (though not necessarily in a diffusible
morphogen) and the successive digits would be specified at differing
thresholds. There would be no obvious correspondence between the set of
positional values and the pattern of digits, the pattern emerging due to the
interpretation of the positional values. By contrast a prepattern mechanism
specifies a wave-like concentration of a chemical and the digits could
correspond to the peaks of the wave. In this way the prepattern would
reflect the overt pattern.

The evidence for a prepattern that would be modified by positional
information is indirect and negative. Negative in the sense that it seems
most unlikely that a mechanism based on positional information alone is
adequate. The most telling observation is the development of cartilaginous
structures resembling digits and more proximal elements in reaggregates of

limb mesenchyme which lack a polarizing region (Pautou, 1973). Our own observations on the development of the humerus when a polarizing region is placed at different positions along the antero-posterior axis are not consistent with a positional signal from the polarizing region specifying the humerus (Wolper and Hornbruch, 1987). Although the humerus is modified by such grafts it is very difficult to either eliminate the humerus or get two, between the host and grafted polarizing region. Yet another difficulty is polydactyly. Polydactyly, which is quite common, could be accounted for by a prepattern mechanism giving an extra "peak" when the bud widens, but is not easily explained in terms of positional information. Again, the large number of phalanges in plesiosaurs argues for some mechanism for generating cartilage elements independent of positional iformation. Even the pattern of muscles might be based on a prepattern mechanism which divides the initial dorsal and ventral blocks first into two followed by further subdivisions. If one thinks of the cartilage element as a spatial pattern of one, followed by two, and then three elements, then the muscle follows a similar pattern but the subdivision takes place with time at the same spatial location. Finally, from an evolutionary point of view it is much easier to imagine the primitive elements of the primitive limb being regenerated by a prepattern mechanism which are then modified by positional information.

A number of models for generating prepattern have been put forward (see Wolpert and Stein, 1984). However, there is not the slightest evidence to support such mechanisms. Reaction diffusion mechanisms (Murray, 1989) could in principle provide just the sort of mechanism required, but again there is virtually no evidence whatsoever for reaction diffusion giving stationary waves of chemical concentration in biological systems. It is striking that the pattern of stripes of gene activity in early Drosophila development does not result from either a prepattern or reaction diffusion but from interacting gradients (Pankratz and Jackloe, 1990). Thus a major problem is the existence and mechanism of prepatterning in the limb. Perhaps the clue will come from understanding how repeated elements - spacing patterns - are generated in other systems. The identification of genes controlling segmentation might be very important.

CELL-TO-CELL INTERACTIONS

The interactions in limb development can be classified in terms of STOP, GO, STAY, and POSITION. This classification serves to emphasis that cell interactions are selective and not instructive (Wolpert, 1990). They select from the few options open to the cells at any stage in their developmental programme. The same signal could, in principle, be used again and again - as seems the case with some growth factors - and yet the response be different because the key feature is the cell's internal programme.

The interaction between the mesenchyme in the progress zone and the overlying ectodermal ridge might be thought of as follows. First the polarizing region specifies position within the progress zone which in turn gives both POSITION in the overlying ectoderm and a STAY signal to the apical ridge. The ridge in turn gives a STAY signal to the progress zone maintaining lability and proliferation. There is evidence that all the cells in the progress zone will form cartilage when removed from the influence of the ectoderm (Cottrill et al, 1987). Thus patterning of the cartilage may essentially involve inhibition of cartilage formation. Thus there is possibly a STOP and GO signal from the ectoderm to the underlying mesoderm, preventing cartilage formation and directing the cells along other pathways. While this is clear in micromass culture (Solursh, 1985) its role in the limb is unclear as complete removal of the dorsal ectoderm has little effect on limb development (Martin and Lewis, 1986). Other interactions include STOP and GO signals to presumptive pigments cells from the ectoderm of early feather buds (Richardson et al, 1990).

None of the signals involved in these interactions have been identified nor are the pathways of the signals known. Some evidence for a role of gap junctions between polarizing region and anterior mesenchyme cells in grafts has been provided by Allen et al, (1990).

REGENERATION

It must be assumed that the mechanisms involved in limb regeneration are based on mechanisms involved in limb development (Brockes, 1989). It is most unlikely that quite new mechanisms should be seen in terms of the retention of the ability to generate new positional values. Nevertheless, it is clear that the size of the blastema, which is ten fold greater than that of the developing bud makes it most unlikely that any of the interactions that operate over distances of 20 to 50 cell diameters as has been suggested for the polarizing region and the apical ectodermal ridge, could play any significant role in regeneration. The blastema is just too big. For this reason mechanisms based on intercalation of positional values are very attractive (Bryant et al, 1981). At the cellular level the mechanism involved in intercalation is not known. However, if it involved diffusible molecules it might turn out to be very similar to those involved in the action of the polarizing region, and so present similar problems. Regeneration seems to provide a very promising system for identifying the molecular basis of positional value.

MORPHOGENESIS

Spatial patterning leads to both differentiation and the development of form. The digits, 2, 3 and 4, have quite distinct shapes and we remain ignorant of the cellular basis of these differences. Early in cartilage development the cells in the future diaphyseal region become oriented at right angles to the long axis of the element, while cells in the region of the future perichondrium become oriented along the long axis (Rooney et al, 1984). The mechanism of these orientations is not known. There is within the long cartilage elements a well defined pattern of cell differentiation and orientations which is symmetrical with respect to the middle of the element. These lead to a typical element with well regionalized diaphyseal and epiphyseal regions and the formation of joints is part of this process. The problem is accentuated by the failure of carpal and tarsals to form such a structure. Yet in the frog, one of the tarsal elements does develop into a long bone element. The mechanism whereby the overall form of the "dumb-bell" shaped cartilage element is generated is not known. Nor is it known how very localized changes in shape that characterise different elements are brought about. What, for example, is the mechanism whereby a lateral extension develops at the distal end of the chick tibia which makes contact and fuses with the distal end of the fibula (Archer et al, 1983). It is not known to what extent the perichondrium, matrix secretion, cell movement, or cell proliferation are involved in this process.

Muscle development involves cell migration and cell contact. Muscle cells migrate in from the somites at a very early stage to form dorsal and ventral aggregations (Chevallier, 1979). Unlike connective tissues which are non-equivalent, muscle cells seem to be equivalent since muscle cells from cervical somites will form a normal pattern (Lewis and Wolpert, 1976). The problem is to identify the molecules in the future muscle connective tissue that "trap" the migrating muscle cells. Splitting of the muscle masses to give rise to individual muscles is not understood, but may involve a change in the pattern of adhesivity and hence migration of muscle cells to new sites - an analogy would be migration of the primary mesenchyme to new

4

sites during sea urchin development (Gustafson and Wolpert, 1967).
patterning of tendons and the orientation of the cells and collage.
presents a further problem.

The overall form of the limb bud presents another problem in for
is attractive to think that the apical ridge acts in a mechanical manr . co
impose a dorso-ventral flattening. When the ridge disappears a local
rounding occurs (Tickle et al, 1989). But even the mechanics of ridge
formation needs investigation - what maintains the long contacts between the
cells and gives them their very columnar morphology? Again there may be an
analogy with sea urchin ectoderm.

Growth is an essential feature of limb development. It seems that the
growth pattern of the individual elements is specified at the same time as
the pattern itself is specified (Wolpert, 1981). However, it is not known
what this means in cellular or molecular terms. The different patterns of
growth of the cartilage elements may require them to be non-equivalent,
having different positional values. These positional values may be involved
in specifying the different growth programmes, which in turn requires an
understanding of how growth plates develop and how their intrinsic
properties are specified. It has been suggested that the length of the
proliferative zone rather than the proliferative rate may be the more
important factor in determining growth rate.

EVOLUTION

Understanding the evolution of the limb and how the different forms of
vertebrate limbs have evolved requires an understanding of limb development.
It could be argued that non-equivalence is an essential prerequisite for
evolutionary diversity (Wolpert, 1983). For only if the limb elements are
non-equivalent can their form be changed locally and independently of the
other elements. Understanding the mechanism of patterning - to what extent
positional information and prepatterning are involved - is fundamental to
understanding how the limb evolves. In addition, only when it is known how
genes control limb development will it be possible to understand how diverse
forms evolved and what developmental constraints on limb form exist. For
example, some changes in form may require so many simultaneous alterations
in genes as to make them virtually impossible.

CONCLUSIONS

We are in a new phase of studying limb development. A few years ago we
would have been overjoyed to identify the expression of any patterns of gene
activity in the limb not directly related to cell differentiation. Now more
than 20 homeobox containing genes are known to be expressed during limb
development. In addition there are reports of graded distributions of
growth factors. The future lies in combining classical techniques of
exerimental manipulation with molecular techniques. The limb is ideal for
such studies. While genetics is lacking it may yet be possible to use
viruses to insert or even delete specific genes in specific regions during
limb development.

It seems not unreasonable to assume that there are three classes of
genes involved in limb development (i) genes controlling spatial
organization, perhaps homeobox containing genes; (ii) genes controlling
cellular differentiation of, for example, muscle, cartilage and tendon
cells, of which Myo-D (Weintraub et al, 1989) is a possible representative;
(iii) genes controlling the form of the limb - such genes could code for
cell adhesion molecules and so possibly specify the long cellular contacts
in the apical ectodermal ridge.

The problem is to identify such genes and show how they control cell behaviour during limb devlopment.

REFERENCES

ALLEN, F., TICKLE, C., and WARNER A., 1990. The role of gap junctions in patterning of the chick limb bud. Development. 108: 623.

ARCHER, C. W., HORNBRUCH, A., and WOLPERT, L., 1983. Growth and morphogenesis of the fibula of the chick embryo. J. Embryol. exp. Morph. 75: 101-116.

BROCKES, J. P., 1989. Retinoids, homeobox genes, and limb morphogenesis. Neuron. 2: 1285-1294.

BRYANT, S., FRENCH, V., and BRYANT, P., 1981. Distal regeneration and symmetry. Science. 212: 993-1002.

CHEVALLIER, A., 1979. Role of somitic mesoderm in the development of the thorax in bird embryos. II. Origin of the thoracic and appendicular musculature. J. Embryol. exp. Morph. 49: 73-88.

COTTRILL, C. P., ARCHER, C. W., HORNBRUCH, A., and WOLPERT, L., 1987. The differentiation of normal and muscle-free distal limb bud mesenchyme in micromass culture. Dev. Biol. 119: 143-151.

DOLLE, P., IZPISUA-BELMONTE, J., FALKENSTEIN, N., RENUCCI, A., and DUBOULE, D., 1989. Coordinate expression of the murine Hox-5 complex homeobox-containing genes during limb pattern formation. Nature. 342: 767-772.

DRIEVER, W., and NUSSLEIN-VOLHARD, C., 1988. The bicoid protein determines position in the Drosophila embryo in a concentration-dependent manner. Cell. 54: 95-104.

GREEN, J. B. A., and SMITH, J. C., 1990. Graded changes in dose of a Xenopus activin A homologue elicit stepwise transitions in embryonic cell surface. Nature. 347: 391-397.

GUSTAFSON, T., and WOLPERT, L., 1967. Cellular movement and contact in sea urchin morphogenesis. Biol. Rev. 42: 442-498.

LEWIS, J., and WOLPERT, L., 1976. The principle of non-equivalence in development. J. Theoret. Biol. 62: 429-490.

MARTIN, P., and LEWIS, J., 1986. Normal development of the skeleton of chick limb buds devoid of dorsal ectoderm. Dev. Biol. 118: 233-246.

MURRAY, J., 1989. "Mathematical biology". Berlin: Springer.

OSTER, G. F., MURRAY, J. D., and MAINI, P. K., 1985. A model for chondrogenic condensations in the developing limb: the role of extracellular matrix and cell tractions. J. Embryol. exp. Morph. 89: 93-112.

PANKRATZ, M. J., and JACKLE, H., 1990. Making stripes in the Drosophila embryo. Trends in Genetics. 6: 287-292.

PAUTOU, M. P., 1973. Analyse de la morphogenese du pied des Oiseaux a l'aide de melange cellulaires interspecifiques. I. Etude morphologique. J. Embryol. exp. Morph. 29: 175-196.

RICHARDSON, M. K., HORNBRUCH, A., and WOLPERT, L., 1990. Mechanisms of pigment pattern formation in the quail embryo. Development. 109: 81-89.

ROONEY, P., ARCHER, C. W., and WOLPERT, L., 1984. Morphogenesis of cartilaginous long bone rudiments. In "The role of the extracellular matrix in development". Ed. R. Trelstad. New York: Alan Liss.

SLACK, J. M. W. and ISAACS, H. V., 1989. Presence of basic fibroblast growth factors in the early Xenopus embryo. Development. 105: 147-153.

SMITH, S. M., PANG, K., SUNDIN, O., WEDDEN, S. E., THALLER, C., and EICHELE, G., 1989. Molecular approaches to vertebrate limb morphogenesis. Development. 107: (Supplement) 121-132.

SOLURSH, M., 1984. Ectoderm as a determinant of early tissue pattern in the limb bud. Cell Differentiation. 15: 17-24.

STRUHL, G., STRUHL, K., and MacDONALD, P. M., 1989. The gradient morphogen bicoid is a concentration-dependent transcription activator. Cell. 57: 1259-1273.

TICKLE, C., CRAWLEY, A., and FARRER, J., 1989. Retinoic acid application to chick wing buds leads to a dose-dependent reorganization of the apical ectodermal ridge that is mediated by the mesenchyme. Development. 106: 691-705.

WEINTRAUB, H., TAPSCOTT, J. J., DAVID, R. L., THAYER, M. J., ADAM, M. A., LARSAR, A. B., and MILLER, A. D., 1989. Activation of muscle-specific genes in pigment, nerve, fat, liver and fibroblast cell lines for forced expression of Myo-D. Proc. Nat. Acad. Sci. 86: 5434-5438.

WOLPERT, L., 1981. Cellular basis of skeletal growth during development. British Medical Bull. 37: 215-219.

WOLPERT, L., 1983 Constancy and change in the development and evolution of pattern. In "Development and evolution". Eds B. C. Goodwin, N. Holder, and C. C. Wylie. Cambridge: Cambridge University Press.

WOLPERT, L., 1989. Positional information revisited. Development. 107: (Supplement) 3-12.

WOLPERT, L., 1990. Signals in limb development: Stop, Go, Stay, and Position. J. Cell Sci. (Supplement) (in press).

WOLPERT, L., and HORNBRUCH, A., 1981. Positional signalling along the antero-posterior axis of the chick wing. The effect of multiple polarizing region grafts. J. Embryo. exp. Morph. 63: 143-159.

WOLPERT, L., and HORNBRUCH, A., 1987. Positional signalling and the development of the humerus in the chick limb bud. Development. 100: 333-338.

WOLPERT, L., and HORNBRUCH, A., 1990. Double anterior chick limb buds and models for cartilage rudiment specification. Development 109: 961-966.

WOLPERT, L., and STEIN, W. D., 1984. Positional information and pattern formation. In "Pattern formation". Eds G. M. Malacinski and S. V. Bryant. New York: Macmillans.

WOLPERT, L., TICKLE, C., and SAMPFORD, M., 1979. The effect of cell killing by X-irradiation on pattern formation in the chick limb. J. Embryol. exp. Morph. 50: 175-185.

THE MOLECULAR BASIS OF LIMB PATTERNING: A REVIEW

Dennis Summerbell

National Institute for Medical Research
The Ridgeway, Mill Hill
London NW7 1AA UK

Ten years ago research into limb development and regeneration had
produced a wealth of material describing interesting biological phenomena. These included the identification of signalling and responding
regions, descriptions of the pattern of changes of cell differentiation
and cell division and the ability to manipulate the embryo so as to
modify these patterns almost at will. The intellectual force behind much
of the research lay in a wealth of theoretical papers discussing possible
mechanisms. These in turn stimulated new waves of experiments. What the
system lacked was any clear evidence supporting a particular molecular
mechanism. However within a very few years the situation has been transformed with the appearance of three distinct lines of enquiry that have
converged to produce a key that may unlock the secrets of development.

THE LIMB BUD

During early development the fertilised egg is transformed into a multicellular layered structure (comprising ectoderm, mesoderm and endoderm)
with a head end, a tail end and bilateral symmetry about the mid line.
This is sometimes known as the formation of the primary body plan. At
this stage we would describe the whole embryo as comprising a single
embryonic field. This first field maps out the positions of secondary
fields and creates the start conditions which control their subsequent
development. The limb bud is one of these secondary fields. It arises
from an oval shaped patch of cells in the lateral plate. It comprises a
thin peripheral layer of ectoderm enclosing a loosely packed mass of
mesoderm. The limb bud mesoderm cells divide faster than the cells of
the surrounding flank tissue so that the limb bud starts to bulge out.
Cell division remains high at the distal tip of the bud but reduces
proximally. Thus the mass of cells is expanding most rapidly at the tip.
The limb unfolds with proximal structures appearing from out of the
undifferentiated tip region before the more distal structures. Wolpert
described the limb as a three dimensional structure mapped along three
orthogonal co-ordinate axes: dorso-ventral, antero-posterior and proximo-
distal. He expected positional information to be provided independently
for each axis so that every cell would know its position within the
solid.

Developmental Patterning of the Vertebrate Limb
Edited by J.R. Hinchliffe *et al.*, Plenum Press, New York, 1991

THE HOMEOBOX

The homeobox was discovered in *Drosophila*, the fruit fly. The adult fly is constructed from a set of prefabricated units that develop in the larva and which are assembled in the pupa. Thus a fly is made out of 6 leg discs, 2 wing discs, 2 antennae discs etc. The name or identity of each disc is determined by the homeotic genes. Expression of a particular subset of the genes causes a disc to develop into an antenna, while expression of a slightly different sub-set results in the development of a leg. If a mutation is induced so that the one gene of the set is deleted then all of the discs that normally require expression of that gene are turned into the appropriate disc type in which that gene is normally not expressed. Thus a simple mutation can convert an antenna into a leg.

The homeotic genes contain a common sequence of about 180 base pairs. This region has been named the homeobox. The homeobox sequence also appears in non-homeotic genes, frequently genes that are thought to be important in controlling development. These included those involved in segmentation of the larva and in the establishment of the early embryonic gradient. Homeobox containing genes code for proteins that act as transcription factors (the protein binds to a specific site on the DNA and regulates transcription of a "responding" gene). The homeobox sequence is the code for the domain that actually binds the transcription factor specifically to its receptor site on the DNA. Homeoboxes are widely seen as being an indicator of genes that are involved in controlling pattern formation rather than genes that respond to the control mechanism. A significant characteristic of homeobox genes is that they have a particular genomic organisation. They tend to lie close together along the chromosomes, forming a number of clusters with genes of similar function being found in the same cluster.

Screening of cDNA libraries from vertebrates soon demonstrated three important correlations with *Drosophila*. 1. Vertebrates contained sequences containing regions with high homology to the homeobox. 2. A gene containing a homeobox could often be clearly related to a homologous (or at least fairly similar) *Drosophila* gene. 3. Vertebrate "homeobox genes" were also frequently found in clusters and the spatial order along the chromosome was identical to the spatial order for the equivalent cluster in *Drosophila*.

It is now clear that the spatial order within the cluster may have functional significance. *In situ* hybridisation of probes to homeobox containing genes shows that the spatial pattern of gene expression is related to the sequence along the chromosome. Thus Hox 2.1[1] is located relatively 3' and is expressed at more cranial positions along the body axis than more 5' located Hox genes. It is indisputable that homeoboxes are of fundamental importance in the control of development in flies, it is natural to assume a similar importance in vertebrates. This assumption was clearly a major premise at the meeting.

---

[1]Nomenclature: Hox 2.1 refers to a single gene. Hox is a species specific title for homeobox containing genes, in this case referring to mouse; the figure "2" refers to a specific cluster, numbered in historical order of discovery; the figure "1" refers to a specific gene, again in historical order of discovery. Unfortunately while Hox is used universally for mice, other species are given differing names by different authors. The historical numbering system is inconsistent and unhelpful, similarly the associated jargon can be obscure. The nomenclature urgently needs revision. 5' and 3' are used to describe relative position along a chromosome or length of DNA. They equate to start and finish. Individual authors may use the words upstream and downstream in the same context.

RETINOIC ACID NUCLEAR RECEPTORS

Retinoic Acid (RA) nuclear receptor proteins (NRPs) are members of the steroid/thyroid hormone nuclear receptor family. Like the homeobox nuclear proteins they are transcription factors. The proteins in this family bind strongly to their specific hormone (or ligand). The resulting ligand-receptor complex is able to bind to a specific response element in DNA. This in turn up- or down-regulates expression of the gene controlled by that response element.

These receptor protein have six regions (or domains), two of which have sequence similarity throughout the family. One is a sequence of about 67 amino acids (40-60% identity) which is characteristic of regions forming zinc-stabilised DNA-binding fingers. The other is a sequence of about 230 amino acids (15-35% identity) which is the ligand binding domain. A concensus sequence for the DNA binding domain of the steroid/thyroid family was used to screen cDNA libraries for other possible family members. One such NRP proved to bind RA. Soon a small family of such retinoic acid receptors (RARs) were identified.

RARs have been identified in several species and many tissues. At the moment three separate genes are known: RAR-α, RAR-ß and RAR-ɣ. The amino acid sequence is highly conserved in the DNA (90-95%) and ligand (85-90%) binding regions. The sequence is also conserved in the B domain (75-85%), but this region differs from the equivalent domain in the steroid /thyroid group. All but a small fragment of RAR-ɣ are very highly conserved between species. Several isoforms with variation in the 5' sequence are known and some isoforms have between species counterparts.

RARs are able to activate transcription of genes normally responding to thyroid hormone but it seems likely that they will have their own specific response elements. So far only one such naturally occurring response element has been detected, the gene for RAR-ß has an RA-responsive element in its promoter region. The three RARs are activated differentially by different retinoids.

RETINOIC ACID AS A MORPHOGEN

Retinoic Acid (RA) is a small water insoluble molecule of 300 Daltons consisting of a hydrocarbon ring with an unsaturated side arm. Together with its close analogs, it is of clinical significance in the treatment of acne, skin cancer and other differentiative diseases of the skin. It is of increasing interest to the cosmetic industry because of supposed anti-aging effects. At the cellular level it is known to control differentiation, eg high levels of RA promote a change from keratinised to mucous epithelia and switches murine F9 teratocarcinoma cells into the parietal endoderm pathway.

Retinoic acid has also been identified as a potential morphogen. It can modify the pattern of digits across the antero-posterior axis of many vertebrate limbs and the entire proximo-distal pattern of amphibian developing and regenerating limbs. Topic application of RA to the anterior margin of the developing chick limb bud causes a local increase in cell division. The additional tissue can develop into perfectly patterned supernumerary limb tissue in mirror image to the original limb. It thus mimics the effect of a ZPA graft. RA can also proximalise an amphibian blastema. A normal blastema produces limb structures in a proximo-distal sequence. RA resets an old blastema so that it starts again from a more proximal level and therefore produces extra proximal structures.

RA has also been detected endogenously in the chick limb. It is asymmetrically distributed with a higher concentration in the posterior (ZPA containing) half than the anterior half. A close analogue 3,4-didehydro-retinoic acid (ddRA) is also present endogenously and also can cause the formation of supernumerary structures. It appears that both ddRA and RA are formed from endogenous retinol, possibly by ZPA cells.

THE MEETING

In an ideal world I would wish to have related the following summary.

> We have now identified a small set of endogenous
> retinoids. They are asymmetrically distributed within
> the limb. They bind differentially to a corresponding
> set of RARs. The RARs each have their own response
> elements but the pattern of expression of the gene
> products is complex because each modifies expression of
> the other members of the family. Apart from the
> interaction within the retinoid family, the main target
> appears to be control of expression of the homeobox
> gene family. The homeobox genes work combinatorially
> so that a particular combination defines the positional
> information of a particular structure. Deleting Hox-
> n.n converts a "little" finger into a "ring" finger.

Alas no such tidy story emerged. The prevailing mood was that things were going to be very complicated and very slow, but that it was also going to be very interesting.

THE PREVAILING THEMES

Research into the developmental significance of homeobox genes is is still very much at a descriptive stage. The Hox-4 cluster has restricted spatial expression in the limb bud. Hox 4.2 appears throughout limb. The others are all present at the tip, but each has a proximal boundary of expression that is progressively more distal than its 3' neighbour (Denis Duboule). Similarly Hox 7 and 8 have an uneven proximo-distal distribution. Hox-7 is expressed preferentially in the Progress Zone and Hox-8 in the Apical Ectodermal Ridge (AER) (Robert Hill, Gary Lyons). The limbless mutant has a defective AER and loses Hox 7 expression at the time that the AER degenerates. It seems likely that an AER product (controlled by Hox-8) is required to maintain Progress Zone as well as distal cell division and that only Progress Zone cells (Hox-7 positive?) can respond to the RA gradient. Mutations of the limb deformity gene (ld) causes loss of the AER (Rolf Zeller). In newt regeneration NvHbox-2 is expressed 3-5 fold higher in proximal as compared to distal regeneration blastemas.

Similarly most of the information about RARs was descriptive, though again there were intriguing correlations with experimental data. RAR-α is expressed in most embryonic tissues. RAR-γ and RAR-ß expression is more restricted being more abundant in head and limb bud (co-expressing with Hox-4.4) while RAR-ß is expressed preferentially in the dorso-proximal part of the limb bud (Denis Duboule), particularly around the prospective shoulder girdle (Annie Rowe, Paul Brickell). It is particularly interesting that there is some RAR-γ expression around Hensen's Node which is clearly an organiser and a possible second source of RA (Hornbruch).

The obvious progression from descriptive data is to look at changing patterns following experimental manipulation. Here there were successes but also disappointments. In chick embryos the limbless mutant described above can be rescued by grafting wild type AER to limbless mesenchyme. The limbless mesenchyme, recovers expression of Hox-7 which had been lost as the limbless AER degenerated (Gary Lyons). An implant of RA stimulates expression of RAR-β in anterior limb mesenchyme (Rowe, Brickell). However when in newts the distal blastema is proximalised using RA the level of NvHbox-2 expression does not change. This result is particularly disappointing. The PD gradient of expression argued a possible role in establishment or maintenance of positional identity but this hypothesis is rather spoilt by the RA experiment. The RA clearly changed the character of the cells to a more proximal positional identity but it did not change the level of expression of NvHbox 2. One would have to argue an effect of RA downstream from the homeobox gene (Jeremy Brockes).

The meeting also discussed the possibility of exciting techniques for the future. At the moment we can modify the genome either by classical genetic manipulation or by making transgenic mice or amphibia. Transgenic chicks may soon be possible. Despite the appearance of these tools it is still a tedious business. It is now feasible to insert genes into chick embryos using both replication competent (thereby infecting the whole embryo), or replication defective (only the progeny of infected cells carry the gene) retroviral vectors (Cliff Tabin). One could therefore manipulate the genetic make up of a whole embryo or of just a small localised region (such as the ZPA) before grafting it into a host embryo.

Perhaps the most surprising outcome of the meeting was a strong debate about the role of retinoic acid. It is now generally supposed that RA is the morphogen produced by the ZPA which controls the pattern across the antero-posterior axis. No less than three papers presented data examining the hypothesis that RA induces anterior limb mesenchyme to form a ZPA (Susan Bryant, Dennis Summerbell, Cheryll Tickle). If an RA implant is made to the anterior margin then the adjacent tissue develops ZPA-like properties after about 12h. However there is still some margin for doubt. Direct measurements of the RA concentration in the grafted tissue show that there are significant levels of RA which may have diffused from the initial implanted source (Dennis Summerbell). Information was also presented about the reaction pathway of RA and of a second endogenous active retinoid, 3,4-didehydro-retinoic acid (ddRA). So far, disappointingly, there are no obvious difference in the effects or properties of ddRA and RA (Reinhold Janocha).

There were other suggestions and queries about the role of RA and/or the ZPA. An anterior half limb deprived of its ZPA fails to develop. It can be reliably rescued using a ZPA graft but not using an RA implant (Sharon Frost). High concentrations of RA promote gap junction formation (Robert Kosher) and increase the number of EGF receptors (Susan Bryant). It is possible that in regeneration RA is involved more in de-differentiation than in axial specification (Malcolm Maden). The ZPA is the source of at least one mitogen, probably not the same signal as that controlling pattern and certainly not RA (John McLachlan). There was conflicting evidence about the mitogenic effects of RA, some workers finding it not to enhance cell division (Susan Bryant, John McLachlan, Cheryll Tickle) while others found a mitogenic effect (Hiroyuki Ide). This disagreement is unlikely to be serious as there seem likely to be clear differences of response in different regions of the limb. Those cells that increase cell division in response to RA also differentiate into the chondrogenic pathway but then lose their ability to respond to RA (Hiroyuki Ide).

A PERSONAL SUMMARY

I left the meeting with three main conclusions. 1. It is by no means clear that RA is the morphogen. Almost all participants at one time or another voiced doubts about the role of RA. Yet it is clear that despite these doubts the concensus position was that RA will have a significant functional role. 2. Unravelling the role of RA, its nuclear receptors, and the homeobox genes is likely to be a long hard slog. However all three will be involved in the control of pattern at a fundamental level and they will provide the key to understanding development. 3. The gene expression people are going to have to look simultaneously in two directions, early and late. At the moment they tend to look for expression at stages which give the best signal-to-noise ratio and not stages when the significant events happen. The stages examined may be uninformative because the pattern is already determined but differentiation has not started.

EXPRESSION OF THE <u>MSH-LIKE</u> HOMEBOX-CONTAINING GENES DURING MOUSE LIMB

DEVELOPMENT

Robert E. Hill and Duncan R. Davidson

MRC Human Genetics Unit
Western General Hospital
Crewe Road
EDINBURGH
Scotland
EH4 2XU, UK

INTRODUCTION

The first reports of the presence of the homeobox domain in *Drosophila* homeotic genes and the subsequent suggestion that this domain was conserved in a wide range of organisms generated a great deal of excitement among researchers interested in mammalian development. This domain contains 61 amino acids and provides the DNA binding function to a larger protein (McGinnis *et al* 1984 a,b; Scott and Weiner 1984). The suggestion that these proteins are transcription factors, now supported by a substantial amount of data, provides an activity consistent with these acting as regulators of developmental processes (Krasnow *et al* 1989; Struhl *et al* 1989; Ohkuma *et al* 1990). The vertebrate developmental biologists using these conserved domains as molecular probes have a direct route to genes which would predictably have similar properties in their chosen organisms of study.

Certain trends have emerged in the study of vertebrate homeobox genes. From molecular analysis of gene structure, and structure of the multigene families and from expression studies it is becoming evident that the vertebrate and *Drosophila* genes share more than sequence similarities (Duboule and Dolle 1989; Graham *et al* 1989; Murphy *et al* 1989, and Wilkinson *et al* 1989). The studies are consistent with the premise that in vertebrate development these genes play a very similar role to their insect counterparts that is, to specify positional identity. (Akam 1987).

The homeobox genes can be sub-divided into different families by sequence homology. The two homeobox-containing genes of concern to this work are homologs of the *Drosophila Msh* (Muscle *segment homeobox*) gene. The mouse *Hox-7* and *Hox-8* genes each show approximately 90% amino acid identity to the Drosophila *Msh* homeobox domain (Hill *et al* 1989, Robert *et al* 1989, and Monaghan *et al,* in press). Comparison of the *Hox-7* and *Hox-8* homeo domains reveal 98% identity; comparison to any other reported homeobox domain is at best 60% identical. *Hox-7* and *Hox-8* are not tightly linked at one chromosomal loci in contrast to the majority of murine homeobox genes which constitute four independent multigene clusters. *Hox-7* is located on mouse chromosome 5; *Hox-8* on chromosome 13 (unpublished data).

We have previously published a preliminary analysis of the expression pattern of *Hox-7* by *in situ* hybridisation (Hill *et al* 1989). In this chapter we focus on the early expression patterns of *Hox-7* and *8* and compare the expression domains of these two genes. In contrast to the genes of the *Hox-1* and *2* multigene clusters which define boundaries along the anterior-posterior axis, these two genes appear to operate along the second embryonic axis, the mediolateral axis. This pattern of expression results in the presence of *Hox-7* and *8* mRNA in the lateral plate mesoderm before limb outgrowth occurs. Subsequently, as the limb develops, the expression of these two genes is modified by the embryonic limb field and marks domains along the limb proximal-distal axis. An intriguing pattern of expression in the 10-day embryo in the distal portion of the limb reveals *Hox-7* expressed in the mesenchyme of the progress zone and Hox-8 in the epithelium of the apical ectodermal ridge (AER).

METHODS

DNA sequence determination and analysis

The *Hox-8* cDNA was sequenced by the chain termination reaction (Sanger *et al* 1987) using Sequenase (US Biochemicals) as previously described for the *Hox-7* cDNA sequence (Hill *et al* 1989). Sequence comparisons were carried out by the GAP and BESTFIT programs in the UWGCG sequence analysis software package.

Preparation of embryo sections and in situ hybridisations

Embryos were obtained from outbred Swiss mice. For staging embryos, midday on the day of detection of a vaginal plug was designated a 0.5-day embryo. Embryos were fixed in 4% paraformaldehyde at 4°C overnight and embedded in parafin wax. 5-7mm sections were cut and floated onto slides treated with TESPA (3-aminopropyltriethoxysilane)(Sigma).

In situ hybridisations were performed as described previously (Davidson *et al* 1988). Radiolabelled ^{35}S RNA probes were generated from the T_7 RNA polymerase promotion of vectors containing inserts specific for either *Hox-7* or *Hox-8*.

RESULTS

Sequence comparison of Hox-7 & 8

Both *Hox-7* and *8* belong to the *Msh* family of homeobox genes. This classification results from comparison of the sequence within the homeobox region of the protein. However, the *Hox-7 and 8* proteins show identity outside the homeobox domain (Fig. 1). The regions directly outside the homeobox also show similarity to the *Drosophila*

Msh gene. A region of significant similarity exists toward the amino-terminal end of the protein starting at amino acid position 14 of *Hox-8* and position 40 of *Hox-7*. A second region of similarity begins at the predicted position of the conserved hexapeptide found in many homeobox genes of which only two and three of these amino acids (starting at position 113 of *Hox-8* and 138 of *Hox-7*) are found in Hox-8 and Hox-7 respectively. These conserved regions may indicate other functional regions of the protein which affect DNA binding specificity, protein - protein interactions, or other transcriptional controls.

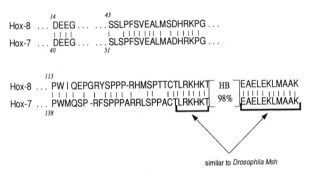

Figure 1. Regions of amino acid identity between *Hox-7* and *Hox-8* outside the homeodomain. The amino acids are presented in the one-letter code; the position of the homeobox (HB) is marked by brackets, and the regions which are similar to the *Drosophila Msh* gene are underlined. The amino acid position within the protein is numbered above the sequence for *Hox-8* and below the sequence for *Hox-7*

Early expression patterns

Both *Hox-7 & 8* are expressed at the time of gastrulation. A stage at 8 days of development near the end of gastrulation has been studied intensively. Adjacent transverse sections through the embryo in posterior to anterior series enable us to build a picture of the expression pattern of *Hox-7 & 8* at progressive developmental stages, since differention following gastrulation proceeds in the anterior to posterior direction. In the primitive streak posterior sections through the presomitic mesoderm anterior to the primitive streak show very broad *Hox-7* expression from the extreme lateral edges of the embryo through the midline. *Hox-8* on the other hand is expressed at a very low level in the midline. Both genes are expressed in all three germ layers the ectoderm, mesoderm and the endoderm. In more anterior regions sections which represent more mature regions of the embryo show that this broad pattern is becoming more restricted. (Fig. 2) *Hox-7* is no longer expressed at a detectable level along the midline and is becoming localized to the lateral tip of the neuroectoderm; ie, the neural crest, as well as the lateral portions of the epithelium

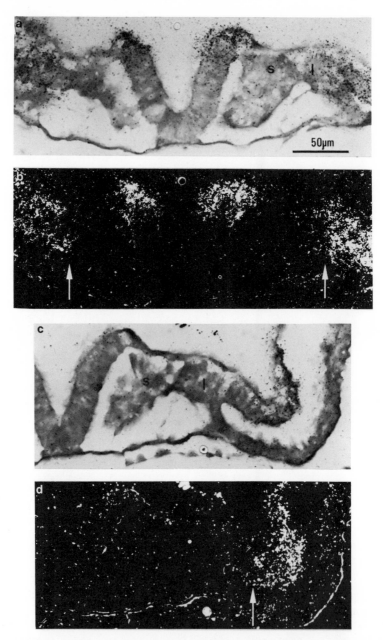

Figure 2. *Hox-7* and *Hox-8* expression in the trunk of the 8.5-day (11-somite) mouse embryo. Transverse sections through an embryo in the region of newly-formed somites. a) and b) : the distribution of *Hox-7* transcripts under a) bright-field and b) dark-field illumination. c) and d): the distribution of *Hox-8* transcripts in a nearby section. Note that the expression of each gene within the mesoderm marks a particular domain on the mediolateral axis of the embryo. The mesial boundary of *Hox-7* expression (arrow) is within the intermediate/lateral-plate mesoderm (l) close to the boundary with the somitic mesoderm (s). The mesial boundary of *Hox-8* expression lies more laterally in the lateral plate (arrow).

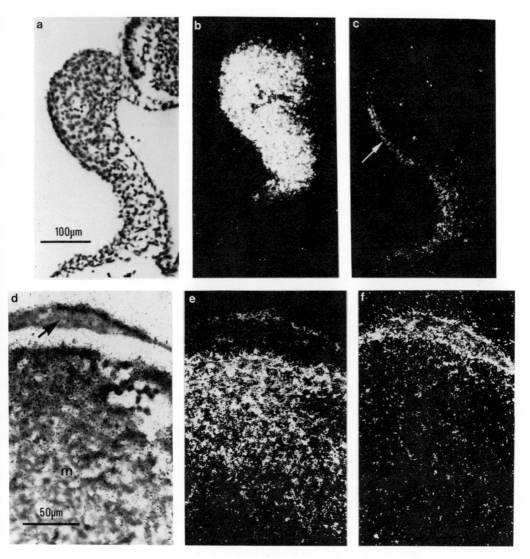

Figure 3. *Hox-7* and *Hox-8* expression in the developing limb.

a)-c): Transverse sections through the fore-limb bud of a 9.5-day mouse embryo. The sections pass through the posterior one-third of the bud. a): bright-field illumination. b): a nearby section showing the distribution of *Hox-7* transcripts throughout the mesoderm and in all but the most dorsal ectoderm of the bud. c): the distribution of *Hox-8* transcripts in the section shown in a). Note that *Hox-8* transcripts are most abundant in the ectoderm (arrow) and in the ventral mesoderm.

d)-f): Sections transverse to the proximodistal axis of the developing hind limb of an 11.5-day mouse embryo. This stage is equivalent to a fore-limb bud at 10.5 days of development. d): bright field (m, mesderm; apical ectodermal ridge arrowed). e) the same section showing the distribution of *Hox-7* transcripts in the mesoderm and in the epithelial cap covering the columnar cells of the apical ectodermal ridge. f): a nearby section showing the distribution of *Hox-8* transcripts in the apical ectodermal ridge and, in much less abundance, in the subadjacent mesoderm.

19

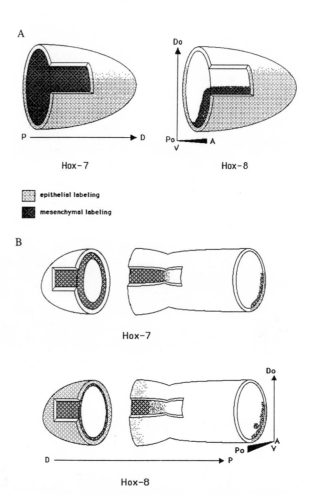

Figure 4. Diagrammatic representation of *Hox-7* and *Hox-8* expression in the developing forelimb at different stages. A.) Representation at 9.5 days of development. B.) Representation at 11.5 days of development. The outer shell represents the epithelium, the inner core the mesenchyme. The expression pattern of each gene is shown separately and designated under the fig. as *Hox-7* or as *Hox-8*. The shaded regions designate the extent of expression predicted by *in situ* hybridization, but not the level of expression. (The level of expression of *Hox-8* is always lower than that of *Hox-7* in the mesenchyme.) The cutaway in each drawing provides a means for examining the extent of mesenchymal expression.

and the lateral plate mesoderm. *Hox-8* is also expressed in the epithelium and in the mesoderm; however, the mesial boundaries of expression lie at a more lateral position than that of *Hox-7* overlapping the *Hox-7* domain of expression. A day later the posterior regions of the embryo will reflect this more restricted pattern. This pattern of expression suggests that the murine *Msh* genes are involved in the initial specification of domains along the mediolateral axis.

Expression in the limb

The expression of *Hox-7* and *8* is set in the flank mesoderm before limb formation. At 9 days in the lateral plate mesoderm *Hox-7* and *8* are expressed in a domain stretching caudally from near the anterior margin of the forelimb buds. The embryo has lost its flatten appearance as the lateral plate folds vertically to yield a cylindrical shaped embryo and the neural tube closes. As a result of these movements, the mediolateral domains marked by *Hox-7* and *Hox-8* expression are now laid out dorsoventrally. Thus the previous mesial boundary of *Hox-7* mesodermal expression now lies dorsal to the boundary of *Hox-8* expression. In the early stages of limb outgrowth *Hox-7* is expressed throughout the entire limb, in the mesenchyme and epithelium, *Hox-8* is predominantly expressed in the anterior and ventral parts of the mesenchyme. (Fig. 3)

These patterns change in the limb with the dynamics of development. In the 10d forelimb bud several major changes occur. Expression of *Hox-7* in the ectoderm is reduced to an almost undetectable level except for a few cells that cap the AER. *Hox-8*, in contrast, continues to be expressed in the distal mesoderm and in the ventral and distal ectoderm including the columnar ectoderm of the AER (Fig. 3). *Hox-7* is eventually excluded from the mesoderm of the proximal core of the limb but is found expressed in the mesoderm at the distal tip in the region referred to as the progress zone. Both genes are expressed in the anterior-ventral region of the limb mesenchyme.

At the time of hand plate formation (11 to 11.5d), the two genes share similar patterns of expression in the mesoderm though each has distinctive features. In the ectoderm, *Hox-8* continues to be expressed and *Hox-7* remains undetectable. At the early stages of cartilage formation in the handplate both genes are expressed in the marginal and interdigital mesenchyme, though by comparison with Hox-7, Hox-8 is more restricted to the superficial mesenchyme. The same patterns of expression of *Hox-7* and *8* are found in forelimbs and hindlimbs. A summary of these expression patterns is shown in Fig. 5.

CONCLUSIONS

Hox-7 and *8* are two closely related genes in the mouse genome. Unlike the *Hox-1*, *Hox-2* and *Hox-4* multigene clusters which within each cluster show coordinated temporal and spatial expression patterns, the *Hox-7* and *8* genes are expressed in a coordinated fashion without the apparent advantage of being closely linked. The early pattern of *Hox-7* and *8* suggests that these genes operate along the medial-lateral axis. The limb field which is generated in the lateral plate begins to grow out to form the limb bud such that the pattern of *Hox-7* and *8* expression that exists along the medial-lateral axis determines the early pattern detected in the limb. It is apparent that this original pattern is further modified as limb development progresses. The resulting pattern at the distal tip of the limb is intriguing. The *Hox-7* gene is expressed in the mesenchyme of the progress zone and *Hox-8* in the AER.

These two adjacent regions interact throughout limb development.(Wolpert, *et al*) As the limb buds out the underlying mesenchyme induces the production of the AER and is subsequently responsible for its maintenance. The interaction is reciprocal, the AER once formed maintains the progress zone, inducing the cells to grow and remain undifferentiated. We suggest that these two homebox-containing genes are involved in this interactive developmental system. Our results are consistent with the suggestion that these genes are involved in specifying regional identity in response to positional information in the limb-bud field. It would therefore be of interest to determine if the expression of these genes in particular cells depends on their position in the limb. Preliminary experiments (manuscript in preparation) with *Hox-7* suggest that indeed this is the case. Nonexpressing mesenchyme from the mouse limb can be induced to express *Hox-7* when transplanted to the distal tip of a similarly stage chicken wingbud. Grafts placed proximally do not express *Hox-7*. Such experiments suggest that *Hox-7* is positionally dependent and introduce the possibility that expression may be induced by the AER. Further evidence, reported elsewhere in this volume (G.Lyons et al.), shows that *Hox-7* expression is not maintained in the distal tip mesenchyme in the limbless mutation in chicken. In these mutants the AER fails to form. Therefore, it appears that at least *Hox-7* is an active participant in this interactive developmental system. These data further suggest that the early pattern along the medial-lateral axis is secondarily modified by the limb field under the influence of the AER.

Hox-7 and *8* are potentially important molecular markers for characterising the positional information involved in limb development. Understanding the expression and function of these genes will aid in determining the nature of the positional information, the interpretation of the positional information by the cell and the effect this information has on limb development in molecular terms.

REFERENCES

Akam, M., 1987. The molecular basis of metameric pattern in the *Drosophila* embryo, *Development* 101, 1-22.
Davidson, D.R., Graham, E., Sime, C. & Hill, R.E. 1988. A gene with sequence similarity to *Drosophila engrailed* is expressed during the development of the neural tube and vertebrae in the mouse. *Development* 104, 305-316.
Duboule, D. & Dolle, P., 1989. The structural and functional organisation of the murine *Hox* gene family resembles that of *Drosophila* homeotic genes. *EMBO J* 5, 1497-1505.
Graham A., Papalopulu, N. & Krumlauf, R., 1989. The murine and *Drosophila* homeobox gene complexes have common features of organisation and expression. *Cell* 57, 367-378

Hill, R.E., Jones, P.F., Rees, A.R., Sime, C.M., Justice, M.J., Copeland, N.G., Jenkins, N.A., Graham, E. & Davidson, D.R. 1989. A new family of mouse homeobox-containing genes; molecular structure, chromosomal location, and developmental expression of Hox 7.1 *Genes & Development;* 3, 26-37.

McGinnis, W., Garber, R.L., Wirz, J., Kuroiwa, A. and Gehring, W. I., 1984a. A homologous protein coding sequence in *Drosophila* homeotic genes and its conservation in metazoans. *Cell* 38, 403-408.

McGinnis, W., Levine, M.S., Hafen, E., Kuroiwa, A., and Gehring, W.J. 1984b. A conserved DNA sequence in homeotic genes of the *Drosophila melanogaster Antennapedia* and *Bithorax* complexes. *Nature* 308, 428-433.

Murphy, P., Davidson, D.R. & Hill, R.E., 1989. Segment specific expression of a homeobox-containing gene in the mouse hindbrain. *Nature* 341, 156-159.

Krasnow, M.A., Saffman, E.E., Kornfeld, K. and Hogness, D.S., 1989. Transcriptional activation and repression by *Ultrabithorax* proteins in cultured Drosophila cells. *Cell* 57, 1031-1043.

Ohkuma, Y., Horikoshi, M., Roeder, R.G. and Desplan, C. 1990. Binding site-dependent direct activation and repression of in vitro transcription by Drosophila homeodomain proteins. *Cell* 61, 475-484.

Robert, D., Sassoon, D., Jacq, B., Gehring, W. and Buckingham, M. 1989. Hox-7, a mouse homeobox gene with a novel pattern of expression during embryogenesis. *EMBO J.* 8, 91-100.

Sanger, F., Nicklen, S. and Coulson, A.R. 1977. DNA sequencing with chain-terminating inhibitors. *Proc. Natl. Acad. Sci.* 74, 5463-5467.

Scott, M.P. and Weiner, A.J., 1984. Structural relationships among genes that control development; Sequence homology between Antennapedia, Ultrabithorax and fushi tarazu loci of *Drosophila*. *Proc. Natl. Acad. Sci. 81*, 4115-4118.

Struhl, G., Struhl, K., and MacDonald, P.M. 1989. The gradient morphogen *bicoid* is a concentration-dependent transcriptional activator. *Cell* 57, 1259-1273.

Wilkinson, D.G., Bhatt, S., Cook, M., Boncinelli, E. & Krumlauf, R. 1989. Segmental expression of Hox 2 homeobox genes in the developing mouse hindbrain. *Nature* 341, 405-409.

Wolpert, L., Lewis, J., and Summerbell, D. 1975. Morphogenesis of the vertebrate limb. in "Cell Patterning", Ciba Symposium 29 (new series). Elsevier, Amsterdam.

THE LIMB DEFORMITY GENE ENCODES EVOLUTIONARILY HIGHLY

CONSERVED PROTEINS

Patricia A Blundell, Jose-Luis de la Pompa,
J. H. Carel Meijers, Andreas Trumpp and Rolf Zeller

EMBL Differentiation Programme
Meyerhofstrasse 1, D-6900 Heidelberg, Germany

INTRODUCTION

Determination and establishment of the two main limb axes is one of the most studied processes in embryonic pattern formation. The molecular signals determining the anteroposterior (ap) and proximodistal (pd) axes seem to be conserved among vertebrates, but none have so far been clearly identified (for review see Tickle 1980). A series of classical experiments led to the proposition that two morphogenetically active regions of the early limb bud are crucial in establishing these axes; the apical ectodermal ridge (AER) appears to be mainly involved in establishing the pd axis, and the zone of polarizing activity (ZPA) is essential for the establishment of ap polarity (for review see Fallon et al 1983, Tickle, 1980). To understand these developmental processes at a molecular level, it is essential to isolate the genes that are involved. Substantial progress in this area has recently been made.

Firstly, many research teams have isolated and analysed the vertebrate cognates of the genes that control early axis formation and segmentation in Drosophila. Analysis of their temporal and spatial expression patterns has suggested homeobox-containing genes as possible key regulators of pattern formation. These genes encode transcription factors which could control major developmental switches. At least one of these genes may be involved in the establishment of the main limb axes (see eg Dolle et al 1989a). Furthermore, genes belonging to the retinoic acid receptor gene family and retinoic acid binding proteins (see eg Dolle et al 1989b) have also been implicated in limb morphogenesis.

Secondly, numerous mutations (mostly in mice) that disrupt limb pattern formation have been described (Gruneberg 1963). Recently, the isolation of one of these genes, the limb deformity gene (ld), has been made possible by chance insertional mutagenesis in transgenic mice (Woychik et al 1985). A review of the genetic, morphological and molecular analysis of the ld gene so far (Woychik et al 1985, Zeller et al 1989, Woychik et al 1990a, 1990b, Maas et al 1990) and a discussion of the results with respect to current models of vertebrate pattern formation is given in this manuscript.

CLONING OF THE CHICKEN ld GENE REVEALS EXTENSIVE EVOLUTIONARY CONSERVATION

Woychik et al (1985) were able to isolate endogenous sequences disrupted by a transgene insertion into the ld locus which created a

Developmental Patterning of the Vertebrate Limb
Edited by J.R. Hinchliffe *et al.*, Plenum Press, New York, 1991

recessive ld phenotype. This transgenic insertional mutation was shown to be allelic to previously characterized ld mutations (see below) and was therefore named ld^{Hd} (Woychik et al 1985).

Within the sequences flanking the transgene insertion site, an exon of an evolutionarily conserved gene was found (Woychik et al 1990a and Fig 1) which is expressed in the developing mouse (Zeller et al 1989, Woychik et al 1990a), the chicken embryo (Fig 1) and specific adult tissues (Woychik et al 1990a).

Detailed molecular analysis eventually led to the identification of the ld gene (Woychik et al 1990a, Maas et al 1990). This gene is expressed as a complex array of low abundance mRNAs created by alternative splicing and differential polydenylation both in the embryo and adult tissues. Such alternatively spliced transcripts seem to encode several related proteins with no significant homology to other known proteins. This group of related ld gene products have been named formins (Woychik et al 1990a). The use of a conserved exon of the mouse ORF (Woychik et al 1990a) as a molecular probe (Fig 1) has led to the cloning of the homologous gene from the chicken. In contrast to previously isolated mouse cDNA clones, which represent ld transcripts present in adult tissues, all the chicken ld cDNAs have been cloned from embryos. Some of these clones should therefore represent ld transcripts relevant to pattern formation (P Blundell, A Trumpp, R Zeller, unpublished). Sequence analysis of these cDNA clones reveals a high degree of evolutionary conservation at the amino acid level.

To study the molecular function of these proteins, a polyclonal antiserum against the large chicken formin was raised. Immunolocalisation studies reveal nuclear localisation of the protein (A Trumpp and R Zeller, unpublished), which indicates that the formins are involved in transcriptional or post-transcriptional control of vertebrate pattern formation.

THE ld GENE IS REQUIRED FOR AER DIFFERENTIATION AND ap LIMB AXIS FORMATION

Woychik et al (1985) were able to show that the recessive ld^{Hd} mutation is allelic to other previously described ld alleles. All four ld alleles isolated so far (ld^J, Cupp 1960, ld^{OR}, Green 1968, ld^{Hd}, Woychik et al 1985, ld^{In2}, Woychik et al 1990b) display a pleiotropic phenotype. The ld phenotype manifests as a prominent synostosis of the long bones of all limbs, and oligodactyly and syndactyly of the bony elements of the hand and foot plates in mutant animals of all four alleles (Woychik et al 1985, Woychik et al 1990b, see also Fig 2). In addition the ld^{Hd}, ld^J and ld^{OR} alleles (Kleinebrecht et al 1982, R Maas and P Leder, personal communication) show a high frequency of uni- and bi-lateral renal aplasias. The kidney phenotype in the ld^{In2} allele is less severe and in most cases is limited to a defect of the urinary tract rather than complete renal aplasias (Woychik et al 1990b). The well established pleiotropic nature of the ld phenotype suggests a more general role for formins during vertebrate pattern formation. This is supported by the finding that both the murine (Zeller et al 1989) and chicken (A Trumpp, R Zeller, unpublished) ld genes are expressed long before the limb pattern is determined.

Figure 1. Upper panel: Comparison of a murine and chicken ld exon to reveal the extent of evolutionary conservation. Lower panel: RNase protection assay using the chicken exon reveals expression of the ld gene during chicken embryonic development. 40 ug of total RNA were used per stage. Asterisks indicate the presence of unspliced precursors in addition to ld mRNAS (149 base fragment).

```
MOUSE     CTTCTGCAG T TTT GAA ACA ACA GTG GGA TAT TTT GGA ATG AAG CCA AAG ACT GGA

CHICK     CCCTCCCAG C --- --- GA- --- --- --- --C --- --G --A --- --- --- C-- --T

MOUSE               F   E   T   T   V   G   Y   F   G   M   K   P   K   T   G

CHICK               -   -   E   -   -   -   -   -   I   -   -   -   P   -


MOUSE     GAG AAG GAG GTC ACC CCC AGC TAT GTG TTT ATG GTG TGG TTT GAG TTC TGC AGT

CHICK     --- --- --- A-- --A --A -A- --- --T --C -C- --- --- -AC --A --- --- ---

MOUSE      E   K   E   V   T   P   S   Y   V   F   M   V   W   F   E   F   C   S

CHICK      -   -   -   I   -   -   N   -   -   -   T   -   -   Y   -   -   -   -


MOUSE     GAC TTC AAG ACC ATT TGG AAG CGG GAG AGT AAG AAC ATA TCT AAA GAA AG GTAA

CHICK     --- --- --- --- --- --- --A --A --- --C --A -G- --T --C --G --G C- GTAA

MOUSE      D   F   K   T   I   W   K   R   E   S   K   N   I   S   K   E

CHICK      -   -   -   -   -   -   -   -   -   -   -   S   -   -   -   -
```

RNA sample

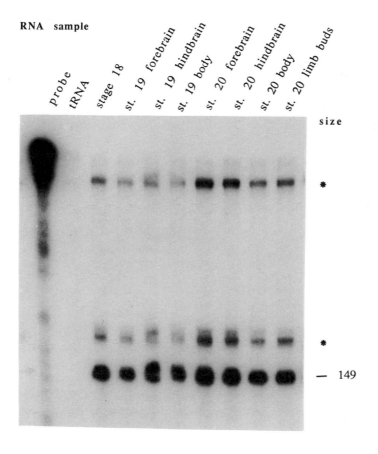

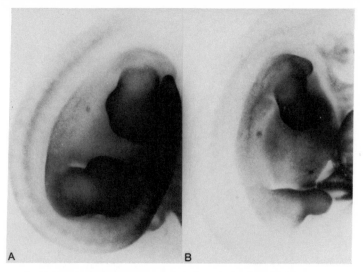

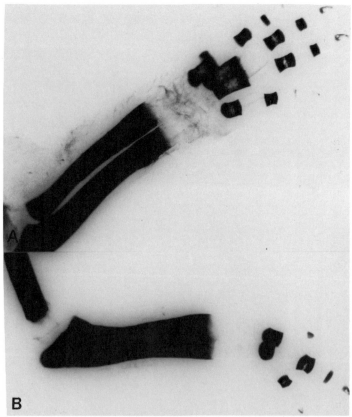

Figure 2. Upper panel: A. Wild type gestational day 11 mouse embryo (stained with hematoxylin) B. Homozygous ld^{Hd}/ld^{Hd} gestational day 11 sibling. Note the characteristic shortening of all ap limb axes (for details see Zeller et al 1989). Lower panel: A. Skeletal stain of an adult wild type murine forelimb. B. Skeletal stain of a homozygous ld^{Hd}/ld^{Hd} forelimb. Note the fusion of ulna and radius and reduction and fusions of digits (for details see Woychik et al 1985).

The nature of the developmental lesion leading to the kidney phenotype remains unknown, but a limited morphological and molecular analysis has been done to determine the role of the ld gene products during limb morphogenesis (Zeller et al 1989). Initial morphological differences between wild type and homozygous ld embryos are apparent during early gestational day 10, a time during which the limb pattern is determined. As a result of a shortened ap axis, the mutant limb bud appears more pointed than its wild type counterpart (Fig 2). This shortening eventually leads to the characteristic truncations affecting the ap but not the pd axis (Fig 2). Furthermore, failure of the AER to differentiate into the characteristic stratified epithelium is apparent in mutant limb buds (for details see Zeller et al 1989). Consistent with this observation, molecular analysis of wild type limb buds reveals an approximately 5-fold higher level of ld transcripts in the ectoderm. Taken together, these data suggest that the formins are required both for differentiation of the AER and ap limb axis formation.

How do these results fit with the current models of limb morphogenesis, which are mostly based on experimental manipulation of the chicken embryo? Wolpert (1969) proposed that ap polarity in the limb mesenchyme is established by a concentration gradient of a morphogen produced by the polarizing region. Retinoic acid (or dihydroretinoic acid) is a candidate molecule for such a morphogen (Tickle et al 1985, Thaller and Eichele 1987, Thaller and Eichele 1990). The morphogen would act by exerting its primary effect on the limb mesenchyme via the polarizing zone (Tickle et al 1989). The role of the AER in ap axis formation is considered to be secondary; it primarily appears to promote outgrowth of the pd axis (for review see Fallon et al 1983). Partial removal of the AER from developing chicken limb buds, however, leads to a loss of specific digits (ie ap truncations, Rowe and Fallon 1981). A series of partial AER removals allowed Rowe and Fallon (1981) to map the regions of the AER which specify individual digits. These studies suggest that local disruption of growth control by the AER probably leads to the observed ap truncations. Such a mechanism can certainly explain the specific effects of ld mutations on AER differentiation and ap axis formation. It is surprising, however, that mutant, non-differentiated AER is still able to support outgrowth and patterning of the pd axis. These findings suggest that mutations in the ld gene lead to a genetic uncoupling of the mechanisms which the AER uses to control pd and ap axis formation.

Now that we have tools available to study the involvement of the different formins in chicken limb morphogenesis, we will hopefully gain more insight into the morphogenetic events that these evolutionarily conserved proteins are involved in. These molecules might help us to unravel how the AER and the limb mesenchyme interact during pattern formation.

ACKNOWLEDGEMENTS

R Zeller would especially like to thank P Leder, R Woychik, R Maas, L Jackson-Grusby and T Vogt for the many hours spent together cloning and analysing the murine ld gene and C Tickle and G Eichele for introducing him to the chicken embryo. We thank M Remy and I Fraignaud for help in preparing this manuscript.

REFERENCES

CUPP, M. B., (1960). In "Mouse News Letter", 22: 50.
DOLLE, P., RUBERTE, E., KASTNER, P., PETKOVICH, M., STONER, C. M., GUDAS, L. J., and CHAMBON, P., (1989a). Differential expression of genes encoding alpha, beta and gamma retinoic acid receptors and CRABP in the developing limbs of the mouse, Nature. 342: 702.

DOLLE, P., IZPISUA-BELMONTE, J.C., FALKENSTEIN, H., RENUCCI, A. and DUBOULE D., (1989b). Coordinate expression of the murine HOX-5 complex homeobox-containing genes during limb pattern formation. Nature. 342: 767.

FALLON, J. F., ROWE, D. A., FREDERICK, J. M. and SIMANDL, B. R., (1983). Studies on epithelial-mesenchymal interactions during limb development. In "Epithelial-mesenchymal interaction in development". R H Sawyer and J F Fallon, New York: Praeger Scientific.

GREEN, M. C., (1968). Mutant genes and linkages. In "Biology of the Laboratory Mouse". Ed E L Green. New York: Dover Publications.

GRUNEBERG, H., (1963). "The pathology of development. A study of inherited skeletal disorders in animals." Oxford: Blackwell Scientific Publ.

KLEINEBRECHT, J. D., SELOV, J. and WINKLER, W., (1982). The mouse mutant limb-deformity (ld). Anat. Anzeiger. 152: 313.

MAAS, R. L., ZELLER, R., WOYCHIK, R. P., VOGT, T. F. AND LEDER, P., (1990). Disruption of formin-encoding transcripts in two mutant limb deformity alleles. Nature. 346: 853.

ROWE, D. A. and FALLON, J. F., (1981). The effect of removing posterior apical ectodermal ridge of the chick wing and leg on pattern formation. J. Embryol. Exp. Morph. 65:309.

THALLER, C. and EICHELE, G., (1987). Identification and spatial distribution of retinoids in the developing chick limb bud. Nature. 327: 625.

THALLER, C. and EICHELE, G., (1990). Isolation of 3,4-didehydroretinoic acid, a novel morphogenetic signal in the chick wing bud. Nature. 345: 815.

TICKLE, C., (1980). The polarizing region and limb development. In "Development in mammals". Vol 4. Ed M. H. Johnson. Amsterdam: Elsevier/North-Holland Biomedical Press.

TICKLE, C., LEE, J., and EICHELE, G., (1985). A quantitative analysis of the effect of all-trans retinoic acid on the pattern of chick wing development. Dev. Biol. 109: 82.

TICKLE, C., CRAWLEY, A., and FARRAR, J., (1989). Retinoic acid application to chick wing buds leads to a dose-dependent reorganisation of the apical ectodermal ridge that is mediated by the mesenchyme. Development. 106: 691.

WOYCHIK, R. P., STEWARD, T. A., DAVIS, L. G., D'EUSTACHIO, P. D., and LEDER, P., (1985). An inherited limb deformity created by insertional mutagenesis in a transgenic mouse. Nature. 318: 36.

WOYCHIK, R. P., MAAS, R. L., ZELLER, R., VOGT, T. F., and LEDER, P., (1990a). Formins: proteins deduced from the alternative transcripts of the limb deformity gene. Nature. 346: 850.

WOYCHIK, R. P., GENEROSO, W. M., RUSSELL, L. B., CAIN, K. T., CACHEIRO, N. L. A., BULTMAN, S. J., SELBY, P. B., DICKINSON, M. E., HOGAN, B. L. M., and RUTLEDGE, J. C., (1990b). Molecular and genetic characterisation of a radiation-induced structural rearrangement in mouse chromosome 2 causing mutations at the limb deformity and agouti loci. Proc. Natl. Acad. Sci. 87: 2588.

ZELLER, R., JACKSON-GRUSBY, L., and LEDER, P., (1989). The limb deformity gene is required for apical ectodermal ridge differentiation and anteroposterior limb pattern formation. Genes Dev. 3: 1481.

THE ROLE OF HOMEOBOX GENES IN AMPHIBIAN LIMB DEVELOPMENT AND

REGENERATION

Robin Brown and Jeremy P. Brockes

Ludwig Institute for Cancer Research
91 Riding House Street
London W1P 8BT

SUMMARY

The newt homeobox gene, NvHbox2, is expressed in the blastema of the
limb and tail after amputation but not in the corresponding normal tissues.
In addition to this restriction of expression to regenerating tissues, the
level of the transcript is 3-5 fold higher in a proximal as compared to a
distal blastema. If a distal blastema is proximalised by treatment with
retinoic acid (RA), the level of the transcript is not affected at two
different times of treatment.

INTRODUCTION

An important group of transcription factors, related by the presence
in their protein products of a homeobox, have been shown to play an important role in a wide range of developmental processes (Scott et al., 1989;
Brockes, 1989). Originally isolated in Drosophila on the basis of sequence
homology with the prototype homeotic gene antennapedia, the more divergent
members of the family are also involved in segmentation (fushi tarazu) and
in the establishment of an early embryonic gradient (bicoid) (Akam, 1987;
Nusslein-Volhard et al., 1987). What makes homeobox genes of greater
interest still is their conservation across species and their remarkably
similar genomic organisation that suggests, at least for those genes present
in the four complexes, an important role in the genetic control of development.

With this background we were encouraged to investigate whether homeobox
genes played a role in regeneration of the urodele limb. We have previously
reported that the regenerating forelimb blastema of the newt Notopthalmus
expresses a homeobox gene NvHbox 1 that is a homolog of the human c8 and
Xenopus XlHbox 1 (Savard et al., 1988). The expression of this gene is not,
however, limited to the limb blastema but is present in the normal limb and
in a number of adult tissues including the liver. Here we report the isolation of another newt homeobox gene, NvHbox 2, whose expression appears
to be limited to regenerating tissue and whose pattern of expression suggests
that the gene may have a more specific role in limb regeneration.

MATERIALS AND METHODS

Details of the materials and methods used in these studies are given
in Savard et al (1988) and Brown and Brockes (Development, submitted).

Developmental Patterning of the Vertebrate Limb
Edited by J.R. Hinchliffe *et al.*, Plenum Press, New York, 1991

RESULTS

In an attempt to identify additional homeobox clones in the limb blastema we screened a lambda gt11 proximal blastema cDNA library in duplicate with a probe to NvHbox1, and with an oligonucleotide fully degenerate to the conserved sequence DRQVKIWFQNRRKEK that is present in a wide range of homeodomain proteins. Clones were isolated from this screen that hybridised with the oligonucleotide alone. Three of the clones isolated in this way were purified and sequenced and shown to be from the same transcript but to differ at their 5' ends.

Translation of the single open reading frame present in the longest of the clones reveals the presence of a homeobox at the c-terminus that is identical to that of the recently isolated human HOX 4f and to the mouse 5.5 gene (Fig 1). Comparison of the sequence of the mouse gene outside of the homeodomain demonstrates that further homology is mostly restricted to the N-terminal 60 amino acids. This suggests that the two genes are closely related although may not be species homologs.

Pattern of Expression

After construction of probes to facilitate RNAse protection analysis the pattern of expression was examined in both regenerating and normal tissues of the newt. These results can be summarised as follows:
1.   The transcript is expressed in both proximal and distal blastemas of the limb and tail. A comparison of the levels of the transcript in a proximal forelimb blastema with the level in a distal reveals the presence of a gradient of expression of 3-5 in favour of the proximal location. Within the forelimb blastema expression is limited to the mesenchyme (Fig 2a).
2.   A tissue survey fails to reveal any detectable expression in adult tissues including liver, heart, kidney, torso, viscera, jaw or normal limb and tail. Transcriptional induction appears to be of the order of 20 fold with the message level in the normal limb being at the limit of detection (Fig 2b).
3.   Treatment of a distal blastema with retinoic acid fails to convert the level of the transcript to that of a proximal blastema (Fig 3). This result was obtained at two different times of regeneration after retinoid treatment. Control animals allowed to regenerate after this treatment demonstrate characteristic duplications, confirming that the distal blastema cells now behave as though they had their origin at a more proximal level.

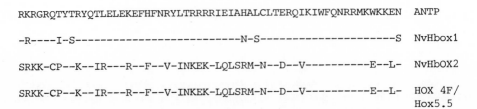

```
RKRGRQTYTYRYQTLELEKEFHFNRYLTRRRRIEIAHALCLTERQIKIWFQNRRMKWKKEN   ANTP

-R----I-S---------------------------N-S--------------------S   NvHbox1

SRKK-CP--K--IR---R--F--V-INKEK-LQLSRM-N--D--V----------E--L-   NvHbOX2

SRKK-CP--K--IR---R--F--V-INKEK-LQLSRM-N--D--V----------E--L-   HOX 4F/
                                                               Hox5.5
```

Figure 1.   Comparison of the homeodomain sequence of the newt NxHbox 1 and NvHbox 2 genes with antennapedia. The sequences of the human HOX-4f and mouse Hox-5.5 homeodomains are identical to NvHbox 2 and show considerable divergence from the antennapedia sequence. A dash represents sequence identity.

DISCUSSION

The two homeobox genes identified so far in the blastema of the newt share a number of features but also differ in at least one key respect. The transcript from NvHbox 2 is not detectable in the normal limb or tail but is induced upon amputation. In contrast NvHbox 1 has a more general role, perhaps being involved in tissue repair as well as regeneration, while the NvHbox 2 protein has a more specific requirement for blastemal cells.

We are left with two apparently incompatible observations that need to be reconciled in our attempt to define the role of homeobox genes in regeneration and the establishment of positional identity. On the one hand the expression of NvHbox 2 shows a gradient on the PD axis and an impressive induction on amputation. By RNAse protection analysis, the transcript has been localised to the mesenchymal cells of the blastema and to show a rapid decline in expression in late regenerates. These results are consistent

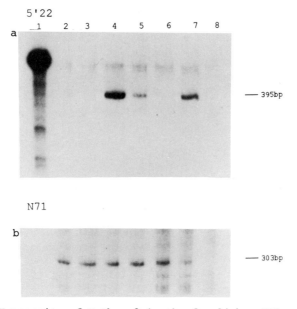

Figure 2. a) Expression of NvHbox 2 in the forelimb. RNAse protection analysis with a 395bp 5' probe (5'22) demonstrates the presence of a proximo-distal gradient and expression in the mesenchymal fraction of the blastema.
Lanes (5ug's each total RNA): 1. input probe 5'22; 2. normal forelimb proximal; 3. normal forelimb distal; 4. proximal blastema; 5. distal blastema; 6. epithelium; 7. mesenchyme; 8. tRNA. The size of the protected fragment is shown.
        b) RNAse protection analysis of the expression of NvHBox 2 in a variety of adult newt tissues. RNA samples were analysed with the probe 5'22 and the newt satellite probe pSP6D6 (Casimir et al., 1988).
Lanes (5ug's total RNA each): 1. proximal forelimb blastema; 2. normal limb; 3. tail blastema; 4. normal tail; 5. belly mesenchyme; 6. whole torso; 7. hand regenerate; 8. jaw; 9. brain; 10. spleen; 11. heart; 12. liver; 13. tRNA (10ug's). The size of the protected fragment is 395bp for 5'22. The satellite probe identifies multiple transcripts in all tissues.

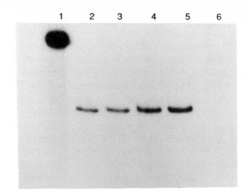

Figure 3.   The effect of RA on NvHbox 2 expression in distal forelimb
blastemas.
Lanes: 1. input probe 5'22; 2. RA treated, early; 2. DMSO treated, early;
3. RA treated, late; 4. DMSO treated, late; 5. tRNA (10ug's). Note that
there is no difference within the two groups but expression increases in
both samples as the later time point.   Early blastemas were harvested
after the retinoic acid induced delay;   late samples were harvested at the
mid-bud stage.

with the hypothesis that NvHbox 2 might be responsible for either the
establishment or maintenance of positional identity. An alternative is
that NvHbox 2 has an important role in some other aspect of blastemal
cell biology not concerned with the establishment of axial specification.
The early determination of the blastemal cells and their commitment to a
differentiation program might be controlled by a gene cascade, of which
NvHbox 2 in particular, and homeobox genes in general, might be members.
In this case the positional value of the blastemal cells might not depend
on the level of the NvHbox 2 transcript.

     The lack of a direct transcriptional response to RA treatment by
NvHbox 2 could be accounted for by suggesting that the gene expression
pathway leading to the establishment of positional identity has a retinoid sensitive step that is downstream of the homeobox gene.   This would
account for the observation that at a time when a PD difference in blastemal expression is observed, the level of the homeobox gene is unaffected
by RA treatment.   This model rests on the assumption that the effect of
RA on the blastemal cells is predominantly transcriptional, and does not
take into account other levels of control such as post-translational
modification either by RA or by an RA-induced mechanism.   It should also
be noted that our studies cannot rule out the unlikely possibility that
there is a transient increase in the expression of either or both of
these genes immediately following RA treatment.

     Evidence for a direct effect of RA on homeobox genes comes from
experiments on tissue culture cells.   In N2T/D1 EC cells transcriptional
activation of genes of the HOX-2 complex occurs in a sequence that reflects the position of the gene within the complex (Simone et al.,1990).
Similarly the human HOX4B gene shows a transcriptionally based response to
RA with increasing concentrations resulting in the induction of two additional transcripts from the gene (Simone et al.,1990).It is known that the
transcripts from this gene show a developmentally distinct pattern relative
to each other (Mavilio et al., 1986 Nature 324,664-668) and so this observation suggests that attractive possibility that homeobox genes are capable
of responding at a transcriptional level to local concentrations of RA.

The co-ordinate expression of homeobox genes is mimicked in the developing limb bud by genes of the Hox-5 complex, members of which show a temporal pattern of expression such that genes most 5' in the complex appear at later stages (Dolle et al., 1989). In the latter case however it is not known whether this effect is due to the action of RA. The position of a gene in its complex would appear from these results to have a predictive value both in terms of timing (and hence position) and RA response, and the observation of a direct transcriptional effect of RA on homeobox genes is intriguing. It is unclear however if such a range of responses would be observed in vivo. An additional caveat is that clones of F9 EC cells that no longer differentiate in the presence of RA continue to show RA-induced expression of the Hox-1.3 gene (Espeseth et al., 1989). The interpretation of these experiments is not straightforward but nevertheless, sequential induction is a striking result and a model whereby the induction of the most responsive homeobox gene in the complex initiates a cascade of expression is an attractive possibility.

Although a considerable weight of descriptive data exists to suggest that homeobox genes play an important role in development and regeneration in vertebrates, it is fair to conclude that their exact role in pattern formation remains unclear. The putative morphogen retinoic acid appears in some culture system to exert an effect on homeobox gene transcription, but the relevance of such effects in vivo is unclear. We hope in the near future to address these problems by the introduction into blastemal cells of recombinant plasmids that would alter the level of expression of NvHbox 2 and to determine the morphogenetic effect of such direct intervention. We also hope to explore the possibility that other families of transcription factors may be more directly linked to the effect of RA and the establishment of the PD axis in the regenerating limb.

REFERENCES

Akam, M.E., 1978, The molecular basis for metameric pattern in the Drosphila embryo, Development, 101:1-22.
Brockes, J.P., 1989, Retinoids, homeobox genes, and limb morphogenesis, Neuron, 2:1285-1294.
Dolle, P., Izposua-Belmonte, J-C., Falkenstein, H., Renucci, A., and Dulboule, D., 1989, Coordinate expression of the murine Hox-5 complex homeobox-containing genes during limb pattern formation, Nature, 342: 767-772.
Espeseth, A.S., Murphy, S.P., and Linney, E., 1989, Retinoic acid receptor expression vector inhibits differentiation of F9 embryonal carcinoma cells, Genes and Development, 3:1647-1656.
Nusslein-Volhard, C., Frohnhofer, H.G., and Lehmann, R., 1987, Determination of anteroposterior polarity in the Drosphila embryo, Science, 238:1675-1681.
Savard, P., Gates, P.B., and Brockes, J.P., 1988, Position dependent expression of a homeobox gene transcript in relation to amphibian limb regeneration, EMBO J., 7:4275-4282.
Scott, M.P., Tamkun, J.W., and Hartzell, G.W., 1989, The structure and function of the homeodomain, BBA, 989:25-48.

HOMEOBOX-CONTAINING GENES AND GAP JUNCTIONAL COMMUNICATION IN PATTERN FORMATION DURING CHICK LIMB DEVELOPMENT

Robert A. Kosher, William B. Upholt, Caroline N. D. Coelho, Kimberly J. Blake, and Lauro Sumoy

Department of Anatomy
University of Connecticut Health Center
Farmington, CT 06030

INTRODUCTION

Elucidating the mechanisms involved in eliciting the formation of the various skeletal elements of the limb in their appropriate position and sequence along the anterior-posterior (A-P) axis is fundamental to understanding normal and abnormal vertebrate limb development. The zone of polarizing activity (ZPA) at the posterior margin of the developing limb bud appears to be the source of a diffusible morphogen, possibly retinoic acid, that becomes distributed in a graded fashion across the A-P axis of the limb, and specifies the A-P positional values of the skeletal elements of the limb according to its local concentration (Brickell and Tickle, 1989; Eichele, 1989). Homeobox-containing genes have been implicated in the regulation of pattern formation during development, and several homeobox genes exhibit spatially-restricted and temporally-regulated patterns of expression during vertebrate limb development (Oliver et al., 1988, 1989; Dolle et al., 1989; Hill et al., 1989; Robert et al., 1989; Wedden et al., 1989). In the present manuscript we describe some of our current studies on the isolation and characterization of several homeobox genes expressed during embryonic chick limb development, and present evidence indicating that some of these genes are expressed in a graded fashion along the A-P axis of the chick limb bud as positional values along the A-P axis are being specified. We also report on studies indicating that intercellular signalling via gap junctions may be involved in specification of pattern along the A-P axis of the chick limb bud.

ISOLATION, CHARACTERIZATION, AND EXPRESSION OF CHICKEN HOMEOBOX-CONTAINING GENES DURING LIMB DEVELOPMENT

A cDNA library prepared from poly(A)+ RNA isolated from stage 22, 26, and 28 embryonic chick limb buds was screened at low stringency with the human homeobox-containing cDNA HHO.c13 (Blake et al., 1990) and with mouse Hox-7.1 (Sumoy et al.,

1990). HHO.c13, which is the human cognate of mouse Hox-4.2
(previously named Hox-5.1) (Duboule et al., 1990), is
transiently expressed during early stages of human limb
development (Mavilio et al., 1986), and Hox-7.1 has been
reported to be expressed in the distal mesenchyme subjacent to
the apical ectodermal ridge (AER) during early stages of mouse
limb development (Hill et al., 1989; Robert et al., 1989).
Three distinct chicken homeobox-containing cDNAs were isolated
by screening with HHO.c13 and two were obtained by screening
with Hox-7.1. Based on their nucleotide and deduced amino
acid sequence similarities to human and mouse homeobox cDNAs,
the chicken homeobox cDNAs isolated with HHO.c13 have been
identified as the chicken cognates of mouse Hox-4.2
(previously named Hox-5.1), Hox-1.4, and Hox-2.2, and,
therefore, the genes encoding these chicken (Gallus) cDNAs
have been designated GHox-4.2, GHox-1.4, and GHox-2.2 (Blake
et al., 1990). One of the chicken homeobox cDNAs isolated
with mouse Hox-7.1 has been designated as GHox-8 (Sumoy et
al., 1990) based on its homology to a recently characterized
mouse homeobox cDNA related to, but distinct from mouse Hox-
7.1 (D. Davidson, personal communication; Hill et al., 1989).
Characterization of the other chicken cDNA isolated with Hox-
7.1 is not yet complete, although partial sequence analysis
suggests it is highly homologous to mouse Hox-7.1. The
homeoboxes and homeodomains encoded by Ghox-4.2, GHox-1.4,
GHox-2.2, and GHox-8 exhibit a striking amount of identity to
those of their cognate mouse and human cDNAs. Each of these
chicken cDNAs also exhibit a high level of sequence identity
to their cognate mouse cDNAs in regions 5' and 3' to the
homeodomain (Blake et al., 1990; Sumoy et al., 1990).

To determine if these chicken homeobox genes are
expressed in a spatial fashion consistent with their
involvement in pattern formation, steady-state levels of mRNAs
for the genes were examined in various regions along the A-P
axis of embryonic chick wing buds at an early stage of
development in which positional values along the A-P axis are
being specified. Ectoderm was removed from stage 20/21 chick
wing buds; the mesoderm was dissected into anterior, middle,
and posterior regions along the A-P axis; and, steady-state
mRNA levels for GHox-4.2, GHox-1.4, and GHox-8 in each region
of the mesoderm were determined by cytoplasmic dot
hybridization (Blake et al., 1990; Coelho et al., 1990). To
ensure specificity, hybridizations were performed under
stringent conditions using regions of the cDNAs 3' to the
homeobox in which there are little or no sequence similarities
among the various homeobox cDNAs (Blake et al., 1990; Coelho
et al., 1990).

We have found that GHox-4.2, GHox-1.4, and GHox-8 are
each expressed in relatively high amounts in the anterior
mesoderm of stage 20/21 wing buds; in considerably lower
amounts in the middle region of the limb; and, in very low
amounts in the posterior mesoderm which contains the ZPA
(Blake et al., 1990; Coelho et al., 1990). This graded
pattern of expression and the complementary relationship of
the expression pattern to the ZPA is suggestive of the
possible involvement of GHox-4.2, GHox-1.4, and GHox-8 in the
specification of positional values along the A-P axis of the
developing limb. In particular, these observations suggest
the possibility that high level expression of these genes may

be involved in the specification of anterior positional values along the A-P axis, and suggest the possibility that the putative diffusible morphogen (possibly retinoic acid) produced by the ZPA may be antagonistic to the expression of these genes. It should be noted that the protein encoded by the Xenopus homeobox gene XlHbox 1 (the Xenopus cognate of mouse Hox-3.3) is also localized predominantly in the anterior compartment of the distal mesoderm of Xenopus, mouse, and chick limb buds (Oliver et al., 1988), whereas the protein encoded by the human Hox-4.4 (previously named Hox-5.2) gene exhibits the complementary pattern of expression, being predominantly localized in the ZPA-containing posterior mesoderm of the limb (Oliver et al., 1989). All of these observations are consistent with the suggestion of Oliver et al. (1989) that interacting opposing gradients of expression of various homeobox genes may be involved in determining positional values during limb development (see Fig. 1).

To more thoroughly define its pattern of expression, the distribution of GHox-8 transcripts has also been examined in various proximal regions of stage 23 and 25 embryonic chick wing buds (Coelho et al., 1990). The proximal portions of stage 23 and 25 wing buds were dissected into five regions, the pre-axial (anterior) periphery, the post-axial (posterior) periphery, the dorsal (myogenic) periphery, the ventral (myogenic) periphery, and the chondrogenic central core; and, the steady-state cytoplasmic levels of GHox-8 mRNA in each of these proximal limb regions were determined (Coelho et al., 1990). These studies indicate that GHox-8 is expressed in high amounts in the proximal pre-axial (anterior) periphery of the chick limb bud. In contrast, only very low levels of GHox-8 transcripts are detectable in the proximal post-axial (posterior) periphery, in the proximal dorsal and ventral (myogenic) peripheries, and in the chondrogenic central core (Coelho et al., 1990). The localization of high levels of GHox-8 expression to the proximal pre-axial periphery of the limb bud is again suggestive of its possible involvement in establishing anterior positional values. It is of particular interest that GHox-8 is not uniformly expressed by mesenchyme subjacent to limb ectoderm, nor is a high level of GHox-8 expression associated with the differentiation of any particular cell type in the limb. Thus, it appears more likely that high GHox-8 expression is involved in specification of polarity or positional values, rather than in the regulation of cytodifferentiation. It should be noted that like GHox-8, the chicken homeobox gene GHox-2.1 is localized in the proximoanterior portion of the chick limb bud (Smith et al., 1989).

GAP JUNCTIONAL COMMUNICATION IN THE REGULATION OF PATTERN FORMATION ALONG THE A-P AXIS OF THE CHICK LIMB BUD

Gap junctions play an important role in mediating cell-cell communication during development, and ultrastructural studies have shown that gap junctions are present between limb mesenchymal cells at early stages of limb development (Kelley and Fallon, 1978; 1983). Allen et al. (1990) have recently demonstrated that grafts comprised of ZPA tissue and anterior limb mesenchymal cells that had been loaded with antibodies against rat liver gap junctional proteins are considerably

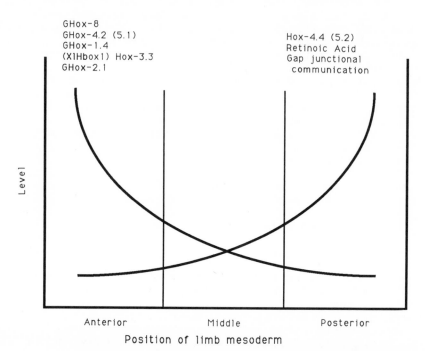

GHox-8
GHox-4.2 (5.1)
GHox-1.4
(XlHbox1) Hox-3.3
GHox-2.1

Hox-4.4 (5.2)
Retinoic Acid
Gap junctional
communication

Level

Anterior   Middle   Posterior

Position of limb mesoderm

Fig. 1. A diagrammatic scheme illustrating the opposing gradients of expression of various molecules across the A-P axis of the chick wing bud at a stage of development during which A-P positional values are being specified. There are several ways in which the opposing gradients might interact in regulating pattern formation. One possibility is that the relatively high concentrations of retinoic acid (RA) in the ZPA-containing posterior region of the limb may promote the extensive gap junctional communication that is occurring in this region, while the lower concentrations of RA in the anterior region of the limb bud may inhibit cell-cell communication (see also Allen et al., 1990). RA may thus generate a gradient of gap junctional communication across the A-P axis that in turn establishes a gradient of an intracellular gap junction-permeable regulatory molecule that specifies positional values. The putative gap junction-permeable regulatory molecule might conceivably specify positional values by regulating the differential pattern of expression of homeobox-containing (HOX) genes along the A-P axis. There are, of course, several other possible scenarios. For example, the graded distribution of a gap junction-permeable regulatory molecule might generate the graded distribution of RA or some other morphogens that regulate HOX gene expression and specify positional values; or, the differential expression of the various HOX genes might regulate the pattern of distribution of RA and/or gap junctional communication. Sorting out these various possibilities will require among other approaches a systematic examination of the effects of the various putative regulatory molecules on one another's expression during limb development.

impaired in their ability to elicit the formation of duplicated skeletal elements when implanted into the anterior margin of limb buds. These observations suggest the possibility that gap junctional communication may be involved in regulating pattern formation.

We have studied gap junctional communication, as assayed by the intercellular transfer of lucifer yellow dye, along the A-P axis of embryonic chick wing buds at an early stage (stage 20/21) of development in which positional values along the A-P axis are being specified (Coelho and Kosher, 1990). We have found that extensive intercellular transfer of lucifer yellow dye via gap junctions occurs among posterior mesenchymal cells of the limb that are adjacent to the ZPA. Considerably less transfer of lucifer yellow occurs among mesenchymal cells in the middle of the limb bud, and little transfer of the dye occurs among mesenchymal cells in the anterior region of the limb bud (Coelho and Kosher, 1990). These observations indicate that there is a gradient of gap junctional communication along the A-P axis of the chick wing bud, with extensive intercellular communication occurring among posterior mesenchymal cells adjacent to the ZPA and progressively less communication occurring among mesenchymal cells in progressively more anterior regions of the limb bud.

The gradient of gap junctional communication along the A-P axis we have observed and its relationship to the ZPA suggests that intercellular signalling via gap junctions may be involved in the specification of pattern along the A-P axis. We suggest that a gradient of gap junctional communication along the A-P axis may generate a graded distribution of a relatively low molecular weight intracellular regulatory molecule involved in specifying A-P positional values (Coelho and Kosher, 1990).

It is of considerable interest that retinoic acid, a putative diffusible morphogen released by the ZPA, is a potent modulator of gap junctional communication in several cell types (Hossain et al., 1989; Mehta et al., 1989). In particular, higher concentrations of retinoic acid enhance gap junctional communication, while lower concentrations inhibit intercellular communication (Hossain et al., 1989; Mehta et al., 1989). We therefore suggest the possibility that the higher concentrations of retinoic acid present in the posterior region of the limb bud (Thaller and Eichele, 1987) may promote the extensive gap junctional communication that is occurring in this region, while the lower concentrations of retinoic acid in the anterior region of the limb bud may be involved in inhibiting cell-cell communication (see also Allen et al., 1990). Retinoic acid may thus generate a gradient of gap junctional communication across the A-P axis that in turn establishes a gradient of a gap junction-permeable regulatory molecule that specifies positional values. Alternately, the graded distribution of a gap junction-permeable regulatory molecule might conceivably generate a graded distribution of retinoic acid along the A-P axis. Similar suggestions concerning the possible relationships between gap junctional communication and retinoic acid in the specification of limb pattern have been discussed by Allen et al. (1990). Finally, it is also possible that differential gap junctional communication may be involved in regulating the differential expression of homeobox genes across the A-P axis of the developing limb. We are currently investigating these and other possibilities.

SUMMARY

The opposing gradients of expression of various molecules across the A-P axis of the developing chick wing bud are diagrammatically illustrated in Fig. 1. It appears likely that interactions between these complementary gradients may be involved in regulating positional values along the A-P axis during embryonic chick limb development.

This research was supported by NIH grants HD22896 and HD 22610.

REFERENCES

Allen, F., Tickle, C., and Warner, A., 1990, The role of gap junctions in patterning of the chick limb bud, Development, 108: 623.

Blake, K. J., Gong, S.-G., Barembaum, M., Paiva-Borduas, J., Kosher, R. A., and Upholt, W. B., 1990, Characterization and expression of chicken homeobox-containing genes during limb development, submitted for publication.

Brickell, P. M. and Tickle, C., 1989, Morphogens in chick limb development, BioEssays, 11: 145.

Coelho, C. N. D. and Kosher, R. A., 1990, A gradient of gap junctional communication along the anterior-posterior axis of the developing chick limb bud, submitted for publication.

Coelho, C. N. D., Sumoy, L., Upholt, W. B., and Kosher, R. A., 1990, Temporal and spatial analysis of the expression of the homeobox-containing gene GHox-8 during embryonic chick limb development, in preparation.

Dolle, P., Izpisua-Belmonte, J.-C., Falkenstein, H., Renucci, A., and Duboule, D., 1989, Coordinate expression of the murine Hox-5 complex homoeobox-containing genes during limb pattern formation, Nature, 342: 767.

Duboule, D., Boncinelli, E., DeRobertis, E., Featherstone, M., Lonai, P., Oliver, G., and Ruddle, F. H., 1990, An update of mouse and human HOX gene nomenclature, Genomics, 7: 458.

Eichele, G., 1989, Retinoids and vertebrate limb pattern formation, Trends in Genetics, 5: 246.

El-Fouly, M. H., Trosko, J. E., and Chang, C.-C., 1987, Scrape-loading and dye transfer. A rapid and simple technique to study gap junctional intercellular communication, Exp. Cell Res., 168: 422-430.

Hill, R. E., Jones, P. F., Rees, A. R., Sime, C. M., Justice, M. J., Copeland, N. G., Jenkins, N. A., Graham, E., and Davidson, D. R., 1989, A family of mouse homeo box-containing genes: molecular structure, chromosomal location, and developmental expression of Hox-7.1, Genes and Dev., 3: 26.

Hossain, M. Z., Wilkens, L. R., Mehta, P. P., Loewenstein, W., and Bertram, J. S., 1989, Enhancement of gap junctional communication by retinoids correlates with their ability to inhibit neoplastic transformation, Carcinogenesis, 10: 1743.

Kelley, R. O. and Fallon, J. F., 1978, Identification and

distribution of gap junctions in the mesoderm of the developing chick limb bud, J. Embryol. Exp. Morph., 46: 99.

Kelley, R. O. and Fallon, J. F., 1983, A freeze-fracture and morphometric analysis of gap junctions of limb bud cells: initial studies on a possible mechanism for morphogenetic signalling during development, in: "Limb Development and Regeneration, Part A," J. F. Fallon and A. I. Caplan, eds., Allen R. Liss, Inc., New York.

Mavilio, F., Simeone, A., Giampaolo, A., Faiella, A., Zappavigna, V., Acampora, D., Poiana, G., Russo, G., Peschle, C., and Boncinelli, E., 1986, Differential and stage-related expression in embryonic tissues of a new human homoeobox gene, Nature, 324: 664.

Mehta, P. P., Bertram, J. S., and Loewenstein, W. R., The actions of retinoids on cellular growth correlate with their actions on gap junctional communication, J. Cell Biol., 108: 1053.

Oliver, G., Wright, C. V. E., Hardwicke, J., and DeRobertis, E. M., 1988, A gradient of homeodomain protein in developing forelimbs of Xenopus and mouse embryos, Cell, 55: 1017.

Oliver, G., Sidell, N., Fiske, W., Heinzmann, C., Mohandas, T., Sparkes, R. S., and DeRobertis, E. M., 1989, Complementary homeo protein gradients in the developing limb, Genes and Dev., 3: 641.

Robert, B., Sassoon, D., Jacq, B., Gehring, W., and Buckingham, M., 1989, Hox-7, a mouse homeobox gene with a novel pattern of expression during embryogenesis, EMBO J., 8: 91.

Smith, S. M., Pang, K., Sundin, O., Wedden, S. E., Thaller, C., and Eichele, G., 1989, Molecular approaches to vertebrate limb morphogenesis, Development, 1989 Supplement: 121.

Sumoy, L., Davidson, D. R., Kosher, R. A., and Upholt, W. B., 1990, Isolation and characterization of GHox-8, a chicken homeobox-containing gene expressed during chick limb development, in preparation.

Thaller, C. and Eichele, G., 1987, Identification and spatial distribution of retinoids in the developing chick limb bud, Nature, 327: 625.

Wedden, S., Pang, K., and Eichele, G., 1989, Expression pattern of homeobox-containing genes during chick embryogenesis, Development, 105: 639.

FUNCTIONAL STUDIES OF GENES IN THE LIMB

Clifford J. Tabin,[1] Bruce Morgan,[1] Hans-Georg Simon,[1] Sara Lazar,[1] Yaoqui Wang,[2] Anuradha Iyer,[1] Julia Yaglom,[1] Changpin Shi,[3] Ken Muneoka[3] and David Sassoon[2]

[1]Department of Genetics, Harvard Medical School, 25 Shattuck Street, Boston, MA 02115

[2]Department of Biochemistry, Boston University School of Medicine, 80 Concord Street, Boston, MA 02118

[3]Department of Cell and Molecular Biology, Tulane University, New Orleans, LA 70118

INTRODUCTION

An increasing number of genes have been implicated by their expression pattern as being important in the developmental control of limb pattern. Such genes, including homeobox-containing genes, genes encoding receptors for potential morphogenic signaling molecules and genes encoding growth factors, are described in detail elsewhere in this volume. A strong case for a role of these genes in controlling limb morphogenesis can be made based on their spatial and temporal expression patterns during normal limb development and in the context of various experimental embryological manipulations. The putative roles of these genes are further supported by analogy to the known function of related genes in other developmental systems.

Nonetheless, the developmental significance of any such gene has not been directly demonstrated. Moreover, while the roles they may play are suggested in general terms, their detailed functions remain to be delineated. To get beyond correlations and to address functional issues, one must have a means of manipulating the genes of interest. We have been developing two approaches to understanding the function of genes in limb morphogenesis: the creation of a limb system which can be manipulated in vitro and the development of retrovirally-mediated gene transfer directly into limb primordia in vivo.

METHODS

In vitro limb cultures are prepared from stage 1 forelimb buds from embryonic day 9.5 mouse embryos (staging according to Wanek et al., 1989). Limb buds are dissected in PBS and dissociated in 0.1% pancreatinase at 4°C. Ectoderms are removed by gentle pipetting. After further digestion, the mesenchyme is dissociated into single cells by triturating with a fine bore pipet. Dissociated mesenchymal cells are plated on glass Lab-TEK slides in medium containing 15% bovine fetal serum (bfs). Cells are allowed to attach (9 hours) and monolayers are treated by adding fresh ectoderm or growth factor (1-100 ng/ml). KP1, a limb mesenchymal cell line (Shi and Muneoka, unpublished), is maintained in medium containing 10% fbs. In situ hybridizations are carried out as previously described (Robert et al., 1989).

Replication competent avian retroviral vector RCAS (Hughes et al., 1987) and replication deficient vector transducing lacZ, RDlac (a gift from A. Stoker) are propogated in vitro, harvested and concentrated as described (Cepko, 1989). Concentrated viral stock (10^7 ffu/ml) is injected into Stage 13 chick embryo presumptive limb mesenchyme (0.05 µl/limb bud). Limb buds are transplanted at Stage 17 as described (Saunders and Reuss, 1974). Viral infection is assayed in cryostat sectioned limb buds either by histological staining for x-gal (Cepko, 1989) or by HRP indirect immunological staining using an antibody, 3C2, directed against the retroviral gag gene product (Potts et al., 1987).

Mice, timed-pregnant CD1 strain (viral free) are obtained from Charles River Lab. Utility-grade and strain 7_2 fertile chicken eggs are obtained from Spafas and are windowed and staged by standard methods (Hamburger and Hamilton, 1951).

RESULTS

In Vitro Hox7 Induction

An in vitro limb system has been developed in which morphologically normal limb buds can be induced to form by placing explanted limb bud ectoderms onto monolayers of dissociated limb mesenchyme cells (Figure 1; Sassoon, unpublished). The development of this in vitro system allows for the study of the cellular and molecular processes underlying early stages of limb morphogenesis. In particular, this system provides a way in which cellular interactions between the ectoderm and the mesenchyme in the initiation of limb bud formation may be studied in vitro. An additional advantage of this approach is our observation that similar in vitro limb buds can be induced when limb ectoderm is placed over monolayers of cells from the KP1 limb mesenchyme cell line. In both cases, limb buds are first observed after 12 hours and reach the equivalent of stage 4 limb bud after 48 hours.

To demonstrate the utility of the in vitro system, we have exploited it to investigate the molecular basis of mesenchymal induction by the apical ectodermal ridge (AER). The role of the ectoderm in the initiation and continued promotion of limb outgrowth is poorly understood. It is clear that the apical ectodermal (AER) is absolutely required for limb outgrowth, however, the molecular basis of any AER derived signal has remained elusive. The AER is thought to affect the subjacent mesenchyme, referred to as the "progress zone," by maintaining progress zone cells in a highly mitotic and positionally labile state (Summerbell et al., 1973). The size of the progress zone is dictated by the AER and as cells leave the influence of the AER at the proximal boundary of the progress zone, their proximal-distal positional value is thought to be fixed. Thus, the identification of any AER derived signal that influences the outgrowth of mesenchymal cells is expected to be important for our understanding of limb patterning mechanisms.

A potential molecular marker for cells in the progress zone has been provided by the isolation of the Hox7 gene (Hill et al., 1989; Robert et al., 1989). Hox7 was initially isolated as a vertebrate homolog of the Drosophila muscle specific homeobox containing the gene Msh. In situ hybridization reveals that Hox7 is specifically expressed in the development of limb mesenchyme at the distal end of the limb bud, just below the AER (Figure 2). When mesenchyme is removed from the influence of the ectoderm, dissociated and plated in vitro, no Hox7 expression is seen. However, when ectoderm explants are placed over the mesenchyme and in vitro limb buds form, Hox7 expression is seen by in situ hybridization only in the appropriate subapical region (Figure 3).

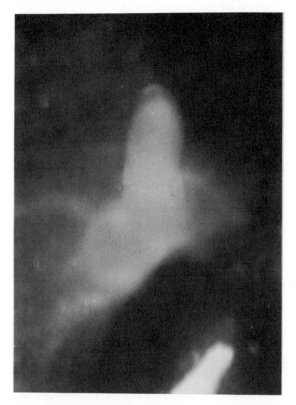

Fig. 1. In vitro limb bud formed by juxtaposing limb ectoderm with dissociated limb mesenchyme.

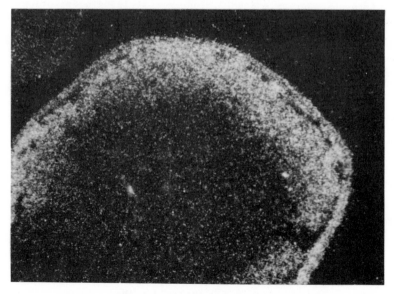

Fig. 2. Expression of Hox7 in the limb bud.

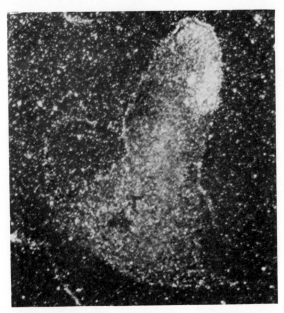

Fig. 3.   Expression of Hox7 in in vitro limbs.

The approach we have taken is to utilize Hox7 expression as an indicator of potential AER derived signals in monolayer cultures of either primary limb bud cells or KP1 cells. Since there is evidence that members of the TGFβ family act as morphogens in early vertebrate development (Smith, 1989) and growth factors related to TGFβ are known to be expressed in the limb bud (Wozney et al., 1988; Lyons et al., 1989), we have initiated a screen of such factors for their ability to cause outgrowth of monolayers of limb mesenchyme and concomitant induction of Hox7 expression. Our preliminary results indicated that TGFβ1, and to a lesser extend BMP2, induce a striking aggregation response analogous to that observed in response to ectoderm explants (Figure 4). We also find that TGFβ1 causes a low level induction of Hox7 expression. Since the Hox7 expression is considerably lower than that induced by limb ectoderm, it may suggest that a related TGFβ-like molecule is the endogenous inducer. The possibility also exists that Hox7 induction results from a combination of inductive signals. In either case, it seems clear that an analysis of Hox7 induction in this in vitro limb system will allow us to rapidly identify potential inducers that can further be tested in vivo.

Retrovirally Mediated Gene Transfer

Gene transfer into limb buds allows the function of cloned genes to be directly tested. We investigated the feasibility of using retroviral vectors for gene transfer into chick limb buds. We used two types of vectors (Figure 5), a replication defective vector, RDlac, carrying the lacZ gene which encodes β-galactosidase, and a replication competent vector RCAS.

RDlac infected cells do not produce virus, thus the RDlac genome is clonally inherited by descendants of infected cells, but no spread occurs to neighboring cells. When limb buds are injected at stage 17 and allowed to develop, histological assay for βgal reveals clones of infected cells in all tissue types of the limb including muscle (Figure 6), ectoderm (Figure 7) and interdigital mesenchyme undergoing programmed cell death (Figure 8).

REPLICATION DEFECTIVE VECTOR

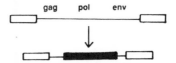

Only spreads when missing viral functions are complemented
Infected cells do not pass virus to neighboring cells
Integrated provirus is clonally inherited

REPLICATION COMPETANT VECTOR

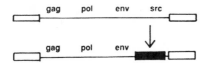

Spreads from infected cells to surrounding mitotic cells

Fig. 4. Limb mesenchymal culture treated with TGFβ.

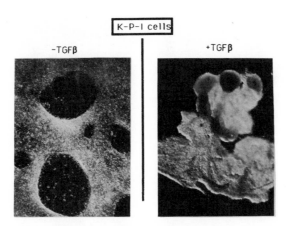

Fig. 5. Retroviral vectors.

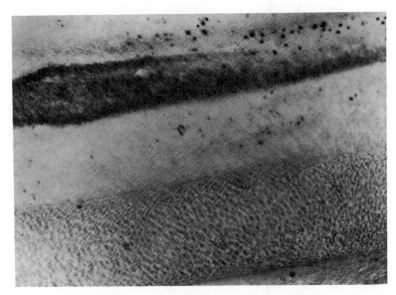

Fig. 6. Clone of muscle cells infected with replication-defective retrovirus transducing lacZ.

Fig. 7. Clone of ectodermal cells infected with replication-defective retrovirus transducing lacz.

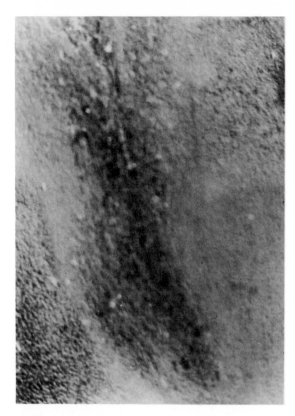

Fig. 8. Clone of interdigital mesenchymal cells infected with
replication-defective retrovirus transducing lacZ.

Fig. 9. Chick with an extra hind limb resulting from transplantation
of early limb bud.

RCAS **Replication Competent Vector** **A/B Tropic**

SPF-11 Outbred Chickens **Infectable by A/B Virus**
7₂ Inbred Chicken Line **Resistant to A,B,E Virus**

Fig. 10. Chick lines resistant and infectable with substrains of retroviruses.

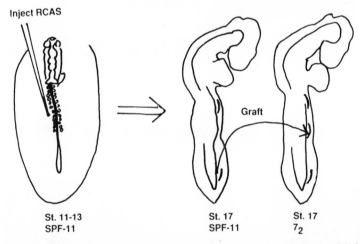

St. 11-13   St. 17   St. 17
SPF-11   SPF-11   7₂

Fig. 11. Experimental protocol for producing transgenic limbs with retroviral vectors.

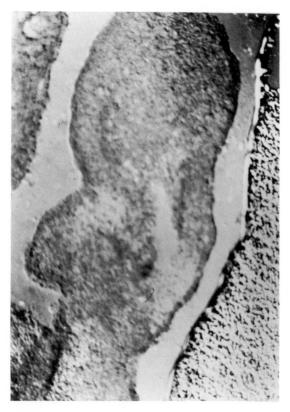

Fig. 12. Transgenic limb produced by infection with replication-
competent retrovirus stained with antisera directed against
retroviral gag determinants.

RCAS is a replication competent vector and thus can spread to all the cells of the limb. However, it can additionally spread to other embryonic tissues. Many genes of interest in the limb have general developmental roles. Thus a misexpression or phenocopy-knockout of these genes may have so many pliotropic effects that their role in limb patterning will be obscured. To overcome this, we are utilizing the classical limb bud transplantation methodology (Figure 9). There are strains of chickens which are highly resistant to infection by certain avian viruses, while other strains are easily infectable (Figure 10). Limb primordia from infectable embryos can be transplanted to the flank of resistant embryos (Figure 11). When chick limb primordia at stage 13 is infected with RCAS, virtually every cell of the limb bud can be seen to be infected using antibodies directed against retroviral gag determinants (Figure 12).

DISCUSSION

The advent of molecular biology has lead to the isolation of a number of genes which are very likely to play important roles in controlling limb morphogenesis. The use of an in vitro system allows

the induction of the expression of these genes to be investigated. Replication-competent retroviruses provide a tool for introducing cloned genes into virtually every cell of the embryonic limb while use of replication-defective vectors allows discrete areas of the limb to be studied. Taken together, these methods hold the promise that the molecular basis for limb pattern formation will be elucidated in future studies.

REFERENCES

Cepko, C., 1979, Lineage analysis and immortalization of neural cells via retroviral vectors, Neuromethods, 16:177.

Hamburger, V., and Hamilton, H.L., 1951, A series of normal stages in the development of the chick embryo, J. Morph., 88:49.

Hill, R.E., et al., 1989, A new family of mouse homeo box-containing genes: molecular structure, chromosomal location, and developmental expression of Hox-7.1, Genes and Devel., 3:26.

Hughes, S.H., et al., 1987, Adaptor plasmids simplify the insertion of foreign DNA into helper-independent retroviral vectors, J. Virol., 61:3004.

Lyons, K.M., et al., 1989, Patterns of expression of murine Vgr-1 and BMP-2a RNA suggest that transforming growth factor-β-like genes coordinately regulate aspects of embryonic development, Genes and Devel., 3:1657.

Potts, W.M., et al., 1987, Epitope mapping of monoclonal antibodies to gag protein p19 of avian sarcoma and leukemia viruses, J. Gen. Virol., 68:3177.

Robert, B. et al., 1989, Hox-7 a mouse homeobox gene with a novel pattern of expression during embryogenesis, EMBO J., 8:91.

Saunders, J.W. and Reuss, C., 1974, Inductive and axial properties of prospective wing-bud mesoderm in the chick embryo, Dev. Biol., 38:41.

Smith, J.C., 1989, Mesoderm induction and mesoderm-inducing factors in early amphibian development. Development 105:665.

Summerbell, D., Lewis, J.H. and Wolpert, L., 1973, Positional information in chick limb morphogenesis. Nature 244:492.

Wanek, N. et al., 1989, A staging system for mouse limb development. J. exp. Zool. 249:41.

Wozney, J.M., et al., 1988, Novel regulators of bone formation: molecular clones and activities, Science, 242:1528.

HOX-7: TEMPORAL PATTERNS OF EXPRESSION IN NORMAL AND *LIMBLESS* CHICK EMBRYOS

G. Lyons*, K. Krabbenhoft°, B.K. Simandl°, M. Buckingham*, J. Fallon°, and B. Robert*

*Pasteur Institute, Paris, France
°University of Wisconsin, Madison, WI, U.S.A.

We have examined by *in situ* hybridization the pattern of expression of the Hox-7 gene in normal and *limbless* (Prahlad et al., 1979) chick embryos. Analysis of the spatial pattern of Hox-7 expression in normal embryos shows that, initially, Hox-7 is expressed in the entire limb field, and, as the limb bud grows out, becomes restricted to the apical ectodermal ridge (AER) and its underlying mesoderm, which has been called the progress zone (Wolpert, 1978). In the chick, this localization occurs by Hamburger-Hamilton stage 21. The AER, which appears shortly after limb bud formation at stage 18, is known to be necessary for normal limb development (Saunders, 1948). *Limbless* mutants do not form an AER and, as a result, exhibit amelia at hatching. However, they do form limb buds at early stages, which regress by stage 23, demonstrating that initial budding is independent of AER activity (Carrington and Fallon, 1988).

In situ hybridization on 5 micron sections of normal and *limbless* embryos was performed using an 35S-labelled antisense cRNA probe which corresponds to the first 100 bp of the conserved 5' half of the homeo box sequence in the mouse Hox-7 gene (Robert et al., 1989; Hill et al., 1989). This analysis reveals that at stages 15-18, both the normal and the limbless embryos express Hox-7 at high levels in newly formed limb buds. At stages 19-20, limb buds in *limbless* begin to necrose and stop expressing Hox-7. Our results demonstrate that the AER is not required for the induction of Hox-7 gene expression at the onset of limb formation, but that it is necessary for the continued expression of Hox-7 in limb buds. The decrease in Hox-7 expression is not associated with necrosis since the mutant limb buds continue to express β-actin mRNAs at high levels at stages 21-23 when Hox-7 transcripts are no longer detectable by *in situ* hybridization. The Hox-7 gene is not directly affected in *limbless* mutants since it is expressed at other locations (e.g. in the neural tube, endocardial cushion, lateral mesoderm, etc.) in these embryos in a temporal pattern identical to that seen in normal embryos. Furthermore, Hox-7 gene expression in *limbless* embryos can be rescued by grafting

Developmental Patterning of the Vertebrate Limb
Edited by J.R. Hinchliffe *et al.*, Plenum Press, New York, 1991

normal ectoderm onto the limb fields of mutants prior to limb bud formation, a manipulation known to restore normal limb development (Carrington and Fallon, 1988).

Grafting experiments show that Hox-7 expression can be reinduced in portions of limb buds which are no longer expressing this gene. *Eudiplopodia* phenocopies (Carrington and Fallon, 1986), in which flank ectoderm is grafted to the dorsal surface of host limb buds, form a supernumerary ectodermal ridge which results in a partial duplication of the limbs. Both outgrowths express Hox-7 in a normal temporal fashion. Grafting a quail AER onto the proximal part of a chick limb bud where Hox-7 is not expressed produces the same partial duplication of limbs, with Hox-7 expression under each AER. These results suggest that there is a diffusible factor transmitted between the AER and the underlying mesenchyme which is responsible for the continued expression of Hox-7.

REFERENCES

Carrington , J., and Fallon, J., 1986. Experimental manipulation leading to induction of dorsal ectodermal ridges on normal limb buds results in a phenocopy of the *Eudiplopodia* chick mutant. Dev. Biol. 116:130-137.

Carrington , J., and Fallon, J., 1988. Initial limb budding is independent of apical ectodermal ridge activity; evidence from a limbless mutant. Development 104:361-367.

Hill, R., Jones, P., Rees, A., Sime, C., Justice, M., Copeland, N., Jenkins, N., Graham, E. and Davidson, D., 1989. A new family of mouse homeo box-containing genes: molecular structure, chromosomal location, and developmental expression of Hox-7.1. Genes Dev. 3:26-37.

Prahlad, K., Skala, G., Jones, D. and Briles, W., 1979. Limbless: A new genetic mutant in the chick. J. Exp. Zool. 209:427-434.

Robert, B., Sassoon, D., Jacq, B., Gehring, W., and Buckingham, M., 1989. Hox-7, a mouse homeobox gene with a novel pattern of expression during embryogenesis. EMBO J. 8:91-100.

Saunders, J., 1948. The proximo-distal sequence of origin of the parts of the chick wing and the role of the ectoderm. J. Exp. Zool. 108:363-404.

Wolpert, L., 1978. Pattern formation in biological development. Scientific American 238:124-136.

G.L. holds an NIH/CNRS fellowship from the Fogarty International Center. This work was supported by grants from the Pasteur Institute, C.N.R.S., I.N.S.E.R.M., A.F.M, N.A.T.O., and A.R.C. to M.B. and by NIH grant P01HD20743 to J.F.

EXPRESSION PROFILE OF TWO HOMEOBOXGENES DURING CHICKEN EMBRYOGENESIS

Berend von Thülen,Reinhold Janocha,
Jürgen Niessing

Institut für Molekularbiologie und
Tumorforschung der Philipps-Universität
Karl-von-Frisch-Str., 3550 Marburg, FRG

INTRODUCTION

Various homeobox genes and zinc finger protein-encoding genes have been isolated from vertebrate organisms ranging from frog to man. In contrast to the most advanced studies on mouse development, very little is known about the structure and expression of developmental control genes during chicken embryogenesis.

For obvious reasons the chicken embryo provides a very advantageous experimental system in which to study morphogenetic events at the molecular level; at all stages of development the embryo is accessible to experimental manipulations which may include grafting experiments, local application of retinoic acid, microinjection and transfection of vectors expressing or suppressing genes involved in the control of developmental events,thereby combining techniques of molecular biology with experimental embryo manipulations.

As a prerequisite for these studies we have cloned and sequenced a number of chicken homologues of Drosophila and mouse homeobox genes, two of which are briefly described in this paper, zinc finger protein - encoding genes as well as conserved sequences of the chicken counterpart to the human retinoic acid receptor gene. In situ hybridization experiments were used to reveal the temporal and spatial distribution of mRNAs transcribed from developmental control genes.

Based on the reference pattern of these gene activities, the aim of our work is to investigate the influence of an experimentally induced disturbed embryonic development on the expression pattern of developmental control genes. These studies are to be complemented by molecular biological experiments directed to influence the activity of these genes in order to determine their functional role in morphogenesis.

RESULTS

CHOX-2

The homeobox of the incomplete cDNA of chox-2 (sequence not shown) exhibits the highest degree of homology to the Hox-2.2 of the mouse (Hart et al., 1987).Using a DNA - fragment 3´ to the homeobox as a template for in situ hybridization of paraffin embedded embryos, a signal can be detected earliest at the caudal end of embryos of 25 hours of incubation in the area of the floor plate of the future neural tube. A more rostral section at the level of the newly formed pair of somites shows the expression to take place in the tissue of the neural fold as well as the splanchnic mesoderm. From this section on, the signal fades along the caudal-rostral axis and disappears at the level of the third pair of somites. By the time of 35 hours of incubation a signal can be detected in the neural tube along the caudal-rostral axis up to the level of the rhombencephalon as well as in the splanchnic mesoderm at the caudal end of the embryo.In embryos of 4 days of incubation, signals can be seen in the neural tube and to a lesser extend in the mesonephros (Fig.1).

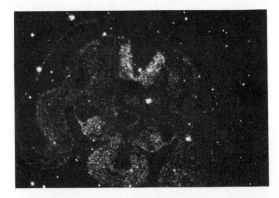

Fig.1   Transvers section of an embryo of 4 days of incubation hybridized with chox-2 antisense RNA. Note the signals in the neural tube and in the mesonephros.

CHOX-3

The homeobox of the incomplete cDNA of chox-3 (sequence not shown) shows the highest degree of homology to the Hox-1.1 sequence of the mouse (Kessler et al, 1987).In situ hybridization experiments show that the first signal is detected at the stage of 25 hours of incubation in the tissue of the segmental plate extending rostrally in the neural fold tissue. As in the case of chox-2, the signal fades and disappears caudal to the level of the newly formed pair of somites. A transverse section at this level reveals the expression in the neural fold to take place at the very dorsal tips of this structure (Data not shown). At the stage of 40 hours of incubation a strong signal is detectable in the caudal half of the neural tube fading rostrally (Fig 2a). By day 4 of incubation the expression domain has extended to the level of the myelencephalon, where a distinct border of expression is seen (Fig 2b + c).

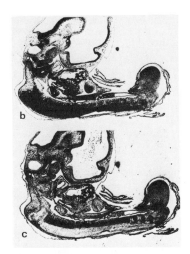

Fig.2   (a)Horizontal section of an embryo of 40 hours of incubation.In situ hybridization was performed with chox-3 antisense RNA. Caudal is left, rostral right. Note the signal in the neural tube, fading from caudal to rostral. (b)Sagital section of an embryo of 4 days of incubation, hybridized with chox-3 antisense RNA. Bright field illumination shows a signal dorsal along the caudal-rostral axis up to the myelencephalon. rostral is left, dorsal is bottom of the illustration. (c)Serial section to b, hybridized with chox-3 sense RNA.

REFERENCE

Hart,C.P.; Fainsod,A.; Ruddle,F.H.;1987; Sequence Analysis of the Murine Hox-2.2, -2.3, and -2.4 Homeo Boxes: Evolutionary and Structural Comparisons; Genomics 1 182-195
Kessler, M.; Schulze, F.; Fibi,M.; Gruss,P.;1987; Primary Structure and nuclear localization of a murine homeodomain protein;Proc.Natl.Acad.Sci.USA 84; 5306-5310

CHARACTERIZATION OF THREE ZEBRAFISH GENES RELATED TO HOX-7

Marie-Andrée Akimenko, Marc Ekker and
Monte Westerfield

Institute of Neuroscience
University of Oregon
Eugene OR 97403, USA

Transcriptional regulators are thought to be involved in
the mechanisms that generate body organization during
vertebrate embryogenesis. Homeobox-containing genes (Hox genes)
may serve this function since they are expressed in specific
regions of the vertebrate embryo. A mouse Hox gene, *Hox-7*, was
recently isolated that shows a unique pattern of expression
compared to the previously described vertebrate Hox genes[1,2].
Hox-7 is expressed in the cephalic neural crest and its
mesectodermal derivatives in the branchial arch region. *Hox-7*
transcripts are also detected in the distal region of mouse
limb buds and, later, in the mesenchymal interdigital
region[1,2].

The zebrafish (*Brachydanio Rerio*) is a simple vertebrate
with many features that facilitate functional studies of Hox
genes. Embryos are easily accessible at all stages of
development and can be produced in large numbers. Their
development is rapid; embryos hatch at two days, and most
morphological features of the adult are visible at seven days.
Most importantly, genetic methods developed to obtain haploid
and gynogenetic offspring[3] and lines of transgenic zebrafish[4]
provide additional advantages for using zebrafish to study
vertebrate development.

Southern analysis using the homeobox region of the mouse
Hox-7 gene revealed that there are at least two other genes in
the mouse genome related to *Hox-7*[1,2] (Fig. 1). Using the same
probe for Southern analysis of zebrafish DNA, we found evidence
for the presence of three genes with sequence similarities
(Fig. 1). We isolated these genes from a cDNA library prepared
from embryos and from a genomic library by hybridization with
the homeobox region of the mouse *Hox-7* gene. The coding regions
of the three genes have been sequenced except for the first
exon of *Hox[zf-71]*.

The homeodomains encoded by the three zebrafish genes are
very similar to the mouse *Hox-7* homeodomain (Fig. 2). The first
gene, *Hox[zf-71]*, encodes a homeodomain identical to that of

Hox-7 . The other two, *Hox[zf-72]* and *Hox[zf-73]*, respectively share 55 and 58, out of 60, amino acid residues of their homeodomains with *Hox[zf-71]*. Sequence comparisons showed that, outside the homeodomain, there is little sequence similarity between the three genes except for the amino acid residues immediately adjacent to the homeodomain and some short motifs such as the tetrapeptide GYSM, shared by *Hox[zf-71]*, *Hox[zf-73]* and *Hox-7*, near the carboxy-terminal end of the protein. A stretch of 26 amino acid residues at the amino-terminal end of the *Hox[zf-73]* protein has a high degree of similarity with the corresponding regions of *Hox-7* and its recently isolated quail homologue (*Quox-7*)[5] (data not shown).

From sequence comparison, it is impossible to determine which of the three zebrafish genes is the homologue of the mouse *Hox-7* gene; sequences outside the homeodomain are too divergent and it is unclear that the few amino acid differences in the homeodomains are important for protein function. The low level of similarity in the regions outside the homeobox enabled us to isolate specific probes for each of the three zebrafish genes. Southern analysis of zebrafish DNA with these probes suggested that the 12.5, 9.4, and the 8.5 kb *Eco*RI fragments detected with the mouse *Hox-7* probe (Fig. 1) contain the homeoboxes of the *Hox[zf-7.2]*, *Hox[zf-7.3]* and *Hox[zf-7.1]* genes, respectively (data not shown).

The patterns of expression of each of the three zebrafish *Hox[zf-7]* genes, which are presently under investigation, should shed light on the role that these genes play in the formation

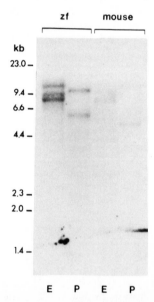

Fig. 1. Southern analysis of zebrafish and mouse DNA using the homeobox of the mouse *Hox-7* gene. A 200 base pair *Sph*I-*Pst*I fragment was ^{32}P-labelled using the random priming procedure. This probe was hybridized, at 65°C, to *Eco*RI (E) and *Pst*I (P) digestions of zebrafish (zf, left) and mouse (right) DNA. Washes were performed in 2x SSC; 0.1% SDS at 65°C. Lambda markers are indicated on the left.

```
         0                                                      49
Hox-7  LRKHKTNRKP RTPFTTAQLL ALERKFRQKQ YLSIAERAEF SSSLSLTETQ  Hox-7
zf-71  ---------- ---------- ---------- ---------- ----------  zf-71
zf-72  ---------- ----S-S--- S--------- ---------- -N--N-----  zf-72
zf-73  ---------- ------S--- ---------- ---------- ----T-----  zf-73


         50                             73
Hox-7  VKIWFQNRRA KAKRLQEAEL EKLK                              Hox-7
zf-71  ---------- ---------- -RF-                              zf-71
zf-72  ---------- ---------- --F-                              zf-72
zf-73  ---------- ---------- ----                              zf-73
```

Fig. 2. Comparisons of the homeodomains of the mouse *Hox-7*
and the three related proteins in zebrafish. The
amino acid sequences deduced from the nucleotide
sequence are shown using the one-letter code. The
zebrafish sequences have been aligned under the
mouse sequence; (-) corresponds to an identity
with the mouse sequence while changes are
indicated by letters. The homeodomain is indicated
by the brackets.

of patterns in the zebrafish embryo. Moreover, if one or more
of these genes is expressed in the developing fins, comparisons
with their putative homologues in tetrapods is likely to
provide information on the evolution of limb development.

ACKNOWLEDGEMENTS

We thank B. Robert and M. Buckingham for the *Hox-7* probe
and A. Molven and A. Fjose for the embryonic cDNA and genomic
libraries. M.-A.A. was supported by the Ministère de la
Recherche et de la Technologie (France). M.E. is a "Centennial"
fellow of the Medical Research Council of Canada. Supported by
grants from the NIH HD22486 and an R.C.D.A. to M.W.

REFERENCES

1. R.E. Hill, P.F. Jones, A.R. Rees, C.M. Sime, M.J.
   Justice, N.J. Copeland, N.A. Jenkins, E. Graham
   and D.R. Davidson, A new family of mouse homeo
   box-containing genes: molecular structure,
   chromosomal location, and developmental expression
   of *Hox-71*, Genes Devel.3:26 (1989).
2. B. Robert, D. Sassoon, B. Jacq, W. Gehring and M.
   Buckingham, *Hox-7*, a mouse homeobox gene with a
   novel pattern of expression during embryogenesis,
   EMBO J. 8:91 (1989).
3. G. Streisinger, C. Walker, N. Dower, D. Knauber and
   F. Singer, Production of clones of homozygous
   diploid zebra fish, Nature 291:293 (1981).
4. G.W. Stuart, J.R. Vielkind, J.V. McMurray and M.
   Westerfield, Stable lines of transgenic zebrafish
   exhibit reproducible patterns of transgene
   expression, Development 109:577 (1990).
5. Y. Takahashi and N. Le Douarin, cDNA cloning of a
   quail homeobox gene and its expression in neural
   crest-derived mesenchyme and lateral plate
   mesoderm, Proc. Natl. Acad. Sci. USA 87:7482
   (1990).

A COMPARISON OF THE EXPRESSION DOMAINS OF THE MURINE HOX-4,

RARs AND CRABP GENES SUGGESTS POSSIBLE FUNCTIONAL

RELATIONSHIPS DURING PATTERNING OF THE VERTEBRATE LIMB [*]

Pascal Dollé[1], Esther Ruberte[2], Juan-Carlos Izpisùa-Belmonte[1],
Hildegard Falkenstein[1], Pierre Chambon[2] and Denis Duboule[1]

[1]EMBL, Meyerhofstr. 1, D-6900 Heidelberg, F.R.G. and [2]LGME du
CNRS, unité INSERM 184, Faculté de Médecine. 11 rue Humann, 67000
Strasbourg, FRANCE

INTRODUCTION

The murine genome contains at least 30 sequences related to the Drosophila
Antennapedia homeobox. The so-called Hox genes are clustered in four complexes
conserved throughout vertebrate evolution[1-5], and accumulating evidence shows that these
genes are expressed during ontogenesis in a coordinated manner[3,6-8] and may be involved
in a regulatory network controlling vertebrate morphogenesis. Retinoic acid (RA) or
related retinoid derivatives are candidates for crucial signalling molecules involved in
vertebrate morphogenetic processes and/or pattern formation as best exemplified by the
developing and regenerating limb system (reviewed in refs. 9,10). RA can interfere very
specifically with anteroposterior (A-P) patterning in the chick wing bud[11] and is believed
to be a natural morphogen released as a concentration gradient from a discrete posterior
area, the zone of polarizing activity (ZPA)[12]. RA can also disturb positional information in
the regenerating amphibian limb[13,14]. The molecular basis of RA activity involves its
binding to a cellular RA binding protein (CRABP) [15] and/or to an appropriate nuclear RA
receptors (RARs). An increasing number of such RA receptors are being characterized and
all of them are ligand-inducible transcription factors belonging to the steroid hormones
receptors family[16-19]. There is increasing evidence that in cultured cells, RA regulates the
steady state level of Hox genes messenger RNAs though the transcriptional or post-
transcriptional nature of this regulation is not yet clearly established (see e.g. ref.20). In
addition, the human Hox-2 genes respond to RA treatment, in teratocarcinoma cell lines, in
a way related to their respective positions within the Hox-2 complex[21]. The molecular
mechanism involved in such a colinear response is not known but probably parallels those
regulating the expression of these genes during ontogenesis. Using the vertebrate limb
development as a model system, we would like to discuss here some circumstantial
evidence that RA, RARs and Hox genes may also be functionally related in vivo, in the
light of recent findings concerning the expression of such genes during mouse
development.

---

[*] Copyright for this chapter is retained by the author.

COORDINATE EXPRESSION OF THE MURINE HOX-4 GENES DURING LIMB DEVELOPMENT

A systematic study of the expression of 5 genes of the HOX-4 (previously named HOX-5) complex, Hox-4.2, -4.4, -4.5, -4.6 and -4.7, has revealed distinct spatially restricted transcript domains in various stages of limb development[22]. The expression of all these genes is cell-lineage specific, since restricted to the mesenchymal cells of the limb buds, and each gene has a distinct proximo-distal (P-D) expression domain within limb mesenchyme. The most 3'-located gene, Hox-4.2, is widely expressed in the limbs and in adjacent axial tissues, whereas other Hox-4 genes display distinct proximal boundaries of expression in the limbs, such that the expression domain of a given gene is always more distally restricted than that of its 3' neighbor. Such defined P-D expression domains are persistent from early stages of limb development up to later stages after cytodifferentiation (Fig.1). These distinct and very specific expression domains could be established through a sequential activation of the Hox-4 genes, progressing 3' to 5', within an interval of

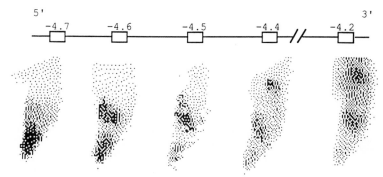

Fig. 1. Spatially restricted expression domains of the Hox-4 genes along the proximo-distal axis of the developing limb. Adjacent sections crossing longitudinaly the whole forelimb of a 12.5 day old fetus were hybridized to antisense RNA probes specific for the genes indicated above. The resulting signal was quantified by computer grain counting. Increasing darkness indicates increasing amount of a given transcript. 3´-located genes (e.g.Hox-4.4) are preferentially expressed in proximal areas whereas more 5´-located genes are more distally expressed.The orientation is: top:proximal; bottom:distal.

several hours during early limb bud outgrowth. Time-delayed appearance of Hox-4 transcripts has been demonstrated for two 5' located genes, Hox-4.6 and Hox-4.7, whose transcripts are initially detected in the same discrete area of the limb buds but at a few hours interval during day 10 of gestation (Fig. 2A,B). These transcripts appear in the most posterior and dorsal cells of the limb buds and remain preferentially located in posterior cells during later stages of development. Hence the expression boundary of a given gene is not orthogonal to the P-D axis of the limbs but extends proximally in posterior areas (Fig. 2B and C). The expression boundaries of the Hox-4 genes can be precisely defined, with a fall of transcripts abundance over a distance of a few cells but we cannot conclude to a regular gradient of transcripts across labelled areas. However, Oliver et al. have reported that the mouse Hox-4.4 protein is indeed distributed as a gradient across the limb bud with maximal amounts in posterior regions[23].

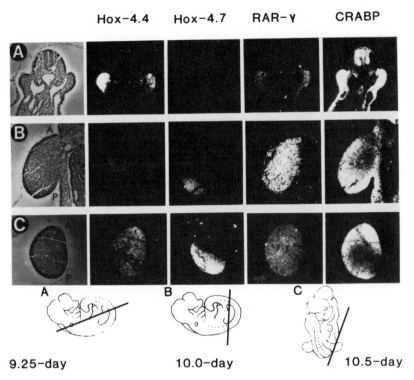

Fig. 2. Comparison of the transcript domains of the Hox-4.4, Hox-4.7, RAR- and CRABP genes in the developing limb buds. Three developmental stages are shown (A,B,C) and the plans of sectionning are diagrammed below. Plan B roughly corresponds to the AP axis of the forelimb whereas plan C is almost orthogonal to the major axis of the limb. In the earliest stage shown (A), Hox-4.7 transcripts are not detected in the forelimb buds. A: Anterior; P: Posterior.

Thus, the expression domain of any Hox-4 gene in the developing limb buds is fully overlapping with those of its 3' neighbor genes and the area where all these domains overlap is located along the postero-distal side of the limb and therefore coincides with the signalling region thought to be involved in patterning the limb AP axis. This region, the ZPA, is present in a number of higher vertebrate species and can alter the A-P pattern when grafted ectopically in chick wing buds[24]. RA (or a dehydrogenated metabolite) is present in higher amounts within the ZPA[12,25] and is able, when locally applied in the anterior part of developing wing buds, to mimic AP alterations induced by ZPA grafting[11]. These two observations have led to the proposal that RA may be a morphogen involved in limb patterning and released from the ZPA. It is therefore an attractive speculation to imagine that a concentration gradient of RA could be translated into more complex and informative Hox protein combinations. Limb patterning could thus be achieved by three or four "molecular steps"; the presence of a gradient of morphogen; the reading and differential translation of this gradient by Hox genes; the differential regulation of target genes (e.g.cytodifferentiation genes) by various Hox protein combinations.

CO-EXPRESSION OF HOX AND RAR GENES DURING LIMB DEVELOPMENT

The RA morphogenetic signalling mechanism is expected to be mediated through the binding of RA to one or several types of RARs, allowing the activated receptor-ligand complex to regulate the transcription of target genes. Thus, a prerequisite for RA-mediated activation of Hox genes is the simultaneous presence of RARs and Hox transcripts in limb bud cells. This has been confirmed by investigating the distribution of transcripts encoded by three RAR genes (RAR-α, −β and -γ) in the mouse fetus[26]. Two types of RAR transcripts, RAR-α and -γ are detected in limb bud cells at the time of pattern formation (gestational day 10-11). Interestingly, RAR-β transcripts are restricted to very proximal and dorsal cells of the limb buds, suggesting that differential expression of RAR genes may account for a distinct response to RA according to cell position. The presence of RAR-α transcripts in the limb buds may be due to the ubiquitous expression of this gene in various embryonic tissues[27]. In contrast, RAR-γ gene expression is regulated in a cell-lineage and tissue specific fashion[26–28]. In day 8-10 embryos, RAR-γ transcripts are specific to some mesenchymal populations, mainly in the head and limb buds. Earlier, RAR-γ is specifically co-expressed with Hox-4.4 in the lateral mesoderm (somatopleure) which includes the presumptive limb fields prior to limb bud outgrowth (day 8-9). In later stages, RAR-γ expression, like that of the Hox-4 genes, is restricted to the limb bud mesenchyme and is not detected in the ectodermal layer or the apical ectodermal ridge. Furthermore, RAR-γ transcripts are homogeneously distributed in the whole limb bud (Fig. 2A-C). Hence, unless some position-dependent translational regulation occurs, RAR-α and RAR-γ would be evenly distributed in the limb mesenchyme. These receptors may thus be activated as a direct function of local RA concentrations and "read" a RA gradient.

The physiological activity of RA can also be modulated by the distribution of CRABP, a cytosolic protein which binds RA with high affinity[15]. CRABP is present at very high concentration in the chick limb bud[29]. However, CRABP transcripts do not appear to be evenly distributed since a distal to proximal gradient across the mesenchyme of the mouse limb bud with maximal amounts in the distal tip and a progressive fall of abundance towards proximal and central regions (Fig.2B,C) was reported[26]. Thus, the localization of CRABP transcripts in limbs is partially overlapping with those of Hox-4 genes and the area of maximal overlap (and probable initial appearance) of Hox-4 transcripts is a region containing the highest amount of CRABP transcripts (Fig. 2). The mesoderm component of the distal tip of the limb bud, or "progress zone", is the area where proliferating undifferentiated mesenchymal cells seem to have the competence to respond to positional information, i.e. the area where the morphogen is expected to be effective[30]. The distribution of CRABP transcripts raises the intriguing possibility that intra-cellular amounts of CRABP could dictate the balance between cell proliferation (progress zone-high amounts) and cell differentiation (low amounts). In addition, a slight A-P gradient of CRABP along the progress zone was reported in the chick wing bud[31] and may help steepen an opposite RA gradient.

Additional similarities in the expression domains of the Hox-4 and RAR-γ genes are seen during later stages of development. Thus, their expression appears to follow two temporal phases. After an initial phase of homogeneous expression of the RAR-γ gene in the undifferentiated mesenchyme, RAR-γ transcripts become specifically restricted to precartilaginous blastema and cartilage forming cells (day 10 to 12 - Fig. 3A) so that the transcripts draw a "preskeletal" pattern. Similarly, Hox-4 transcripts, after a rather widespread distribution, become restricted to some central condensations too. However, instead of being systematically associated with chondrifying areas, Hox-4 transcripts are found only around the extremities of bone models and in joint regions, i.e. in areas of preferential growth of bone models (Fig. 3A, B). Hox-4 and RAR-γ late expression domains are thus only partially overlapping since Hox-4 transcripts appear to mark a limited time interval in the differentiation of mesenchymal cells into chondroblasts. Simultaneously, CRABP transcripts become restricted to areas exclusive for RAR-γ transcripts, with a minimal overlap around chondrifying bone models (Fig. 3A). A similar relationship between RAR-γ, CRABP and Hox gene expression can be extrapolated to the axial skeleton, namely the developing ribs and vertebral column, where other members of various Hox complexes are expressed (see e.g. ref.6). For example, in the sclerotome-derived prevertebrae, the Hox-1.4 homeogene is expressed preferentially in the ventral and lateral regions of the prevertebrae, whereas RAR-γ transcripts, more limited to the dorsal chondrifying centers, are closely exclusive from those of CRABP (Fig. 3C).

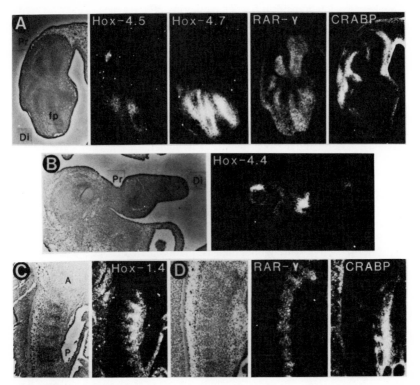

Fig. 3. Comparison of the expression domains of some Hox genes with those of the RAR- and CRABP genes. A: Section through the hindfoot plate of a 12.5 day-old fetus. B: Section through the proximo-distal axis of a 12.5 day old forelimb. The Hox-4.4 homeogene is specifically expressed in peripheral cells of the humeral condensation (left) and the radius blastema. C-D: Sagittal sections through the cervical prevertebrae of a 12.5 day-old fetus.

The two-step expression patterns of the Hox-4, RAR-γ and CRABP genes in developing limb mesenchyme may be related to the dual effects of RA on this tissue both in vitro and in vivo. In micromass cultures, a different responsiveness to RA is observed between distal and proximal limb mesenchymal cells. At low concentrations, RA has a growth-promoting effect and induces chondrogenesis only on distal mesenchyme of early limb buds (stage 20-24). Proximal mesenchyme and cells from late developmental stages are not responsive whereas higher RA concentrations are even inhibitory for chondrogenesis and myogenesis[32,33]. In stage 23-24 embryos, RA treatment can lead to the typical AP mirror-image duplication of the chicken wing but only when applied to distal (undifferentiated) regions, whereas proximal application produces local deletions of structures[34]. This latter teratological effect is similar to the effect of systemic administration of RA in mouse embryos[35] which produces morphological anomalies such as shortening of skeletal elements with deletion of bone extremities and fusion of some joints.

A UNIFORMITY OF PATTERN FORMATION MECHANISMS ALONG EMBRYONIC AXES ?

Hox genes are expressed in a set of spatially restricted and partially overlapping domains along the A-P axis of the mouse embryo. The organization of these expression domains is subject to the rule of colinearity which says that the spatial sequence of the developmental expression domains reflects the ordering of the genes in their respective complexes (3´-anterior; 5´-posterior)[3,6-8]. This concept, originally proposed by Ed Lewis studying the genetics of Drosophila homeotic genes from the Bithorax complex (BX-C),[36] is in vertebrates, not restricted to the trunk axis. The coordinate expression of the Hox-4 genes in developing limbs demonstrates that it also stands true for the limb axis, which is formed later and is distinct from the A-P body axis. Consequently, the same set of regulatory proteins may provide positional cues along independent axes. The appearance of Hox-4 transcripts is an early event during limb growth. Similarly, Hox transcripts are generally earliest detected during the gastrulation phase, at the time when the A-P axis of the embryo is determined, and in domains which are already restricted spatially along the undifferentiated ectoderm and mesoderm[e.g.37]. Altogether, this suggests that a common mechanism is used to set up the coordinate expression of the Hox genes along the limbs and the body axis.

Some experimental evidence supports an involvement of RA in A-P pattern specification. RA treatment of Xenopus embryos results in abnormal A-P differentiation in the central nervous system and instructs anterior neural tissue with a posterior specification[38]. In the chick embryo, Hensen's node, the anterior extremity of the primitive streak, has polarizing activity throughout the gastrulation period (stage 4 to 10) and later, some activity is still found posterior to the node[39]. Thus, an area related to A-P axis specification has the same polarizing activity as the limb ZPA. Grafting of these regions in the anterior margin of the wing bud give rise to comparable patterns of digit duplications. In later stages of chick embryogenesis (stage 16-17) polarizing activity is characteristic of two midline structures, the notochord and the ventral-most region of the neural tube or floor plate, at the time of regional differentiation in the neuroectoderm. The latter area has been shown to synthesize morphogenetically active retinoids[40]. Hox genes are widely expressed in the neural tube throughout the period of neurodifferentiation, and are expressed differentially along the A-P and dorso-ventral (D-V) axis of the neural tube e.g.[41]. The neural tube is also a region of specific expression of RARs and CRABP (E. R and P. C; unpublished results). Although direct proof of a D-V retinoid gradient along the neural tube is missing, it is intriguing that polarizing activity in the neural tube does not extend anterior to the floor plate, which terminates in the midbrain. Similarly, the expression boundaries of the most "anterior" Hox genes do not extend more anteriorly than the hindbrain-midbrain junction[3,42].

Although a polarizing activity does not indict Hensen's node as a source of retinoids, it suggests that a gradient of signalling molecule could be released by the node region throughout gastrulation, concomitantly to the lateral and anterior migration of newly formed mesodermal cells. There is no direct proof that Hensen's node is the local source of retinoids but it is worth mentioning that RAR-γ transcripts are specifically expressed during such early developmental stages, in posterior areas of the embryo where the neural plate has not yet closed. In these areas, RAR-γ expression is ubiquitous in all germ layers and furthermore, becomes progressively restricted in an A-P direction roughly following somite appearance and neural tube closure[28]. Hence, as in the limb buds, the dynamic pattern of RAR-γ expression in areas which have recently gastrulated makes this receptor a possible candidate to mediate a putative RA signal. Thus, patterning along the limb and trunk axes could be achieved by similar mechanisms involving retinoids, Hox genes and further "downstream" cytodifferentiation genes. The logic of this process would be to translate a rather primitive signal (e.g. a monomolecular gradient of retinoid) into a more complicated and informative network of multiple homeoproteins allowing a fine tuning in the determination of the patterns. The particular structural organization of the HOX complexes allows, in principle, autoregulatory mechanisms which could prevent all the Hox genes from responding at the same time to the same dose of the morphogen. For example, the presence of "proximal" homeoproteins could be required, in the developing limb, for a correct response of the more "distal" genes to the morphogen thus giving rise to the partially overlapping expression domains observed both in the trunk and limbs (discussed in ref.22).

There is presently only circumstantial evidence that active retinoids could direct pattern formation via homeogene products. Undoubtedly, this attractive functional relationship will be clarified in the near future through both biochemical and biological approaches. The combination of in vitro studies addressing the potential regulation of Hox gene expression by RARs with in vivo work using organisms such as the chick embryo which are easily amenable to experimental manipulation should help to better understand the functions of homeogenes and retinoids during morphogenesis.

ACKNOWLEDGEMENTS

We thank C. Ovitt for her helpful suggestions.

REFERENCES

1. A. M. Colberg-Poley, S. D. Voss, K. Chowdhury, C. L. Stewart, E. F. Wagner, and P. Gruss, Clustered homeo boxes are differentially expressed during murine development, Cell 43:39 (1987).
2. C. P. Hart, A. Fainsod, and F. H. Ruddle, Sequence analysis of the murine Hox-2.2, -2.3, and -2.4 homeo-boxes: evolutionary and structural comparisons, Genomics 1:182 (1989).
3. D. Duboule and P. Dollé, The structural and functional organization of the murine HOX gene family resembles that of Drosophila homeotic genes, EMBO J. 8:1497 (1989).
4. K. Schughart, C. Kappen, and F. H. Ruddle, Mammalian homeobox-containing genes: Genome organization, structure, expression and evolution, Br. J. Cancer 58:9 (1988).
5. K. Schughart, C. Kappen and F. H. Ruddle, Duplication of large genomic regions during the evolution of vertebrate homeobox genes, Proc. Natl. Acad. Sci. USA 86:7067 (1989).
6. S. J. Gaunt, P.T. Sharpe and D. Duboule, Spatially restricted domains of homeo-gene transcripts in mouse embryos: relation to a segmented body plan, Development 104:71 (1988).
7. A. Graham, N. Papalopulu, J. Lorimer, J. -H. McVey, E. G. J. Tuddenham and R. Krumlauf, The murine and drosophila homeobox gene complexes have common features of organization and expression, Cell 57:367 (1989).
8. M. Akam, Hox and HOM, homologous gene clusters in insects and vertebrates, Cell 57:347 (1989).
9. J. P. Brockes, Retinoids, homeobox genes and limb morphogenesis, Neuron 2:1285 (1989).
10. G. Eichele, Retinoids and vertebrate limb pattern formation, TIG 5:246 (1989).
11. C. Tickle, B. Alberts, L. Wolpert, and J. Lee, Local application of retinoic acid to the limb bud mimics the action of the polarizing region, Nature 296:564 (1982).
12. C. Thaller, and G. Eichele, Identification and spatial distribution of retinoids in the developing chick limb bud, Nature 327: 625 (1987).
13. M. Maden, Vitamin A and pattern formation in the regenerating limb, Nature 295:672 (1982).
14. S. D. Thoms, and D. L. Stocum, Retinoic acid-induced pattern duplication in regenerating urodele limbs, Dev. Biol. 103:319 (1984).
15. D. E. Ong, and G. Chytil, Cellular retinoic acid-binding protein from rat testis, J. Biol. Chem. 253:4551 (1978b).
16. M. Pektovich, N. J. Brand, A. Krust and P. Chambon, A retinoic acid receptor which belongs to the family of nuclear receptors, Nature 330, 444-450 (1987).
17. V. Giguère, S. E.Ong, P. Segui, and R. M. Evans, Identification of a receptor for the morphogen retinoic acid, Nature 330:624 (1987).
18. A. Zelent, A. Krust, M. Pektovich, P.Kastner, and P.Chambon, Cloning of murine a and b retinoic acid receptors and a novel receptor g predominantly expressed in skin, Nature 339:714 (1989).

19. R. M. Evans, The steroid and thyroid hormone receptor superfamily, Science 240:889 (1988).
20. F. Mavilio, A. Simeone, E. Boncinelli, P.W. Andrews, Activation of four homeobox gene clusters in human embryonal carcinoma cells induced to differentiate by retinoic acid, Differentiation 37:73 (1988).
21. A. Simeone, D. Acamora, L. Arcioni, P. W. Andrews, E. Boncinelli, and F. Mavilio, Human Hox2 Homeobox genes are sequentially activated by retinoic acid in embryonal carcinoma cells, Nature In Press (1990).
22. P. Dollé, J.-C. Izpisuá-Belmonte, H. Falkenstein, A. Renucci, and D. Duboule, Coordinate expression of the murine Hox-5 complex homeobox-containing genes during limb pattern formation, Nature, 342:767 (1989).
23. G. Oliver, N. Sidell, W. Fiske, C. Heinjmann, T. Mohandas, R. S. Sparkes, and E. M. DeRobertis, Complementary homeoprotein gradients in developing limb buds, Genes and Dev., 3:641 (1989).
24. C. Tickle, G.Shilswell, A. Crawley, and L. Wolpert, Positional signalling by mouse limb polarizing region in the chick wing bud, Nature 259:396 (1976).
25. C. Thaller, and G. Eichele, Isolation of 3,4-didehydroretinoic acid, a novel morphogenetic signal in the chick wing bud, Nature 345:815 (1990)
26. P. Dollé, E. Ruberte, P.Kastner, M. Petkovich, C. M. Stoner, L.Gudas, and P. Chambon, Differential expression of genes encoding a, b and g retinoic acid receptors and CRABP in the developing limbs of the mouse, Nature 342:702 (1989).
27. P. Dolle, E. Ruberte, P. Leroy, G. M. Morris-Kay, and P.Chambon, Retinoic acid receptors and cellular retinoid binding proteins I: a systematic study of their differential pattern of transcription during mouse organogenesis. Development In Press (1990).
28. E. Ruberte, P. Dollé, A. Krust, A. Zelent, G. Morriss-Kay, and P. Chambon, Specific spatial and temporal distribution of retinoic acid receptor gamma transcripts during mouse embryogenesis, Development 108:213 (1990).
29. M. Maden, and D. Summerbell, Retinoic acid-binding protein in the chick limb bud: identification at developmental stages and binding affinities of various retinoids, J. Embryol. exp. Morph. 97: 239 (1986).
30. D. Summerbell, J. H. Lewis, and L. Wolpert, Positional information in chick limb morphogenesis, Nature 244:492 (1973).
31. M. Maden, D. E. Ong, D. Summerbell, and F. Chytil, Spatial distribution of cellular protein-binding to retinoic acid in the chick limb bud, Nature 335:733 (1988).
32. H. Ide, and H. Aono, Retinoic acid promotes proliferation and chondrogenesis in the distal mesodermal cells of chick limb bud, Developmental Biol. 130:767 (1988).
33. D. F. Paulsen, R. M.Langille, V. Dress, and M. Solursh, Selective stimulation of in vitro limb-bud chondrogenesis by retinoic acid, Differentiation 39:123 (1988).
34. C. Tickle, and A. Crawley, The effect of local application of retinoids to different positions along the proximo-distal axis of embryonic chick wings, Roux´Árch.Dev.Biol. 197:27 (1988).
35. D. M. Kochhar, J. D. Penner, and C. Tellone, Comparative teratogenic activities of two retinoids: Effects on palate and limb development. Teratg. Carcinog. Mutagen. 4:377 (1984).
36. E. B. Lewis, A gene complex controlling segmentation in Drosophila, Nature 276: 556 (1978).
37. S. J. Gaunt, Homeo-box gene Hox-1.5 expression in mouse embryos: earliest detection by in situ hybridization is during gastrulation, Development 101:51 (1987).
38. A. J. Durston, J. P. M. Timmermans, W. J. Hage, H. F. J. Hendriks, N. J. De Vires, M. Heideveld, and P. D.Nieuwkoop, Retinoic acid causes an anteposterior transformation in the developing central nervous system, Nature 340:140 (1989).
39. A. Hornbuch, and L. Wolpert, L., Positional signalling by Hensen's node when grafted to the chick limb bud, J. Embryol. exp. Morph 94:257 (1986).
40. M. Wagner, C. Thaller, T. Jessell, and G. Eichele, Polarizing activity and retinoid synthesis in the floor plate of the neural tube, Nature, 345:819 (1990).

41. S. J. Gaunt, P. L. Coletta, D. Pravtcheva, and P. T. Sharpe, Mouse Hox-3.4: homeobox sequence and embryonic expression patterns compared with other members of the Hox gene network, <u>Development</u> 109:329 (1990).
42. D. G. Wilkinson, S. Bhatt, M. Cook, E. Boncinelli, and R. Krumlauf, Segmental expression of Hox-2 homeobox-containing genes in the developing mouse hindbrain, <u>Nature</u> 341:405 (1989).

RETINOIC ACID NUCLEAR RECEPTORS

P. Kastner, N. Brand, A. Krust, P. Leroy, C. Mendelsohn, M.
Petkovich, A. Zelent and P. Chambon

Laboratoire de Génétique Moléculaire des Eucaryotes du CNRS
Unité 184 de Biologie Moléculaire et de Génie Génétique de
l'INSERM, Institut de Chimie Biologique, Faculté de
Médecine, 11, Rue Humann, 67085 STRASBOURG Cédex, France

INTRODUCTION

The pleiotropic effects that RA exerts during vertebrate
development have been studied in a number of experimental systems. For
example, retinoic acid (RA) is thought to be the morphogen released by
the zone of polarizing activity (ZPA) in the developing chick limb bud
(see Smith et al., 1989) for review, and also several other articles in
the present book). RA has also been implicated in the patterning of the
antero/posterior axis of the body in Xenopus (Durston et al., 1989;
Sive et al., 1990), and of the developing CNS in chicken (Wagner et
al., 1990). It appears therefore that RA might have been selected as a
molecule which plays a crucial role in directing pattern formation in
vertebrate development. In order to undestand better the molecular
mechanism for RA action, it is important to characterize the multiple
components involved in the interpretation of the signal provided by RA
and/or its plausible concentration gradients.

The cellular retinoic acid binding proteins (CRABPs), which are
encoded by at least two closely related genes (Giguère et al. 1990a;
Blomhoff et al., 1990) are unlikely to be the final transducers of the
RA signal, because they are not present in all RA responsive systems,
e.g. the human promyelocytic leukemia cell line HL60. Several small
hydrophobic molecules like the steroids, thyroid hormone, and vitamin
D3 exert their biological activity via nuclear receptors. These receptors are ligand-inducible transcriptional transactivators (or repress-
ors in some cases) which act by binding to specific responsive elements
of target genes (termed enhancers in the case of transactivation) (see
Green and Chambon, 1988; Evans, 1990; Beato, 1989; Diamond et al.,
1990). These receptors form a superfamily of genes (Green and Chambon,
1988 and see below). That the effect of RA at the gene expression level

may be mediated by steroid receptor-like proteins has been suspected for a long time (Lotan, 1980). Nuclear RA binding proteins, different from CRABP, have been detected in the past (Daly and Redfern, 1987; Hashimoto et al., 1988), and, indeed, in the last few years, our group, as well as others, has cloned the cDNAS encoding several RA receptors (RARs) which were found to belong to the same superfamily of genes as the steroid/thyroid hormone receptors. In this paper, we describe the structure of the different RARs and show that they possess specific patterns of expression during embryogenesis and in the adult. In this light we discuss the possible functional significance of these multiple RAR genes and their isoforms.

THREE RETINOIC ACID RECEPTOR (RAR) GENES

How many RAR genes?

The first human RAR (named hRAR-α) cDNA was cloned by using a consensus oligonucleotide probe derived from the sequence of several members of the steroid/thyroid hormone receptor family (Petkovich et al., 1987). An identical human RAR was also identified by Giguère et al., 1987. Subsequently, a gene within which integration of the hepatitis B virus occured in a single hepatocarcinoma (Dejean et al., 1986; de Thé et al., 1987) was demonstrated to encode a second RAR, hRAR-β (Brand et al., 1988). A third human retinoic acid receptor cDNA, hRAR-γ, has also been isolated together with its mouse counterpart (mRAR-γ) by cross-hybridization with hRAR-α and hRAR-β probes (Krust et al., 1989; Zelent et al., 1989). The corresponding mouse hRAR-α and β cDNAs have also been cloned (Zelent et al., 1989, see also Fig. 1). No additional RAR gene could be detected by low stringency Southern blot analysis of mouse and human genomic DNA with probes corresponding to a conserved part of the RA binding domain (region E, see Fig. 1) (our unpublished data). Therefore, the RAR-α, β and γ genes appear to represent all of the members of the RAR subfamily of nuclear receptors in mouse and man. Furthermore, three RARs homologous to the mammalian α, β and γ genes have recently been identified in the newt [Ragsdale et al., 1989; Giguère et al., 1989; the newt RAR named RAR-δ appears to be the homologue of mammalian RAR-γ (J. Brockes, personal communication)] and chicken (our unpublished results), suggesting that 3 RAR genes exist in all higher vertebrates.

Recently, a member of another subfamily of nuclear receptors, the RXRs (Mangelsdorf et al., 1990), has been shown to function as a RA dependent transcriptional enhancer factor. However, the weak apparent affinity of this receptor for RA, as well as its inability to respond to the synthetic retinoid TTNPB suggest that RA may not be the physiological ligand for this receptor.

Domain structure of the RARs

Comparison of the amino acid sequence of the 3 RARs with each other, as well as with other members of the nuclear receptor super-family revealed that the RAR primary sequence consists of 6 regions (A to F) which show differential degrees of conservation (see Fig.1). The most highly conserved is the C region (93% to 95% identity between the 3 RARs, see Fig. 2) comprised of 66 amino acids, which are conserved throughout the superfamily of nuclear receptors and contains 2 putative "Zinc fingers". Region C has been shown to belong to the DNA binding domain of the nuclear receptors (see Green and Chambon, 1988 for review).

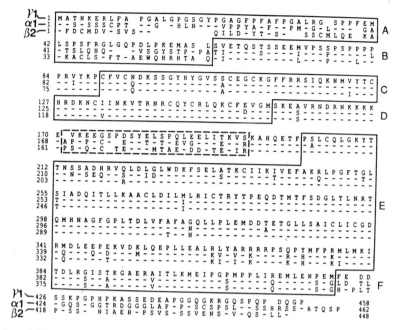

Fig. 1 . Alignment of murine RAR-γ1, α1 and β2. Regions A, B, C, D, E and F are indicated. Region D2 is encased in dashed line.

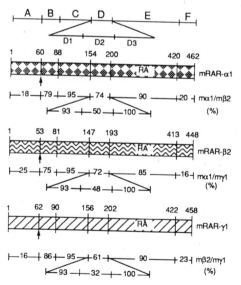

Fig. 2 . Homology comparison of the murine RAR subtypes with each other. The arrow corresponds to the splicing junction between the A and B regions.

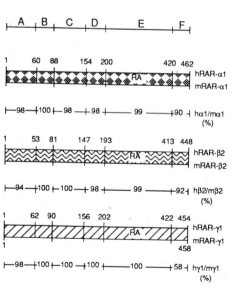

Fig. 3 . Homology comparison of each RAR subtype between mouse and man.

Region E which is 85-90% conserved between the RARs (see Fig. 2) corresponds to the RA binding domain. In addition, this region has been shown for other nuclear receptors to contain a ligand-inducible transcription activation function (Webster et al., 1988; O'Donnel and Koenig, 1990). This is also the case for the RARs (M. Saunders and P. Chambon, unpublished). The B region is also well conserved between the 3 RARs (75 to 86%). Although encoded in the same exon as the first "zinc finger" of region C, its amino acid sequence is unrelated to that found at a similar position in other nuclear receptors. It may, therefore, perform a RAR specific function. It is worth noting that this region contains a strong consensus recognition sequence for casein kinase II (T Q S S/T S S E E; all the serine and threonine residues would be phosphorylated), and may therefore be a target for phosphorylation. Whether this is indeed the case is presently unknown.

The D region, which in the oestrogen receptor is a less conserved hinge region, has been further subdivided into D1, D2 and D3 sub-regions, where D1 and D3 are highly conserved and D2 is not conserved between the three RARs (see Fig. 1). Region D1 is highly basic and may contain a nuclear localization signal which is found at a similar position in other nuclear receptors. The A as well as F regions differ markedly among the three receptors within a given species (Fig. 2). In the case of other nuclear receptors the A/B region has been shown to contain a cell- and promoter-specific activation function (Tora et al., 1988; Tora et al., 1989, see below). However no function has yet been ascribed to the F region. Nevertheless the functional importance of RAR regions A, D2 and F is suggested by their nearly complete amino acid sequence conservation between mouse and man, for a given RAR subtype (either α, β or γ) (Fig. 3). The F region of RAR-γ is an apparent exception to that rule, due to a frameshift in the coding sequence which affects only the C-terminal amino acids (Krust et al., 1989).

This strong evolutionary conservation for a given RAR subtype is also seen with respect to sequences of the B, C and E regions (Fig. 3), thus suggesting that even these conserved domains may possess some degree of functional specificity for each RAR subtype.

MURINE ISOFORMS OF RETINOIC ACID RECEPTORS

Isolation of several mouse RAR cDNA isoforms

In the process of cloning the human RAR-γ, several cDNA clones were isolated with diverging 5' sequence (Krust et al., 1989). This finding led to a systematic search for 5' variants of all 3 mouse RAR genes. By a combination of cDNA library screening and anchored PCR analysis, we found 7 RAR-α, 3 RAR-β and 7 RAR-γ cDNA isoforms (Kastner et al., 1990; Leroy et al., 1990; Zelent et al., 1990). These isoforms were all found to differ in their sequence upstream of the point corresponding to the 5' end of the exon encoding the B region (see Figs 4-6).

The B to F region sequences appear to be common to all isoforms of a given RAR-subtype, as was confirmed by PCR using isoform-specific 5' primers and common 3'-untranslated region (3'-UTR) primers. Analysis of genomic DNA revealed that the specific sequences of all the isoforms are encoded in separate exons present in the 5' region of their corresponding genes (see Fig. 4b-6b). Several of these isoforms encode

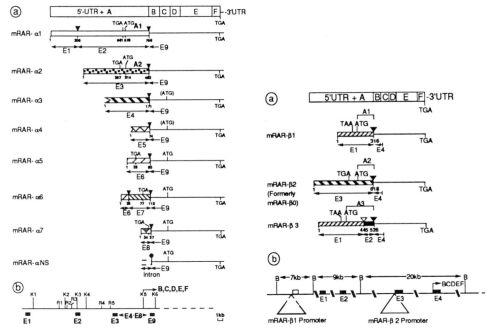

Fig. 4 . Murine RAR-α isoforms
a) schematic drawing of mRAR-α cDNA isoforms. b) exonic organization in the 5' part of the mouse RAR-α gene (see Leroy et al., 1990, for further details).

Fig. 5 . Murine RAR-β isoforms
a) schematic drawing of mRAR-β cDNA isoforms. b) structure of the mRAR-β gene in its 5' region (see Zelent et al.,1990, for further details).

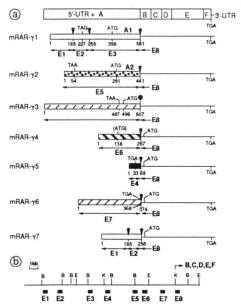

Fig. 6 . Murine RAR-γ isoforms a) schematic drawing of mRAR-γ cDNA isoforms b) exonic organization in the 5' part of the mRAR-γ gene (see Kastner et al., 1990, for further details).

RAR proteins with distinct A regions: RAR-α1 (previously mRAR-α0), mRAR-α2, mRAR-β1, mRAR-β2 (previously mRAR-β0), mRAR-β3, mRAR-γ1 (previously mRAR-γ0) and mRAR-γ2. Note that mRAR-β3 is identical to RAR-β1, except for an additional 27 amino acid-encoding exon present upstream of the B region encoding exon (Fig. 5).

The reading frame of mRAR-α3 and -α4 remains open up to the 5' end of the known sequence and it is presently unknown whether these two isoforms also encode alternative A regions. Due to a variability in the length of a poly-dC stretch (see Kastner et al., 1990), the structure of the possible mRAR-γ4 protein is also unclear. mRAR-α5 to 7, and mRAR-γ5 and 6, possess in-frame stop codons upstream of their B region-encoding sequence. Thus, RAR-α5 to 7 would yield N-terminally truncated receptors lacking the whole B region and the first finger (CI) of the C region (see Fig. 4), since their first AUG codon would correspond to Met 163 (Fig. 1). This lack of the CI finger was previously found for one of the isoforms of the Drosophila "orphan" receptor E75 (Segrave and Hogness, 1990). On the other hand, mRAR-γ5 and mRAR-γ6, as well as RAR-γ7, which is an alternatively spliced form of RAR-γ1 (see Fig 6), would encode truncated receptor proteins whose initiation codon would be Met 73 in the B region (Fig. 1). However, it cannot be excluded at the present time that some of these isoforms may have an A region initiated at a non-AUG codon.

RAR isoform conservation between mouse and man

The sequences of RAR-α1, RAR-β2, RAR-γ1, -γ2 and -γ3 are known in both mouse and man. RAR-γ3 appears to be a partially processed (spliced) primary transcript without obvious conservation between the 2 species (Kastner et al., 1990). In contrast, the other isoforms exhibit a remarkable degree of conservation, both in their A region-coding sequence, and, surprisingly, in their 5' untranslated region (5'-UTR) sequences, suggesting that these regions perform a specific function. In the case of the 5'-UTRs, it could be related to regulation of translation, mRNA stabilization, or both (see Zelent et al., 1990; Leroy et al., 1990 for further discussion of this point).

The existence of human counterparts to mRAR-α2, mRAR-β1/β3, mRAR-γ4 and mRAR-γ6 is suggested by the fact that specific sequences corresponding to these isoforms can be readily detected by cross-hybridization in Southern blot analysis of human genomic DNA (Leroy et al., 1990; Zelent et al., 1990; Kastner et al., 1990). Again, note that RAR-γ6 and possibly RAR-γ4 5'-conserved sequences, would be untranslated sequences.

Specific expression pattern of some RAR isoforms

Northern Blot analysis performed with the total cDNA of each RAR subtype showed that the 3 mRAR genes have different pattern of expression in adult mouse tissues. Most notably, mRAR-γ was expressed almost exclusively in skin (Zelent et al., 1989). Furthermore, the three genes were found to be differentially expressed at various stages of mouse development (for a detailed analysis of the RAR subtype mRNA distribution in the mouse embryo, see Ruberte et al., 1990a and b; Dollé et al., 1989 and 1990).

In order to investigate the expression pattern of each isoform, we carried out Northern blot analysis with isoform specific probes. Only mRAR-α1/α2, mRAR-β1/β2/β3 and mRAR-γ1/γ2 could be readily detected by this assay. Note, however, that the sequences specific to RAR-α3-7 and RAR-γ5 are short and cannot generate probes with high sensitivities.

All 7 mRAR isoforms whose expression could be detected were expressed during embryogenesis, suggesting that they all mediate some of the effects of RA during development (see Figs. 7,8,9). In addition, each of these 7 isoforms exhibits a specific pattern of transcript distribution (Figs. 7,8,9). mRAR-α1 shows a relatively ubiquitous expression pattern with, however, some variation in the absolute transcript level among various tissues. For example, the mRAR-α1 expression is particularly high in the skin. The broad spectrum of mRAR-α1 expression is reminiscent of that of a housekeeping gene. Interestingly, the hRAR-α1 promoter (Brand et al., 1990) exhibits all the features of a housekeeping gene promoter, such as multiple SP1 binding sites, high G/C content and lack of TATA box. This ubiquitous expression of RAR-α1 suggest that the presence of a RA responsive machinery may be required by most cell types.

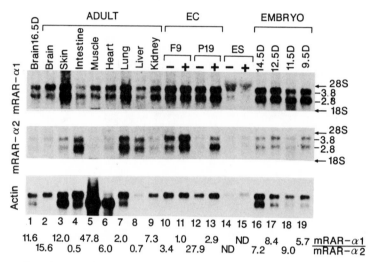

Fig. 7 . Northern blot analysis of mRAR-α1 and mRAR-α2 transcript distribution. Values at the bottom represent the ratio of mRAR-α1 to mRAR-α2 transcript levels, as determined by densitometric scanning (see Leroy et al. for further details).

The expression of RAR-α2 is more restricted. In the adult, its mRNA levels are highest in intestine, lung and liver and very low in brain, muscle, heart and kidney. Interestingly, the RAR-α2 mRNA is induced by approximately 11-fold in the P19 embryonal carcinoma (EC) cell line, and 3 to 5-fold in the F9 EC cell line after 24 hr of RA treatment, whereas the levels of mRAR-α1 are unaffected by RA.

The expression of RAR-β1 and RAR-β3 (which differ by the presence in RAR-β3 of an additional exon encoding 27 amino acids upstream of the B region, see Fig. 5) has been analysed by Northern blot and PCR (see Zelent et al., 1990). mRAR-β1 and β3 are particularly abundant in adult and embryonic brain, and may therefore play some specific role in the development of the central nervous system (Fig. 8). mRAR-β1/β3 (particularly, mRAR-β3) are also relatively abundant in skin and lung. Interestingly mRAR-β1/β3 messages are induced by RA treatment of P19 and F9 EC cells, and RAR-β1 transcripts are present at a much higher level than mRAR-β3 transcript in these cells (see Zelent et al., 1990). mRAR-β2 is the only mRAR-β isoform detectable in kidney, heart, liver and muscle. The expression of this isoform is also very strongly induced by RA in P19 and F9 EC cells.

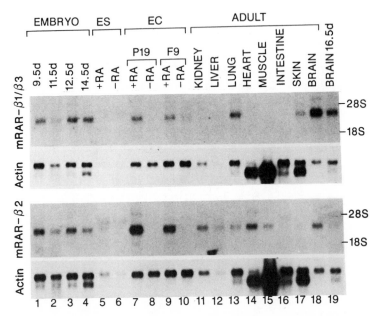

Fig. 8 . Northern blot analysis of mRAR-β1/β3 and mRAR-β2 expression. The probe used in the top panel did not distinguish between the β1 or β3 isoforms (see Zelent et al., 1990, for further details).

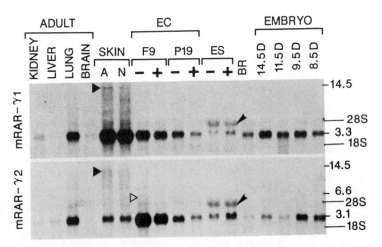

Fig. 9 . Northern blot analysis of mRAR-γ1 and mRAR-γ2 expression (see Kastner et al. 1990 for details).

All mRAR-β isoforms were expressed in the embryo at day 9.5, 11.5, 12.5 and 14.5 post-coitum; however, the relative amounts of mRAR-β2 and mRAR-β1/β3 transcripts decreased and increased, respectively, during the course of embryogenesis (Fig. 8).

mRAR-γ1, in the adult is expressed predominantly in skin, where it accounts by far for the majority of mRAR-γ transcripts (Fig. 9). Both mRAR-γ1 and mRAR-γ2 isoforms are expressed in adult lung, mRAR-γ2 being the predominant form in this tissue. Both isoforms are also expressed in EC cells and embryonic stem cells, where RAR-γ2 is the predominant mRAR-γ isoform (80% of the transcripts). Both mRAR-γ1 and mRAR-γ2 are expressed in the embryo, but there is a relative decrease in mRAR-γ2 transcripts abundance during the course of embryogenesis, with a concomitant increase in the level of mRAR-γ1 transcripts (Fig. 9; see also Kastner et al., 1990).

Similar organization of the three RAR genes

The specific exons encoding the isoforms type 1 (RAR-α1, RAR-β1, RAR-γ1) lie upstream of the exons encoding the isoforms type 2 (mRAR-α2, mRAR-β2, mRAR-γ2) in all three mouse RAR genes (see Fig. 4b, 5b, 6b). Since a functional promoter has been found upstream of both the human (de Thé et al., 1990) and mouse (Sucov et al., 1990; CM and PC, in preparation) RAR-β2 specific exon, there must be another promoter in the RAR-β gene which is necessary for transcription of the RAR-β1 and RAR-β3 isoforms. The situation appears to be very similar in the case of the mRAR-α gene in which a RA-responsive promoter would be responsible for mRAR-α2 transcription (our unpublished results). Whether mRAR-γ1 and mRAR-γ2 expression is also controlled by two promoters remains to be seen (see below).

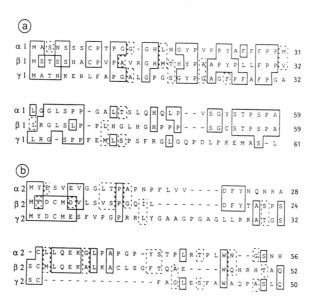

Fig. 10 . Sequence comparisons of the A regions of type 1 and type 2 mouse isoforms
a) alignment of A region sequences of mRAR-α1, -β1 and -γ1
b) alignment of A region sequences of mRAR-α2, -β2 and γ2
Conserved residues are boxed. Conservative replacement are boxed with dashed lines. Gaps (dashes) have been introduced for optimal alignment.

In addition to their relative position within the gene, the α, β and γ isoforms type 1 display some common features: (i) there are clear similarities between their A region amino acid sequences (Fig. 10a); (ii) their specific 5'- UTR and A region sequences are encoded in several exons (2 exons for RAR-α1, at least 2 exons for RAR-β1, and 3 exons for RAR-γ1); (iii) they display some common expression features. For example the RAR isoforms type 1 are predominantly expressed in the skin, when compared with the isoforms type 2. The temporal expression of mRAR-β1 and mRAR-γ1 during the course of embryogenesis is also similar, these isoforms being predominantly expressed at late stages of development.

Similarly, the RAR isoforms type 2 display also some common features: (i) there are clear sequence similarities between the amino acid sequences of their A regions (see Fig. 10b); (ii) all isoforms type 2 specific sequences seem to be encoded in a single exon; (iii) In contrast to mRAR-γ1 and mRAR-β1 isoforms, both mRAR-γ2 and mRAR-β2 are expressed at higher levels at early stages of development.

All of the above observations suggest that all three RAR genes have evolved by duplication of an ancestral gene possessing two promoters controlling the transcription of two major isoforms type 1 and 2. However, the exact exon pattern may not be identical for the 5'-region of the three RARs. For instance, the RAR-β3 additional exon does not appear to have any counterpart in the RAR-α and RAR-γ genes.

At the present time, the origin of the other RAR-α isoforms and of RAR-γ4 to γ6 is unclear, although the conservation of some of them between mouse and man suggest that they could be functionnally important. It remains to be seen whether they are transcribed from a separate promoter or produced by alternative splicing from a primary transcript originating from one of the 2 promoters described above. At the present time it cannot be excluded that the isoforms which are expressed at a low level correspond to physiologically insignificant splicing products, occurring by "chance".

WHY MULTIPLE RARs?

Their conservation through evolution, and also their specific pattern of expression support strongly the idea that the multiple RAR subtype isoforms perform specific functions. In principle, these specific functions could be exerted at three levels: ligand binding, binding to RA-responsive elements of target genes, and transcriptional activation.

The three RAR subtypes appear to be activated differentially by retinoic acid and/or synthetic retinoids (Brand et al., 1988; Giguère et al., 1990b; Ishikawa et al., 1990; Aström et al., 1990; Hashimoto et al., 1990). These observations are in keeping with the almost complete conservation (99-100%) of the E region across species for a given mRAR subtype, whereas the E region of the three RARs within a given species is 85-90% conserved. Thus, it is possible that the three RAR subtypes respond differentially to the presently identified natural retinoids [all trans-RA, 13-cis-RA, 3,4-didehydroretinoic acid (Thaller and Eichele, 1990) and to those which may not yet have been found. Such differential responses of the three RARs would obviously increase the combinatorial possibilities for controlling the transcriptional trans-activation of retinoid responsive genes.

Although the C region which belongs to the DNA binding domain is highly conserved between the 3 RAR subtypes, each of them possesses some specific residues in this region (Fig.1), and some of these residues are conserved through evolution. For example, the serine residue 106 (mouse sequence as reference) is RAR-γ specific and conserved in mouse, man, newt (Ragsdale et al., 1989) and chicken (our unpublished result).These subtle conserved differences suggest that the responsive element recognition may differ between the 3 RAR subtypes. Only 2 natural RARE have been characterized to date, in the laminin B1 promoter (Vasios et al., 1989) and in the RAR-β2 promoter (De Thé et al., 1990; Sucov et al., 1990). All 3 RAR types appear to bind equally well to the RAR-β2 RARE (S. Mader and P. Chambon, unpublished results), and no data is yet available concerning the laminin B1 RARE. As more RAREs will be characterized, RAR subtype-specific responsive elements may be uncovered.

The amino acid sequence of the different receptors and of their isoforms diverge strikingly in their N-terminal A region. Given the evolutionary conservation of this region for a given RAR-isoform, it is highly probable that this region performs an important and receptor isoform-specific function. Although no function has been found yet for these regions in the case of RARs, it has been shown that the N-terminal regions of the estrogen, progesterone and glucocorticoid receptors contain transcriptional activation functions (TAFs) which exhibit cell type- and promoter-specificity (Tora et al., 1988; Tora et al., 1989; Bocquel et al., 1989). Thus, each RAR A region may correspond to a cell type- and promoter-specific TAF. This possibility is particularly attractive since, together with the possible ligand- and RA-responsive element specificity of the three RAR subtypes, it would generate the functional diversity which is required to account at the molecular level for the pleiotropic effect of RA during development and in the adult.

ACKNOWLEDGEMENTS

We thank C. Werlé, B. Boulay, J.M. Lafontaine and the secretarial staff for help in preparing the manuscript. We also thank C. Reibel and J.M. Garnier for technical assistance, A. Staub and F. Ruffenach for oligonucleotide synthesis.

REFERENCES

A. Aström, U. Pettersson, A. Krust, P. Chambon, and J.J. Voorhees, 1990, Retinoic acid and synthetic analogs differentially activate retinoic acid receptor dependent transcription, Biochm. Biophys. Res. Comm., in press.

M. Beato, 1989, Gene regulation by steroid hormones, Cell, 56:335.

R. Blomhoff, M.H. Green, T. Berg, and K.R. Norum, 1990, Transport and storage of vitamin A, Science, 250:399.

M.T. Bocquel, V. Kumar, C. Stricker, P. Chambon, and H. Gronemeyer, 1989, The contribution of the N- and C-terminal regions of steroid receptors to activation of transcription is both receptor and cell-specific, Nucl. Acids Res., 17:2581.

N. Brand, M. Petkovich, A. Krust, P. Chambon, H. de Thé, A. Marchio, P. Tiolais, and A. Dejean, 1988, Identification of a second human retinoic acid receptor, Nature, 332:850.

N. Brand, M. Petkovich, and P. Chambon, 1990, Characterization of a
    functional promoter for the human retinoic-acid receptor-alpha
    (hRAR-α), Nucl. Acids Res., in press.
A.K. Daly, and C.P.F. Redfern, 1987, Characterisation of a retinoic-
    acid-binding component from F9 embryonal-carcinoma-cell nuclei,
    Eur. J. Biochem., 168:133-139.
A. Dejean, L. Bougueleret, K.H. Grzeschick, and P. Tiollais, 1986,
    Hepatitis B virus DNA integration in a sequence homologous to
    v-erb-A and steroid receptor genes in a hepatocellular
    carcinoma, Nature, 322:70.
H. de Thé, A. Marchio, P. Tiollais, and A. Dejean, 1987, A novel
    steroid thyroid hormone receptor related gene inappropriately
    expressed in human hepatocellular carcinoma, Nature, 330:667.
H. de Thé, M. del Mar Vivanco-Ruiz, P. Tiollais, H. Stunnenberg, and A.
    Dejean, 1990, Identification of a retinoic acid responsive
    element in the retinoic acid receptor β gene, Nature,
    343:177-180.
M.I. Diamond, J.N. Miner, S.K. Yoshinaga, and K.R. Yamamoto, 1990,
    Transcription factor interactions: selectors of positive or
    negative regulation from a single DNA element, Science,
    249:1266.
P. Dollé, E. Ruberté, P. Kastner, M. Petkovich, C.M. Stoner, L. Gudas,
    and P. Chambon, 1989, Differential expression of the genes
    encoding the retinoic acid receptors α, β, γ and CRABP in the
    developing limbs of the mouse Nature, 342:702.
P. Dollé, E. Ruberté, P. Kastner, G. Morris-Kay, and P. Chambon, 1990,
    Retinoic acid receptors and cellular retinoid binding proteins.
    I. A systematic study of their differential pattern of
    transcription during mouse organogenesis, Develop., in press.
A.J. Durston, J.P.M. Timmermans, W.J. Hage, H.F.J. Hendriks, N.J. de
    Vries, M. Heideveld, and P.D. Nieuwkoop, 1989, Retinoic acid
    causes an anteroposterior transformation in the developing
    central nervous system, Nature, 340:140.
R. Evans, 1988, The steroid and thyroid hormone receptor family,
    Science, 240:889.
V. Giguère, E.S. Ong, P. Segui, and R.M. Evans, 1987, Identification of
    a receptor for the morphogen retinoic acid, Nature,
    330:624-629.
V. Giguère, E.S. Ong, R.M. Evans, and C.J. Tabin, 1989, Spatial and
    temporal expression of the retinoic acid receptor in the
    regenerating amphibian limb, Nature, 337:566.
V. Giguère, S. Lyn, P. Yip, C.-H. Siu, and S. Amin, 1990a, Molecular
    cloning of cDNA encoding a second cellular retinoic acid-binding
    protein, Proc. Natl. Acad. Sci. USA, 87:6233-6237.
V. Giguère, M. Shago, R. Zirngibl, P. Tate, J. Rossant, and S. Varmuza,
    1990b, Identification of a novel isoform of the retinoic acid
    receptor γ expressed in the mouse embryo, Mol. Cell. Biol.,
    10:2335.
S. Green, and P. Chambon, 1988, Nuclear receptors enhance our
    understanding of transcription regulation, Trends Genet.,
    4:309-314.
Y. Hashimoto, H. Kagechika, E. Kawachi, and K. Shudo, 1988, Specific
    uptake of retinoids into human promyeolocytic leukemia cells
    HL-60 by retinoid-specific bind proteins: possibly the true
    retinoid receptor. Jpn. J. Cancer Res. (Gann), 79:473-483.
Y. Hashimoto, H. Kagechika, and K. Shudo, 1990, Expression of retinoic
    acid receptor genes and the ligand-binding selectivity of
    retinoic acid receptors (RAR's), Biochm. Biophys. Res. Comm.,
    166:1300.

T. Ishikawa, K. Umesono, D.J. Mangelsdorf, H. Aburatani, B.Z. Stanger, Y. Shibasaki, M. Imawari, R.M. Evans, and F. Takaru, 1990, A functional retinoic acid receptor encoded by the gene on human chromosome 12, Mol. Endocrinology, 4:837.

P. Kastner, A. Krust, C. Mendelsohn, J.M. Garnier, A. Zelent, P. Leroy, A.Staub, and P. Chambon, 1990, Murine isoforms of retinoic acid receptor-γ with specific patterns of expression, Proc. Natl. Acad. Sci. USA, 87:2700.

A. Krust, P. Kastner, M. Petkovich, A. Zelent, and P. Chambon, 1989, A third human retinoic acid receptor, hRAR-γ, Proc. Natl. Acad. Sci. USA, 86:5310.

P. Leroy, A. Krust, A. Zelent, C. Mendelsohn, J.-M. Garnier, P. Kastner, A. Dierich, and P. Chambon, Multiple isoforms of the mouse retinoic acid receptor alpha are generated by alternative splicing and differential induction by retinoic acid, EMBO J., in press.

J.R. Lotan, 1980, Effects of vitamin A and its analogs (retinoids) on normal and neoplastic cells, Biochem. Biophys. Acta, 605:33.

D.J. Mangelsdorf, E.S. Ong, J.A. Dyck, and R.M. Evans, 1990, Nuclear receptor that identifies a novel retinoic acid response pathway, Nature, 345:224.

A.L. O'Donnell, and R.J. Koenig, 1990, Mutational analysis identifies a new functional domain of the thyroid hormone receptor, Mol. Endocrinology, 4:715.

M. Petkovich, N.J. Brand, A. Krust, and P. Chambon, 1987, A human retinoic acid receptor which belongs to the family of nuclear receptors, Nature, 330:444.

C.W. Ragsdale Jr., M. Petkovich, P.B. Gates, P. Chambon, and J.P. Brockes, 1989, Identification of a novel retinoic acid receptor in regenerative tissues of the newt, Nature, 341:654.

E. Ruberte, P. Dollé, P. Chambon, and G. Morris-Kay, 1990a, Retinoic acid receptors and cellular retinoic binding proteins: II. Their differential pattern of transcription during early morphogenesis in mouse embryos, Develop., in press.

E. Ruberte, P. Dollé, A. Krust, A. Zelent, G. Morris-Kay, and P. Chambon, 1990b, Specific spatial and temporal distribution of retinoic acid receptor gamma transcrits during mouse embryogenesis, Develop., 108:213.

W.A. Segraves, and D.S. Hogness, 1990, The E75 ecdysone-inducible gene responsible for the 75B early puff in Drosophila encodes two new members of the steroid receptor superfamily, Genes Develop., 4:204.

H.L. Sive, B.W. Draper, R.M. Harland, and H. Weintraub, 1990, Identification of a retinoic acid-sensitive period during primary axis formation in Xenopus laevis, Genes Develop., 4:932.

S.M. Smith, K. Pang, O. Sundin, S.E. Wedden C. Thaller, and G. Eichele, 1989, Molecular approaches to vertebrate limb morphogenesis, Develop., supplement:121.

H.M. Sucov, K.K. Murakami, and R.M. Evans, 1990, Characterization of an autoregulated response element in the mouse retinoic acid receptor type β gene, Proc. Natl. Acad. Sci. USA, 87:5392.

C. Thaller, and G. Eichele, 1990, Isolation of 3,4-didehydroretinoic acid, a novel morphogenetic signal in the chick wing bud, Nature, 345:815-819.

L. Tora, H. Gronemeyer, B. Turcotte, M.-P. Gaub, and P. Chambon, 1988, The N-terminal region of the chicken progesterone receptor specifies target gene activation, Nature, 333:185-188.

L. Tora, J. White, C. Brou, D. Tasset, N. Webster, E. Scheer, and P. Chambon, 1989, The human estrogen receptor has two independent nonacidic transcriptional activation functions, Cell, 59:477.

G.W. Vasios, J.D. Gold, M. Petkovich, P. Chambon, 1989, A retinoic acid-responsive element is present in the 5' flanking region of the laminin B1 gene, Proc. Natl. Acad. Sci. USA, 86:9099-9103.

M. Wagner, C. Thaller, T. Jessel, and G. Eichele, 1990, Polarizing activity and retinoid synthesis in the floor plate of the neural tube, Nature, 345:819.

N.J.G. Webster, S. Green, J.R. Jin, and P. Chambon, 1988, The hormone-binding domains of the estrogen and glucocorticoid receptors contain an inducible transcription activation function, Cell, 54:199.

A. Zelent, A. Krust, M. Petkovich, P. Kastner, and P. Chambon, 1989, Cloning of murine α and β retinoic acid receptors and a novel receptor γ predominantly expressed in skin, Nature, 339:714.

A. Zelent, C. Mendelsohn, P. Kastner, J.-M. Garnier, F. Ruffenach, P. Leroy, and P. Chambon, Differentially expressed isoforms of the mouse retinoic acid receptor beta are generated by usage of two promoters and alternative splicing, EMBO J., in press.

THE ROLE OF RETINOIC ACID AND CELLULAR RETINOIC ACID-BINDING PROTEIN IN

THE REGENERATING AMPHIBIAN LIMB

Malcolm Maden[1], Nick Waterson[2], Dennis Summerbell[2], Jean
Maignon[3], Michel Darmon[4], Braham Shroot[4]

[1] Anatomy & Human Biology Group, Kings College, Strand
London WC2R 2LS, UK. [2] Division of Physical Biochemistry
National Institute for Medical Research, Mill Hill, London
NW7 1AA, UK. [3] Laboratoires de Recherche L'Oreal, Aulnay-
Sous-Bois, France. [4] Centre International de Recherches
Dermatologiques Galderma, Sophia Antipolis, Valbonne, France

INTRODUCTION

We describe here two aspects of our work on retinoic acid (RA) and
pattern formation in the regenerating limb. The first concerns the presence
of endogenous RA in the regenerating limb and the second concerns
experiments on the role of cellular retinoic acid-binding protein (CRABP)
in the process of pattern respecification.

The effects of exogenous administration of RA in respecifying pattern
formation in the regenerating limb are now well-documented and need no
repeating here. They immediately raise the question of whether this is
just a random interference with the process of pattern formation which
happens to result in duplications or whether it is truly a reenactment of
the original cellular and molecular mechanisms which established the initial
pattern. In the developing chick limb bud, crucial evidence that the
latter is the case came from the discovery of endogenous RA in the bud
(Thaller & Eichele, 1987). Using HPLC Thaller & Eichele found that whereas
retinol, the metabolic precursor of RA was found in equal amounts in the
anterior vs the posterior part of the limb bud, RA itself was concentrated
in the posterior part in a 2.5-fold excess over the anterior part. This is
precisely the behaviour we would expect of a morphogen released by the zone
of polarizing activity (ZPA) which diffuses out of the ZPA and across the
anteroposterior (AP) axis of the bud thereby generating positional
information (Tickle, Summerbell & Wolpert, 1975). There is some doubt as to
whether a 2.5-fold gradient is steep enough to specify the requisite number
of thresholds across the AP axis. In combination with our data on the
distribution of cellular retinoic acid-binding protein (Maden et al., 1988)
Smith et al. (1989) have calculated that a 10-fold gradient of free RA can,
in fact, be generated. Most recently, an additional retinoid which also
causes duplications, 3, 4-didehydroretinoic acid, has been identified in
the chick limb bud at a tissue concentration about six times greater than
that of all-trans-RA (Thaller & Eichele, 1990) although its distribution
across the AP axis is not yet known.

In the regenerating amphibian limb we obviously would like to know whether the same principles are true and we describe here our preliminary evidence to suggest that RA is indeed an endogenous component of the regenerating limb.

If RA is a morphogen we need to understand its cellular mechanism of action. RA interacts with two classes of proteins within the cell, one class in the cytoplasm and one class in the nucleus. The cytoplasmic proteins are the cellular retinoic acid-binding protein I and II (Bailey & Siu, 1988; Kitamoto et al., 1988; Giguere et al., 1990) which are present in high concentrations in the cells of the developing and regenerating limb (Keeble & Maden, 1986; Maden & Summerbell, 1986) and which although exhibiting a high degree of specificity for RA (Keeble & Maden, 1989; Maden & Summerbell, 1989) has a relatively low dissociation constant. The nuclear proteins are the retinoic acid receptors (RARs and RXRs) (Zelent et al., 1989; Mangelsdorf et al., 1990) of which there are now many variants. They have both a high affinity ligand binding site and a DNA binding site and are thus involved in the ligand dependent control of DNA transcription by RA.

Our previous work on the role of CRABP in the process of respecification in the limb has provided several suggestive correlations. The levels of CRABP in the regenerating limb rise after amputation and reach a peak during the cone stage when blastemas are most receptive to respecification by RA (Keeble & Maden, 1986). CRABP is present in the progress zone of the chick limb bud as shown by immunocytochemistry and is distributed across the AP axis with the opposite polarity to that of endogenous RA (Maden et al., 1988). Finally, the potency of a range of natural and some synthetic retinoids at respecifying the limb correlated well with their ability to bind to CRABP (Keeble & Maden, 1989; Maden & Summerbell, 1989). Here we continue these studies with a new type of synthetic retinoid which has the property of binding to the RARs, but not CRABP (Darmon et al., 1988). Thus we can test whether binding to CRABP is an obligatory step during limb respecification.

MATERIALS AND METHODS

Extraction of retinoids from normal and regenerating axolotl limbs which had been amputated through the mid-stylopodium was performed exactly as described in Thaller & Eichele (1987). Briefly, normal limbs and cone stage blastemas from 12cm axolotls were removed, sonicated in stabilising buffer and extracted 3 times with 8:1 ethyl acetate/methyl acetate containing BHT. The organic phases were pooled and dried down over nitrogen and 100µl methanol added for HPLC analysis. As an internal radioactive standard 250pg ^{3}H all-trans-RA was added to the sonicated tissues in some experiments. A Beckman Gold system was used with reverse phase chromatography. An XL ODS C column was used with methanol/acetonitrile/water (40/40/20) pH 6 as the solvent.

For testing the biological effects of the new retinoid, termed CD-564 or 6-(5,6,7,8-tetrahydro-5,5,8,8-tetramethyl-2-naphthoyl)-2-naphthoic acid, we have used our standard limb regeneration assay system. The compound was mixed with silastin and small pieces of silastin inserted under the skin adjacent to the distal blastemas as previously described (Maden, Keeble & Cox, 1985). The resulting limbs were stained with Victoria blue and scored according to the duplication index.

To confirm that CD-564 did not bind to axolotl blastemal CRABP, mid-cone stage limb regeneration blastemas from 10-15 cm axolotls were collected, homogenised in PBS and high-speed supernatants were prepared as previously described (Keeble & Maden, 1986). 1mg aliquots of these cytosols were incubated with 40 pmole of ^{3}H-RA plus the relevant cold retinoid analogue. Samples were dialysed overnight, spun overnight on 5-20% sucrose gradients, fractionated and counted.

RESULTS

HPLC Analysis

Extracts of whole, unamputated axolotl limbs showed only one small peak of absorbance at 350nm on reverse phase HPLC (Fig. 1). By the criteria of addition of non-radioactive standards to parallel extracts and co-elution with radioactive standards this peak corresponds to all-trans-retinol. When blastemal extracts were chromatographed a completely different picture emerged (Fig. 2). Four peaks appeared at the typical elution times for retinoids and they were identified by the criteria described above as all-trans-RA (arrow 1), all-trans-retinol (arrow 2) and 13-cis-RA (arrow 3). The additional peaks which were rapidly eluted from the column (arrow 4) may correspond to the newly described 3,4-didehydro-retinoids (Thaller & Eichele, 1990). Thus RA seems to be present in the blastema, but not in the normal unamputated limb.

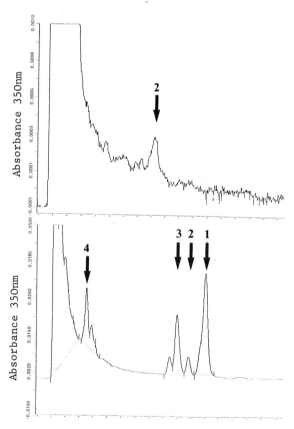

Fig. 1 (upper) and Fig. 2 (lower). HPLC chromatograms of unamputated axolotl limbs (upper) and blastemas (lower). See text for details of the retinoids identified in arrows.

CRABP Binding Assay

On fractionated sucrose gradients a single peak of radioactivity was detected after incubation with ^{3}H-RA at the expected position for the molecular weight of CRABP (Fig. 3). The addition of a 100-fold excess of RA, as expected, abolished the peak completely. In contrast a 100-fold excess of CD-564 had no effect on the radioactive peak (Fig. 3). For comparison we also performed incubations in the presence or retinol which at the same molar excess did not abolish the radioactive peak either. Since we know that retinol does not bind to CRABP we may conclude that CD-564 also does not bind to axolotl CRABP.

Potency of CD-564 in the Regenerating Limb

We began by establishing a baseline figure for RA to compare with CD-564. At a concentration of 0.3M in the silastin block, RA gave a range of

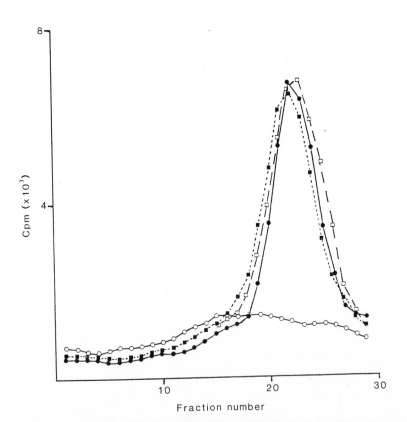

Fig. 3.   Sucrose density gradient fractionation of cyto-
          plasmic protein preparations from cone stage
          axolotl blastemas incubated with 40pmole of ^{3}H-RA
          in the presence or absence of 100-fold molar
          excess of retinoid analogues. The peak appears
          in the 2S (17KD) region of the gradient.
          ●———● ^{3}H-RA alone.   o———o RA.
          ■------■ CD-564.   □— —□ retinol.

92

duplications ranging from normal (4 cases) to completely duplicated limbs (2 cases). This resulted in a duplication index of 2 (Fig. 4). This is of the same order, although slightly lower, than that recorded in a previous study of retinoid analogues (Keeble & Maden, 1989).

Using CD-564 at a concentration of 0.3M in the silastin blocks, all the limbs were inhibited from regenerating, giving a duplication index of 6 (Fig. 4). Not only did CD-564 affect regeneration, but the whole body showed dramatic changes despite the tiny silastin block being implanted into the forearm (or shank). The animals ceased growing and the epidermis began to be sloughed off. At the end of the experiment (12 weeks after amputation) these animals were at least 50% smaller than the RA treated animals. CD-564 was thus toxic and although none of the animals died they remained in suspended animation, apparently permanently.

At a 10-fold lower concentration (0.03M) the majority of limbs did regenerate (10 out of 14) and every one of these regenerates produced a partially duplicated humerus giving a duplication index of 4.6 (Fig. 4). This remarkable uniformity of response has not been seen before, normally a range of duplications are seen as described above for RA. CD-564 was thus about 20x more potent than RA at inducing duplications during regeneration.

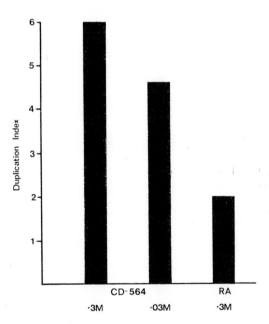

Fig. 4.   Histogram of the results of administering
         CD-564 or RA to regenerating limbs. The
         concentration of the retinoid is shown
         below the columns. The duplication index
         is a measure of the amount of extra tissue
         regenerated in the PD axis and is described
         more fully in Materials and Methods.

DISCUSSION

The identification of endogenous RA in the developing chick limb bud (Thaller & Eichele, 1987) was of major importance in establishing that this compound plays a crucial role in the generation of pattern in the limb, although this picture is now somewhat more complicated by the presence of an additional class of retinoids, the 3,4-didehydroretinoids (Thaller & Eichele, 1990). We have shown here that RA (and probably the didehydro-retinoids too) is present endogenously in the other retinoid responsive system, the regenerating amphibian limb. It seems to be specifically switched on in the tissues of the blastema because it could not be detected in the unamputated limb. Obviously the metabolism of a blastemal cell is completely different from that of its fully differentiated precursor and one of these changes is presumably the up-regulation of the enzymes involved in the metabolism of RA from retinol.

These observations raise several questions of interest about the location and role of RA in the regenerating limb. Is it involved in the generation of AP pattern as in the chick limb bud or proximodistal (PD) pattern? From the initial reports of the effects of exogenous administration of RA (Maden, 1982) we would presume that it is involved in PD pattern formation, but since then it has been revealed that RA can also respecify the AP axis (Maden, 1983; Kim & Stocum, 1986a) and the dorsoventral (DV) axis as well (Ludolph et al., 1990). So how can we say that RA is respon-sible for specification in any one axis? It must be involved in all three. More likely, RA may be used in an earlier event in regeneration such as dedifferentiation (Brockes, pers. comm.). It has been noted before that the increasing PD duplications induced by increasing doses of RA involve an increased amount of dedifferentiation of the stump tissues (Kim & Stocum, 1986b). There is a good deal of experimental data which can be mustered in support of this contention since the original observations of Fell & Mellanby (1952) that vitamin A caused the resorption of the cartilage matrix of cultured mouse and chick embryonic limb bones. It is important to note that this effect is not simply a degenerative phenomenon, but an active one involving specific changes in RNA, protein and glycoprotein synthesis since inhibitors of these processes prevent matrix resorption (Kistler, 1978; 1982). Most interestingly, these inhibitors (Actinomycin D, cyclo-hexamide and tunicamycin) also prevent vitamin A indiced respecification of the regenerating limb (Scadding, 1988) thereby adding weight to the idea that RA may be involved in the dedifferentiation and establishment of the blastema. Our further characterisation of endogenous retinoids in the limb will have to consider such questions as variations in concentration at different stages, tissues of origin and variations in concentrations at different PD levels, AP locations and DV locations.

The other experiments reported here concern the role of CRABP in the respecification of pattern (and by implication the initial establishment of pattern) by RA. We had previously shown that there was a good correlation between biological efficacy (ability to respecify pattern) of a range of natural and synthetic retinoids and their ability to bind to CRABP (Keeble & Maden, 1989). Those compounds that did not bind to CRABP were inactive, such as retinyl palmitate and retinol, whereas those compounds that were potent did bind to CRABP such as RA and the synthetic analogue TTNPB. Here we have used a new synthetic retinoid CD-564 that does not bind to CRABP (Fig. 2), but does bind to the RARs, at least in F9 cells (Darmon et al., 1988), yet is 20x more potent than RA at respecifying pattern (Fig. 3). We are currently testing other retinoids of the same family for similar behaviour. However, we may conclude from these initial experiments that binding to CRABP is not an obligatory step in the mechanism of respecification.

REFERENCES

Bailey, J. S., and Siu, C-H., 1988, Purification and partial characterisation of a novel binding protein for retinoic acid from neonatal rat. J. Biol. Chem., 263:9326.

Darmon, M., Rocher, M., Cavey, M-T., Martin, B., Rabilloud, T., Delecluse, C., and Shroot, B., 1988, Biological activity of retinoids correlates with affinity for nuclear receptors but not for cytosolic binding protein. Skin Pharmacol., 1:161.

Fell, H. B., and Mellanby, E., 1952, The effect of hypervitaminosis A on embryonic limb-bones cultivated in vitro. J. Physiol., 116:320.

Giguere, V., Lyn, S., Yip, P., Siu, C-H., and Amin, S., 1990, Molecular cloning of cDNA encoding a second cellular retinoic acid-binding protein. Proc. Natl. Acad. Sci. USA, 87:6233.

Keeble, S., and Maden, M., 1986, Retinoic acid-binding protein in the axolotl: Distribution in mature tissues and time of appearance during limb regeneration. Dev. Biol.,117:435.

Keeble, S., and Maden, M., 1989, The relationship among retinoid structure, affinity for retinoic acid-binding protein, and ability to respecify pattern in the regenerating axolotl limb. Dev. Biol., 132:26.

Kim, W-S., and Stocum, D.L., 1986a, Retinoic acid modifies positional memory in the anteroposterior axis of regenerating limbs. Dev. Biol., 114:170.

Kim, W-S., and Stocum, D.L., 1986b, Effects of retinoic acid on regenerating normal and double half limbs of axolotls. Histological studies. Roux's Arch. Dev. Biol., 195:243.

Kistler, A., 1978, Inhibition of vitamin A action in rat bone cultures by inhibitors of RNA and protein synthesis. Experientia, 34:1159.

Kistler, A., 1982, Retinoic acid-indiced cartilage resorption. Induction of specific changes in protein synthesis and inhibition by tunicamycin. Differentiation, 21:168.

Kitamoto, T., Momoi, T., and Momoi, M., 1988, The presence of a novel cellular retinoic acid-binding protein in chick embryos: purification and partial characterisation. Biochem. Biophys. Res. Comm., 157:1302.

Ludolph, D. C., Cameron, J. A., and Stocum, D. L., 1990, The effect of retinoic acid on positional memory in the dorsoventral axis of regenerating axolotl limbs. Dev. Biol., 140:41.

Maden, M., 1982, Vitamin A and pattern formation in the regenerating limb. Nature, 295:672.

Maden, M., 1983, The effect of vitamin A on limb regeneration in Rana temporaria. Dev. Biol., 98:409.

Maden, M., and Summerbell, D., 1986, Retinoic acid-binding protein in the chick limb bud: Identification at various developmental stages and binding affinities of various retinoids. J. Embryol. Exp. Morph., 97:239.

Maden, M., and Summerbell, D., 1989, Biochemical pathways involved in the respecification of pattern by retinoic acid. NATO Advanced Research Workshop on Recent Trends in Regeneration Research, V. Kiortsis, S. Koussoulakos, and H. Wallace, eds., pp 313, Plenum Press, New York.

Maden, M., Keeble, S., and Cox, R. A., 1985, The characteristics of local application of retinoic acid to the regenerating axolotl limb. Roux's Arch. Dev. Biol., 194:228.

Maden, M., Ong, D. E., Summerbell, D., and Chytil, F., 1988, Spatial distribution of cellular protein binding to retinoic acid in the chick limb bud. Nature, 335:733.

Mangelsdorf, D. J., Ong, E. S., Dyck, J. A., and Evans, R. M., 1990, Nuclear receptor that identifies a novel retinoic acid response pathway. Nature, 345:224.

Scadding, S. R., 1988, Actinomycin D, cyclohexamide, and tunicamycin inhibit
    vitamin A indiced proximodistal duplication during limb regeneration
    in the axolotl Ambystoma mexicanum. Can. J. Zool., 66:879.
Smith, S. M., Pang, K., Sundin, O., Wedden, S. E., Thaller, C., and Eichele,
    G., 1989, Molecular approaches to vertebrate limb morphogenesis. Development
    (suppl.) 121.
Thaller, C., and Eichele, G., 1987, Identification and spatial distribution
    of retinoids in the developing chick limb bud. Nature, 336:775.
Thaller, C., and Eichele, G., 1990, Isolation of 3,4-didehydroretinoic acid,
    a novel morphogenetic signal in the chick wing bud. Nature, 345:815.
Tickle, C., Summerbell, D., and Wolpert, L., 1975, Positional signalling
    and specification of digits in chick limb morphogenesis. Nature,
    254:199.
Zelent, A., Krust, A., Petkovich, M., Kastner, P., and Chambon, P., 1989,
    Cloning of murine α and β retinoic acid receptors and a novel
    receptor γ predeominantly expressed in skin. Nature, 339:714.

RETINOIC ACID TREATMENT ALTERS THE PATTERN OF RETINOIC ACID RECEPTOR BETA

EXPRESSION IN THE EMBRYONIC CHICK LIMB

Annie Rowe[1,2], Joy M. Richman[2], James O. Ochanda[1]
and Paul M. Brickell[1]

[1]The Medical Molecular Biology Unit, Dept. of Biochemistry
[2]Dept. of Anatomy and Developmental Biology
University College and Middlesex School of Medicine
The Windeyer Building, Cleveland Street, London W1P 6DB, U.K.

INTRODUCTION

Local application of retinoic acid (RA) to the anterior margin of the chick wing bud can affect patterning in both the developing limb and face, and there is evidence that RA may be a natural signalling substance in the limb (reviewed in Eichele, 1989; Brickell and Tickle, 1990, and Wedden et al, 1988). In the limb, local application of RA produces a mirror image duplication of structures across its anterior-posterior (A-P) axis. A second effect of RA application to the bud is inhibition of outgrowth of the upper beak primordium (the frontonasal mass), resulting in clefting of the primary palate.

The nuclear retinoic acid receptors (RARs) provide a means by which presence of RA in a cell can be interpreted in terms of changes in gene expression (reviewed in Evans, 1988 and Green and Chambon, 1988). In a number of cell lines in culture, transcription of the RAR-beta gene itself is stimulated by RA acting through both RAR-beta and RAR-alpha (de The et al, 1990). RAR-beta could therefore act as a feedback system through which the effects of RA on gene expression could be amplified.

To investigate the role of RARs in normal chick morphogenesis, and in the generation of RA-induced defects, we have isolated chick RAR-beta cDNA clones and used them to analyse RAR-beta expression by both Northern blotting and in situ hybridisation.

ISOLATION OF CHICK RAR-beta cDNA CLONES

We screened a 10 day (stage 36) chick embryo cDNA library with a fragment of the human RAR-alpha cDNA clone p63 (Petkovich et al, 1987) which contained sequences encoding the A, B and C domains. The inserts from two strongly-hybridising lambda gt11 clones, RAR1 and JO5, were subcloned, restriction mapped and sequenced.

The nucleotide sequences of RAR1 and JO5 showed that they were overlapping cDNA clones derived from the same mRNA species. The composite nucleotide and predicted amino acid sequences of the two clones were compared with those of human and mouse RAR-alpha, -beta and -gamma. The extent of the homologies demonstrated clearly that RAR1 and JO5 were derived from chicken RAR-beta mRNA.

RAR-beta TRANSCRIPTS IN THE CHICK EMBRYO

The whole RAR1 insert was used as a template for synthesis of RNA probes for hybridisation to Northern blots and tissue sections. Northern blot analyses revealed polyadenylated transcripts of 2.8 and 3.5kb in a range of embryonic chick tissues including limb bud and head during the times at which RA treatment affects their development (Fig. 1). RAR-beta transcripts were also present in adult tissues, but at much lower levels.

Hybridisation of Northern blots with an oligonucleotide probe corresponding to sequences encoding part of the A domain, which were specific to RAR-beta, identified the same two transcripts as the RAR1 probe. This showed that both transcripts were derived from the RAR-beta gene and that the RAR1 probe was specific for RAR-beta transcripts.

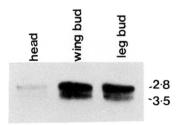

Fig. 1. Northern blot of total RNA from stage 22 chick embryo tissues, hybridised with the chick RAR-beta probe RAR1. The sizes of the transcripts are shown in kilobases.

IN SITU HYBRIDISATION TO NORMAL AND RA-TREATED WING BUDS

We investigated the effect of RA treatment on RAR-beta expression in the wing bud by in situ hybridisation with ^{35}S-labelled RNA probes synthesised using the same template as for the Northern hybridisation probe. RA soaked beads were grafted to the anterior margin of one wing bud of stage 20 embryos. 24 hours later (stage 24), beads were removed and embryos were prepared for in situ hybridisation. Each embryo had one treated and one untreated bud, so that RAR-beta expression could be compared between the treated and normal buds of individual embryos.

In the normal bud at stage 24, RAR-beta transcripts were detected in the proximal mesenchyme (i.e. in the region of the shoulder girdle), extending into the anterior of the bud, and in the flanking mesenchyme around the bud (Fig. 2). Local application of RA resulted in a strong and specific local increase in RAR-beta transcripts around the site of the bead, so that the levels were increased anteriorly and somewhat distally as compared to the normal pattern (Fig. 2).

The RAR-beta gene is the only RAR whose expression has been shown to be directly stimulated by RA application in vitro, and our results show that its expression is also affected in vivo. The change in RAR-beta transcript distribution around the bead suggests that RA could be involved in the reprogramming of gene expression in cells in this area. However, it is unclear how this local effect could be related to the interpretation of changes in a gradient of RA across the A-P axis of the bud.

At stage 24, RAR-beta transcripts in the wing bud were confined to the shoulder girdle region, and the morphology of the bud was already changing as a result of RA treatment. Further investigation will be necessary to establish if there is any real association between the changes in RAR-beta expression and the effects of RA on limb bud morphogenesis. In particular, examination of the distribution of RAR-beta transcripts during the 18 hours after bead implantation, when RA is required to obtain duplications, may be informative.

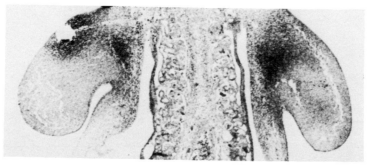

Fig. 2. In situ hybridisation of the chick RAR-beta probe RAR1 to a frontal section of a stage 24 embryo. The hole in the wing bud on the left was the location of the RA-soaked bead. Sections hybridised with a sense RNA control probe (not shown) showed low uniform levels of signal which were unchanged by RA treatment.

ACKNOWLEDGEMENTS

We thank Cheryll Tickle and Lewis Wolpert for advice and helpful discussions, and also Cheryll Tickle for grafting RA-soaked beads. We are grateful to Pierre Chambon for the gift of the human RAR-alpha cDNA clone p63. This work was supported by the Medical Research Council of Great Britain (AR), the Medical Research Council of Canada (JMR), and the British Council (JOO).

REFERENCES

Brickell, P. M. and Tickle, C., 1990, Morphogens in chick limb development, Bioessays, 11:145.
de The, H., del Mar Vivanco-Ruiz, M., Tiollais, P., Stunnenberg, H. and Dejean, A., 1990, Identification of a retinoic acid responsive element in the retinoic acid receptor beta gene, Nature, 343:177.
Eichele, G., 1989, Retinoids and vertebrate limb pattern formation, Trends in Genetics, 5:246.

Evans, R. M., 1988, The steroid and thyroid hormone receptor superfamily, <u>Science</u>, 240:889.

Green, S. and Chambon, P., 1988, Nuclear receptors enhance our understanding of transcription regulation, <u>Trends in Genetics</u>, 4:309.

Petkovich, M.,Brand, N. J., Krust, A. and Chambon, P., 1987, A human retinoic acid receptor which belongs to the family of nuclear receptors, <u>Nature</u>, 330:444.

Wedden, S. E., Ralphs, J. R. and Tickle, C., 1988, Pattern formation in the facial primordia, <u>Development</u>, 103 Suppl.:31.

EXPRESSION PATTERN OF AN RXR NUCLEAR RECEPTOR GENE

IN THE CHICK EMBRYO

Annie Rowe[1,2], Nicholas S. C. Eager[1], Melanie Saville[1],
Lewis Wolpert[2] and Paul M. Brickell[1]

[1] The Medical Molecular Biology Unit, Department of
Biochemistry

[2] Department of Anatomy and Developmental Biology

University College and Middlesex School of Medicine
The Windeyer Building
Cleveland Street
London W1P 6DB

INTRODUCTION

Retinoic Acid (RA) affects differentiation and morphogenesis in
various developmental systems, including the vertebrate limb (Brickell
and Tickle, 1989; Brockes, 1989), and acts by binding to nuclear RA
receptors which are members of the steroid/thyroid hormone family of
ligand-binding transcription factors (Green and Chambon, 1988; de The et
al, 1990). Three closely-related RA receptors (RAR-alpha, RAR-beta and
RAR-gamma), encoded by distinct genes and with distinct patterns of
expression (Ruberte et al, 1990), have been identified in humans and
mice, and a fourth RA receptor (hRXR-alpha) was discovered recently in
humans (Mangelsdorf et al, 1990). Whilst still a member of the
steroid/thyroid hormone receptor family, hRXR-alpha shares no significant
homology with the RA-binding domain of RAR-alpha, RAR-beta or RAR-gamma
and appears to represent an evolutionarily distinct RA response pathway
(Mangelsdorf et al, 1990). We have isolated a cDNA clone encoding a
chicken homologue of hRXR-alpha (cRXR) and have used this to examine the
distribution of cRXR transcripts in chick embryos.

ISOLATION OF A cDNA CLONE ENCODING A CHICKEN HOMOLOGUE OF hRXR-alpha

A stage 36 (10 day) chick embryo cDNA library was screened at low
stringency with a radiolabelled probe derived from the human RAR-alpha
cDNA clone p63 (Petkovich et al, 1987). Amongst the clones isolated was
R2. The nucleotide sequence of the insert of R2 and the predicted amino
acid sequence of the protein which it would encode (termed cRXR) were
determined. The predicted amino acid sequence of cRXR was strongly
homologous to that of hRXR-alpha (Mangelsdorf et al, 1990), not only in
the putative DNA-binding domain, but also in the putative ligand-binding
domain (Fig. 1). There was much lower homology to the DNA-binding
domains of RAR-alpha, RAR-beta and RAR-alpha, and no significant homology

Developmental Patterning of the Vertebrate Limb
Edited by J.R. Hinchliffe *et al.*, Plenum Press, New York, 1991

to the ligand-binding domains of these proteins or to those of any other
known members of the steroid/thyroid hormone receptor family (data not
shown). The cDNA clone R2 therefore appears to have been derived from an
mRNA encoding the chicken homologue of hRXR-alpha, or a closely-related
protein.

NORTHERN BLOTTING ANALYSIS OF cRXR GENE EXPRESSION

A fragment of R2 was used as a template for synthesis of a
^{32}P-labelled antisense RNA probe for cRXR transcripts. This probe was
complementary only to sequences encoding the putative ligand-binding
domain of cRXR (Fig. 1), which are not conserved amongst members of the
steroid/thyroid hormone receptor family. Northern blotting analysis
identified a single cRXR transcript of approximately 2.5kb in chick limb
buds, heads and bodies at stages 22 and 25. This transcript was also
found at low levels in a range of adult chicken tissues, and at high
levels in adult liver.

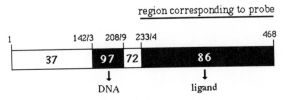

Fig. 1. Predicted domain structure of cRXR protein. DNA- and
ligand-binding domains are indicated. Amino acid residues at
domain boundaries are numbered. Percentage amino acid identity
between cRXR and hRXR-alpha (Mangelsdorf et al, 1990) is shown
in each domain. The region corresponding to the probe used in
expression studies is indicated.

DISTRIBUTION OF cRXR TRANSCRIPTS IN STAGE 24 CHICK EMBRYOS

To determine the distribution of cRXR transcripts, in situ
hybridisation to tissue sections from stage 24 chick embryos was
performed using a ^{35}S-labelled RNA probe synthesised from the same
template as the Northern blotting probe. As shown in Fig. 2, the probe
hybridised strongly to liver and to elements of the peripheral nervous
system, including the dorsal root ganglion, the ventral root, the spinal
nerve and the sympathetic chain. Adjacent sections showed strong
hybridisation to the dorsal root. There was no hybridisation to the
neural tube or developing ventral horn. There was punctate hybridisation
to the gut wall, consistent with hybridisation to cells of the developing
enteric ganglia. There was hybridisation to the developing innervation
of the wing buds.

Examination of frontal and sagittal sections of stage 24 chick
embryos demonstrated that cRXR transcripts were present in all of the
dorsal root ganglia along the body axis, and in the spinal

nerves associated with them.   Transcripts were also present within the trigeminal ganglion.   Since there was hybridisation to nerve roots, cRXR transcripts must be expressed by peripheral glial cells.   However, neurons may also express cRXR transcripts.   in either case, expression of cRXR transcripts within nervous tissue is restricted to cells derived from the neural crest (Le Douarin and Smith, 1988).

DISCUSSION

The distribution of cRXR transcripts is quite different to those of RAR-alpha, RAR-beta and RAR-gamma transcripts.   In particular, RAR-alpha, RAR-beta and RAR-gamma transcripts have not been found in the peripheral nervous sytem at any stage of development (Ruberte et al, 1990).   This indicates that the biological roles of the RXR-like and RAR-like RA receptors are different.   The distribution of cRXR transcripts is similar in some respects to that of the cellular RA-binding protein   (Maden et al, 1990).   However, the distribution of cRXR transcripts correlates more closely with neural crest origin than does cellular RA-binding protein expression, and cRXR provides a marker for a specific population of neural crest-derived cells.

Mangelsdorf et al (1990) showed by Northern blotting that high levels of RXR transcripts were present in adult liver, and suggested a role for hRXR-alpha as a regulator of vitamin A metabolism by this organ. Our data support this idea, but also indicate a role for cRXR as a regulator of gene expression in the neural crest-derived cells of the peripheral nervous system.   The precise nature of this role is unclear, particularly since there is no well-documented function for RA in the peripheral nervous system.   Retinoids do induce changes in

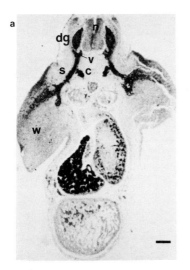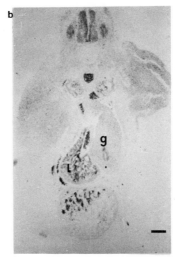

Fig. 2.   Adjacent transverse sections of stage 24 chick embryo hybridised with an antisense RNA probe for cRXR transcripts (a) or a sense RNA probe as a negative control (b).   Scale bars: 200um.   c, sympathetic chain; dg, dorsal root ganglion; g, gut; l, liver; n, neural tube; s, spinal nerve; v, ventral root; w, wing bud. Sections were stained with malachite green

cell-substratum adhesion in cranial and trunk neural crest cells in vitro, and it has been suggested that this inhibits neural crest migration in vivo. However, retinoids probably exert these effects on neural crest cells in vitro by disrupting cell membrane structure rather than by inducing changes in gene expression (Smith-Thomas et al, 1987). It is also possible that a metabolite related to RA, rather than RA itself, is the natural ligand of cRXR (Mangelsdorf et al, 1990). As with the RAR family of RA receptors, a fuller understanding of function will require the identification of target genes regulated in vivo.

ACKNOWLEDGMENTS

We thank Pierre Chambon for the gift of the human RAR-alpha cDNA clone p63 and are grateful to Cheryll Tickle for her comments. This work was supported by the Medical Research Council.

REFERENCES

Brickell, P.M. and Tickle, C., 1990, Morphogens in chick limb development, BioEssays, 11:145.
Brockes, J.P., 1989, Retinoids, homeobox genes and limb morphogenesis, Neuron, 2:1285.
de The, H., del Mar Vivanco-Ruiz, M., Tiollais, P., Stunnenberg, H. and Dejean, A., 1990, Identification of a retinoic acid responsive element in the retinoic acid receptor beta gene, Nature, 343:177.
Green, S. and Chambon, P., 1988, Nuclear receptors enhance our understanding of transcription regulation, Trends Genet., 4:309.
Le Douarin, N.M. and Smith, J., 1988, Development of the peripheral nervous system from the neural crest, Ann. Rev. Cell Biol., 4:375.
Maden, M., Ong, D.E. and Chytil, F., 1990, Retinoid-binding protein distribution in the developing mammalian nervous system, Development, 109:75.
Mangelsdorf, D.J., Ong, E.S., Dyck, J.A. and Evans, R.M., 1990, Nuclear receptor that identifies a novel retinoic acid response pathway, Nature, 345:224.
Petkovich, M., Brand, N.J., Krust, A. and Chambon, P., 1987, A human retinoic acid receptor which belongs to the family of nuclear receptors, Nature, 330:444.
Ruberte, E., Dolle, P., Krust, A., Zelent, A., Morris-Kay, G. and Chambon, P., 1990, Specific spatial and temporal distribution of retinoic acid receptor gamma transcripts during mouse embryogenesis, Development, 108:213.
Smith-Thomas, L., Lott, I. and Bronner-Fraser, M., 1987, Effects of isotretinoin on the behaviour of neural crest cells in vitro, Devl. Biol., 123:276.

RETINOIC ACID AND LIMB PATTERN FORMATION IN CELL CULTURE

Hiroyuki Ide[1], Yasushi Ohkubo[1], Sumihare Noji[2], and Shigehiko Taniguchi[2]

[1]Biological Institute, Tohoku University, Sendai and [2]Department of Biochemistry, Okayama University Dental School, Okayama, JAPAN

INTRODUCTION

It is well known that retinoic acid (RA) applied to the anterior margin of chick limb buds induces antero-posterior duplicate formation as ZPA (zone of polarizing activity) does, and the degree of duplication depends on the concentration of applied RA. RA is considered to be formed in ZPA and affects limb bud cells as a morphogen (Wolpert, 1990). The limb bud cells recognize their positions with respect to antero-posterior axis with the concentration of RA and differentiates to form cartilage pattern. However, it is possible as originally demonstrated by Summerbell and Harvey (1983) that RA may convert adjacent mesodermal cells to ZPA and the ZPA may affect the cells of progress zone by releasing some other molecule(s) with morphogen-property or through the process of local interactions between the tissues with different positional values.

Thus, although it remains uncertain whether RA is a morphogen or inducer of ZPA, an important problem concerning the action of RA is what is the first molecular change in the process of duplicate formation. The expression pattern of chick homeobox gene changed after the application of RA into the anterior margin of limb bud (Yokouchi and Kuroiwa, 1990; Koyama et al., 1990). These results are very important for the analysis of RA-induced duplicate formation at molecular level since we can see regional changes of these expression pattern directly in the limb bud. However, it is difficult to analyze the quantitative changes of other unidentified molecules after the application of RA. Thus, we have established a cell culture system of progress zone to elucidate the molecular changes of RA-induced duplicate formation in the chick limb bud, although it is uncertain whether the changes observed in the cell culture accurately correspond to the changes essential for the duplicate formation. We have already reported that RA promotes cell proliferation and chondrogenesis in the distal mesodermal cells of chick limb bud in cell culture (Ide and Aono, 1988) and compared the effects of RA in culture with those of duplicate formation (Ide, 1990). In this paper, we have examined the spatial and temporal changes in the responsibility to RA in detail and analyzed the molecular changes induced in RA-treated limb bud cell cultures.

Developmental Patterning of the Vertebrate Limb
Edited by J.R. Hinchliffe *et al.*, Plenum Press, New York, 1991

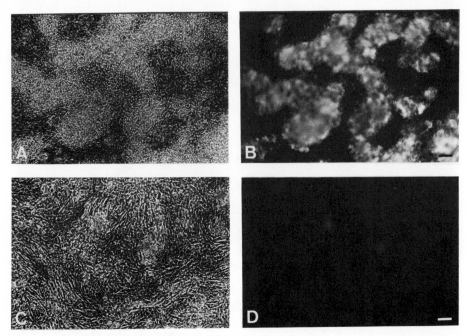

Fig.1 Chondrogenesis in the cultured mesodermal cells of anterior
half(A, B) and posterior half(C, D) of proximal limb budsat
stage 24. Proximal half of limb bud was excised and dissected
into anterior and posterior halves. Mesodermal cells were
isolated from these fragments after Hattori and Ide (1984) and
cultured for 5 days in Ham's F12 medium containing 1% fetal calf
serum with a micro-culture system (Hattori and Ide, 1984). A
and C, phase contrast. B and D, The cultures were fixed in PLP
and stained with rabbit anti-chick proteoglycan H (PGH) antibody
(cartilage specific, gifted from Dr. K. Kimata) and FITC-labeled
anti-rabbit IgG. Bars, 100u.

CHONDROGENIC DIFFERENTIATION IN RA-FREE MEDIUM

Distal cells

Distal mesodermal cells (cells of progress zone) of stages 20-26
limb buds formed a uniform sheet and gradually differentiated into
chondrocyte. Almost all the cells differentiated into the chondrocytes
as a sheet and few regional differences like formation of cartilage
nodules were observed in the culture. However, a tendency that distal
cells of young limb buds (stages 21-23) showedhigher chondrogenic
activity than those of old limb buds (stages 24-26) was observed.

Proximal cells

It is well known that when whole mesodermal cells of proximal limb
bud region are cultured at high cell density, cartilage nodule formation
occurs within 3 days. The cells of whole proximal half of young limb
buds (stages 21-22) showed the capability of nodule formation. However,
at stages 23-26, differences in the cell proliferation and chondrogenic
activity between the anterior and posterior halves of the proximal limb
bud was observed (Fig. 1). High chondrogenic activity was observed

Table 1. Chondrogenesis and Proliferation
of Proximal Limb Bud Cells

Region[a]	Chondrogenesis[b] (A_{600})	Proliferation[c] (Cell No. $X10^5$)
Most anterior	0.103	7.06
Central anterior	0.020	3.52
Central posterior	0.025	3.01
Most posterior	0.020	1.59

[a]Proximal mesoderm was dissected into four pieces in the anteroposterior sequence, and dissociated cells of each region were used after 5 days in culture for assay of chondrogenic and growth activity.
[b]Alcian blue staining after Wedden et al.(1987). The results are means of 2-3 samples.
[c]The results are means of two samples.

only in the anterior half, especially anterior 1/4 at stage 24 (Table 1). The cell proliferation was also high in the anterior 1/4 (Table 1). This localization of high chondrogenic activity in the anterior region may correspond to the regional difference in cell proliferation demonstrated by cell culture experiments (Aono and Ide, 1988). The cells isolated from the most anterior region responded most actively to co-cultured ZPA and fibroblast growth factor to proliferate, although these cells did not proliferate in 0.1% serum condition. It is likely that some growth factors contained sufficiently in the medium with 1 % serum stimulate the proliferation of the cells from the most anterior regions and the resultant high cell density promoted chondrogenesis in the culture.

CHONDROGENIC DIFFERENTIATION IN THE MEDIUM CONTAINING RA

Distal cells

RA at 10-25 ng/ml rapidly promoted proliferation and chondrogenesis in distal mesodermal cells (Ide and Aono, 1988). The stimulation of chodrogenesis was confirmed by the immunofluorescence staining with

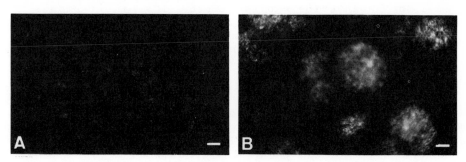

Fig. 2 Chondrogenesis induced by RA. Distal mesodermal cells from
stage 21 limb buds were cultured for 5 days in the medium
containing 0 (A) and 10 (B) ng/ml. Immunofluorescence
staining with anti PGH antibody. Bars, 100 u.

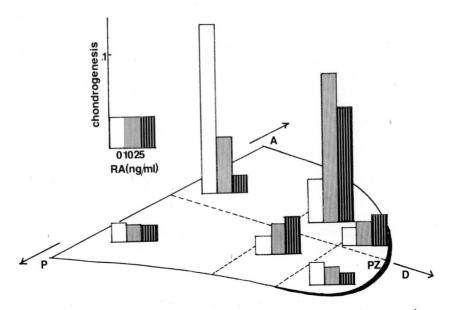

Fig. 3 RA-promoted chondrogenesis in the cells of various regions of
stage 24 limb buds. The cells were cultured for 5 days and
chondrogenic activity was measured with alcian blue staining.
The values were means of 2-3 samples.

antibody to PGH (Fig. 2).   As discussed previously, RA promoted cell
proliferation and chondrogenesis in vitro showed marked resemblance to
RA-induced duplicate formation (Ide, 1990). One difference is that both
the cells of anterior and posterior progress zone were responsive to RA,
although, in duplicated digit formation, only RA applied to the anterior
progress zone was effective. In the present experiment, a difference in
RA-promoted chondrogenic activity between the anterior and posterior
region was found. In the young progress zone (stage 20-23), no
difference in RA-promoted chondrogenic activity between the anterior and
posterior half was observed, but in the old progress zone (stage 24-25),
the responsiveness to RA disappeared from the posterior half and only
the cells of anterior progress zone were responsive to RA (Fig. 3). This
seems to correspond to the lack of excess cartilage formation when RA is
applied to the posterior progress zone. Further, as limb bud development
proceeds, the most responsive region moved from the anterior progress
zone to the anterior subdistal zone (Fig. 3). No promotion in
chondrogenesis was observed when the cells of anterior distal region and
of anterior proximal region were mixed and cultured, indicating that the
responsiveness to RA was remaining only in the anterior subdistal
region and the responsiveness in the subdistal region was not caused by
the interaction between the cells of distal and proximal regions. This
change in localization of responsiveness to RA may have occurred already
in the young progress zone but not detected since the width of the young
progress zone is narrow. The subdistal localization of responsiveness to
RA may be concerned with the effective region for implantation of beads
containing RA for duplicate formation since the beads remain in the
subdistal region as limb bud development proceeds.

Proximal cells

     All the cells of proximal limb buds, including the region of
posterior half, showed the inhibitory response of chondrogenesis after
the RA treatment.

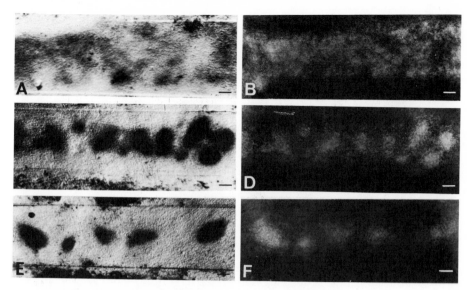

Fig. 4 RA-induced cartilage pattern formation in the cultures of distal
cells surrounded by retinal pigment cells. In the medium without RA
(A-B), chondrogenesis occurred as a sheet. In the medium with 10
ng/ml RA (C-D), chondrogenesis occurred as a cluster of nodules. In
the medium with 25 ng/ml RA (E-F), chondrogenesis occurred as a
spaced nodules. A, C and E: alcian blue staining. B, D and F:
anti-PGH immunofluorescence. Bar: 100 u.

EFFECTS OF RETINAL PIGMENT CELLS ON CARTILAGE PATTERN FORMATION

Limb pattern formation occurs during the process of epithelio-
mesenchymal interaction. Non-ridge ectodermal cells are known to inhibit
chondrogenesis and this may cause the formation of fibrous connective
tissues (dermis) in the peripheral area of the limb (Solursh, 1983). The
retinal pigment cells (ectodermal cells) of chick embryo also inhibited
chondrogenic differentiation (Solursh et al., 1981). Thus, we co-
cultured the retinal pigment cells with the culture of progress zone
cells to mimic the epithelio-mesenchymal interaction in the limb buds
(Ide, 1990). Spatially ordered aggregate formation of the mesodermal
cells occurred in the center of rectangular culture area and these
aggregates developed to cartilage nodules, which may correspond to
cartilage elements of cross-sectioned forelimb at radius-ulna or digit
level. Internodule space was occupied by fibroblastic cells. The nodule
formation occurred only in the culture with RA (Fig. 4). Without RA,
almost all the mesodermal cells differentiated to form cartilage sheet,
although the degree of the chondrogenesis was poor. RA may affect to
form these structures from homogeneous cell population as a
morphogenetic substance in the distal limb bud region.

MOLECULAR CHANGES IN RA-TREATED CULTURES

It remains uncertain what is the first step in the process of
chondrogenesis induced by RA. Using 2D-electrophoresis, we have analyzed
protein synthesis in cultured cells of distal mesoderm after RA-
treatment. The synthesis of two proteins significantly changed. Within 6
hours after the addition of 10 ng/ml RA, the synthesis of one protein

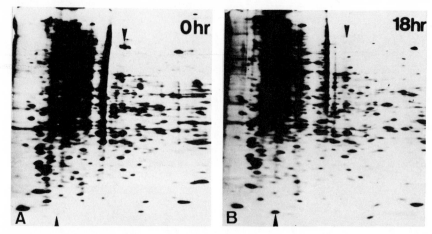

Fig. 5 Two dimensional gels of protein synthesized by cultured limb
bud distal mesodermal cells. The cells were treated with
10 ng/ml RA for 0 (left) and 18 (right) hours. The cells
were labeled with ^{35}S-methionine for 3 hours. Molecular
weight of upper and lower spots indicated by arrow heads was
200kd and 18kd, respectively.

(Mr. 18kd) was promoted. The synthesis was maximal at 18 hours (Fig. 5)
and decreased after 36 hours. The synthesis of the other protein
(Mr.200kd) was inhibited rapidly by RA and stopped after 6 hours.
Although the roles of these proteins in RA-induced chondrogenesis and
cell proliferation and in duplicate formation, they may be concerned
with the early responses of distal limb bud cells to the RA as a
morphogen or ZPA-inducer.

RA is known to bind to RAR (retinoic acid receptor) to control gene
expression. The expression of RARbeta, a subfamily of RAR, is up-
regulated by the RA (de The et al., 1989). Thus, it is likely that the
expression of RARbeta gene in the limb bud cells is one of early
responses after RA treatment. In developing mouse limb bud, RARbeta
gene was expressed in the proximal region (Dolle et al., 1989). Thus,
we have examined the expression of RARbeta gene in the culture with or
without RA. Within 12 hours after the addition of RA, the expression of
RARbeta increased in the culture of distal and subdistal mesodermal
cells, and at 24 hours, intense signal was observed in the peripheral
region of aggregates of subdistal mesodermal cells (Fig. 6). This type
of expression pattern was also observed after 48 and 96 hours. The
promotion of RARbeta gene expression suggests the possibility that RA
proximalizes the distal limb bud cells as in the case of amphibian limb
regeneration (Stocum and Crawford, 1987).

RA AND LIMB PATTERN FORMATION

In cell culture, the distal mesodermal cells constituting progress
zone responded to RA by promoting the proliferation and chondrogenic
differentiation. The effective concentration of RA in cell culture was
almost similar to the concentration of RA measured in the limb buds. The
responding cells were localized in the distal region and not in the

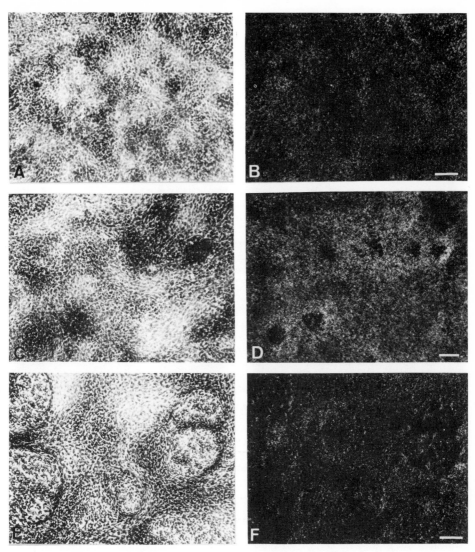

Fig. 6 Expression patterns of chick RARbeta gene in the subdistal
    limb bud cells cultured with 0 (A, B) and 10 (C, D) ng/ml RA
    for 24 hours or with 10 ng/ml RA for 96 hours (E, F). The
    procedures for in situ hybridization were described
    previously (Noji et al., 1989). The method of the
    preparation of antisense probe and the base sequence of the
    chick RARbeta cDNA will be described elsewhere (Noji et al.,
    submitted). Bright- (left) and dark- (right) views.
    Bar, 100 u.

proximal region. Further, at least in stage 24-26 limb buds, the antero-
posterior difference in the responsiveness to RA was observed as in the
case of grafting beads containing RA. These results suggest that the
cellular and molecular changes observed in RA-treated culture reflects
at least partially the process of RA-induced duplicate formation.
    One remarkable feature of the response to RA is the promotion of
cell proliferation, which will correspond to the active growth

preceding the duplicate formation (Cook and Summerbell). The promotion of chondrogenic differentiation by RA is not merely the result of promotion of cell proliferation (Ide and Aono, 1988). RA promotes differentiation of the cells of cartilage and probably of fibrous connective tissues directly.

The distribution of the competence to RA is a crucial factor for changing limb pattern. RA must be applied at the anterior margin for duplicate formation in chick limb buds. In salamander regenerating limb, the region duplicating along proximo-distal axis is confined to anterior half of the blastema (Wigmore, 1990).

In the present experiments, it was found that the region responsive to RA confined to anterior region and changed from distal to subdistal region as the limb bud development proceeds. Gradual decrease of the responsiveness to RA in the posterior region may be caused by some diffusible factor(s) from ZPA. RA itself will be a candidate for the factor. Chondrogenic differentiation occurs from posterior to anterior region at least in chick limb bud (Hinchliffe, 1977). RA or some other ZPA factors may promote differentiation in this direction and uncommitted cells may remain in the anterior region (Ide, 1990). On the other hand, cell lineage analysis revealed the gradual translocation of descendant of apex mesodermal cells to anterior region by the active proliferation of posterior distal cells (Bowen et al., 1989). Hence, the descendant of distal mesodermal cells which are responsive to RA may translocate to the anterior region as development proceeds. The distribution of RA-responsive cells in the subdistal region of old limb buds may have occurred already in young limb bud as described above. If so, the most distal cells of limb bud are unresponsive to RA, and the descendants of these cells may acquire the responsiveness when they moved to the region of proximal progress zone. The other possibility is that the responsiveness of the distal cells is actually lost in old limb buds and the responsiveness remained only in the subdistal cells, which may be concerned with the change of AER activity. Although somewhat different in time and place, the expression pattern of AV-1 (a position-specific antigen of chick limb buds) showed similar changes (Ohsugi et al., 1988), suggesting some relationships between the cell's state for the responsiveness to RA and the expression of the AV-1 antigen.

Acknowledgments. The original research in this chaptor was supported by grants-in-aid from the Ministry of Education, Science and Culture of Japan, and grant from the Mitsubishi Foundation.

REFERENCES

Aono, H. and Ide, H., 1988, A gradient of responsiveness to the growth-promoting activity of ZPA (Zone of polarizing activity) in the chick limb bud. Dev. Biol., 128: 136-141.

Bowen, J. Hinchliffe, J. R. Horder, T. J. and Reeve, A. M. F., 1989, The fate map of the chick forelimb-bud and its bearing on hypothesized developmental control mechanisms, Acta Embryol., 179: 269-283.

Cook, J. and D. Summerbell, 1980, Cell cycle and experimental pattern duplication in the chick wing during embryonic development, Nature, 287: 697-701.

Dolle, P., Ruberte, E., Kastner, P., Petkovich, M., Stoner, C. M., Gudas, L. J. and Chambon, P., 1989, Differential expression of genes encoding alpha, beta and gamma retinoic acid receptors and CRABP in the developing limbs of the mouse, Nature, 342: 702-705.

Hattori, T. and Ide, H., 1984, Limb bud chondrogenesis in cell culture, with particular reference to serum concentration in the culture medium, Exp. Cell Res., 150: 338-346.

Hinchliffe, J. R. , 1977, The chondrogenic pattern in chick limb morphogenesis: A problem of development and evolution. in "Vertebrate Limb and Somite Morphogenesis" D. A. Ede, J. R. Hinchliffe, and M. Balls, eds., Cambridge Univ. Press, Cambridge.

Ide, H, 1990, Growth and differentiation of limb bud cells in vitro: Implications for limb pattern formation, Develop., Growth & Differ., 32:1-8.

Ide, H. and Aono, H., 1988, Retinoic acid promotes proliferation and chondrogenesis in the distal mesodermal cells of chick limb bud, Dev. Biol., 130: 767-773.

Koyama, E., Oyama, K., Noji, S., Taniguchi, S., Nohno, T., Saito, T. and Kuroiwa, A., 1990, Identification of chick Chox 5 gene cluster and its expression in chick limb buds, Develop. Growth & Differ., 32: 422

Noji, S., Yamaai, T., Koyama, E. Nohno, T., and Taniguchi, S., 1989, FEBS lett., 257: 93-96.

Ohsugi, K., Ide, H. and Momoi, T., 1988, Temporal and spatial expression of a position specific antigen, AV-1, in chick limb buds. Dev. Biol., 130: 454-463.

Solursh, M., 1983, Cell interactions during in vitro chondrogenesis. in "Limb Development and Regeneration" Part B R. O. Kelley, P. F. Goentinck and J. A. MacCabe eds. Alan R. Liss, New York.

Solursh, M., Singley, C. T. and Reiter, R. S., 1981, The influence of epithelia on cartilage and loose connective tissue formation by limb mesenchyme cultures, Dev. Biol., 86, 471-482.

Stocum, D. L. and Crawford, K., 1987, Use of retinoids to analyze the cellular basis of positional memory in regenerating amphibian limbs. Biochem. Cell. Biol., 65, 750-761.

Summerbell, D. and Harvey, F., 1983, Vitamin A and control of pattern in developing limbs. in "Limb Development and Regeneration " Part A J. F. Fallon and A. I. Caplan eds. Alan R. Liss, New York.

Wedden, S. E., Lewin-Smith, M. R., and Tickle, C., 1987, The effects of retinoids on cartilage differentiation in micromass cultures of chick facial primordia and the relationship to a specific facial defects, Dev. Biol., 122: 78-89.

Wigmore, P, 1990, Serially duplicated regenerates from the anterior half of the axolotl limb after retinoic acid treatment. Roux's Arch. Dev. Biol., 198: 252-256.

Wolpert, L., 1990, Positional information revisited, Development, 107 supplement: 3-12.

Yokouchi, Y. and Kuroiwa, A., 1990, Up and down regulation of region specific homeobox gene expression by retinoic acid in developing limb bud, Develop. Growth & Differ., 32: 435.

GROWTH FACTORS PRODUCED BY THE POLARISING ZONE –

A COMPLEMENT TO THE RETINOIC ACID SYSTEM

John C. McLachlan

Department of Biology and Pre-Clinical
Medicine, University of St. Andrews
Fife KY16 9TS, SCOTLAND

INTRODUCTION

After a ZPA graft to the limb anterior margin, the
limb bud becomes wider, and the resulting reduplicated limb
contains many more cells than a comparable control (Smith and
Wolpert, 1981). This indicates that the ZPA has initiated
growth in the host limb. This effect can also be observed at
the cellular level, in that cells close to a ZPA graft
show an increased labelling index compared to controls
(Cooke and Summerbell, 1980). Such a growth promoting effect
might be at least formally separable from the reduplicating
effect of the ZPA, and it is certainly more amenable to
quantitative analysis *in vitro*. We have therefore been
engaged in investigating the mitogenic effect of ZPAs in
culture by a number of different techniques, including time
lapse video recording, and uptake of tritiated thymidine as
detected both by scintillation counting and by autoradiography. Although each of these methods may present
special problems, in general we detect good concordance
between the different techniques.

Unless otherwise stated, NIH-3T3 cells were employed in
the assays. These were preferred to primary chick limb cells
on the grounds that they are homogenous, do not differentiate during the experimental period, are resistent to
low serum conditions, and have well established cell
kinetics.

The amount of tissue, and medium conditioned by the
presence of tissue, that can be obtained directly from the
embryo is very small in absolute terms. Conventional
biochemical techniques for the identification of unknown
growth factors are not feasible. Instead we have relied upon
sensitive bio-assays to detect mitogenic effects from limb
tissues. Once such effects were well characterised, the
physical nature of the underlying mitogen was explored.
Next, in order to attempt to identify the class of molecule

Developmental Patterning of the Vertebrate Limb
Edited by J.R. Hinchliffe *et al.*, Plenum Press, New York, 1991

involved, a panel of cell types with known differential growth factor responses was employed. The response of the panel of cell types to a number of test growth factors was compared to its response to the ZPA mitogen. In addition, Anchorage Independent Growth and macrophage colony stimulating assays were carried out.

Some of these results have been published previously, and are summarised here for the sake of completeness.

METHODS

Tissue Preparation

Tissue pieces for assay were removed from limb buds at Hamburger and Hamilton (1951) stages 19-22, unless otherwise stated. Cubes of side 200 um, as measured by a photo-graphically reduced grid placed underneath the culture dish, were isolated from the region of maximum ZPA activity (Honig and Summerbell, 1985) using electrolytically sharpened tungsten needles. Control pieces were removed from the anterior margin of the limb bud. Tissue cubes were washed at least twice in serum-free medium before assay. Conditioned medium was prepared by incubating ZPAs or control limb cubes in serum free medium for 24 hours. The medium was then removed using disposable syringes and needles - use of glass pipettes decreased the mitogenic activity.

Mitogenesis Assays

In general, sparse suspensions of the test cells were allowed to attach to the substrate in 5% FCS medium for four hours. Then the cells were switched to 0.5% FCS medium for 18-24 hours to reduce the number of background divisions. (This regime did not induce full quiescence, in that a few divisions were observed in the absence of any additives). Then test materials, such as (i) tissue pieces (ii) medium conditioned by the presence of tissue pieces (iii) known growth factors, were added for at least another 24 hours. Full details of the cell culture and tritiated thymidine regimes are contained in Bell and McLachlan (1985), Bell (1986) and Smith and McLachlan (1990)

In time lapse video recording studies, sparse 3T3 cells were placed in a drop of 5% FCS medium near the neck of a Falcon 25 cm^2 tissue culture flask. Additional drops of sterile distilled water were spotted around the medium to prevent evaporation. After 4 hours, the medium was replaced with 0.5% FCS medium, and the flask gassed with 5% CO_2 and sealed. After 24 hours further incubation, 24 ZPAs or anterior margin pieces (prepared as described below) were added to the drop of medium. The flasks were regassed and sealed, and transfered to a 39°C hot room. A field of view containing 10-20 cells (but no tissue pieces) was selected, and recording was carried out continuously over the next 42 hours using a National Panasonic NV8030 time lapse video recorder, attached to a Hitachi CCTV camera and a Zeiss Inverted Phase microscope. Tapes were coded for blind analyses: cells were traced on the monitor screen using a transparent overlay, and 1st and 2nd cell division times were noted.

Physical Assays

Conditioned medium was passed through Centricon molecular weight exclusion filters prior to assay, to obtain an indication of the dimensions of the ZPA mitogen. The nature of the mitogen was explored by co-incubating conditioned medium with various Sephadex and Sepharose products. Phenyl Sepharose binds hydrophobic molecules, while Heparin Sepharose binds FGF and related growth factors. DAEA-Sephadex binds negatively charged molecules; CM-Sephadex binds positively charged molecules, and Con-A Sephadex binds glycoproteins.

Cellular Assays

Details of the panel of cell types employed, and their various culture regimes, are contained in Bell (1986). A positive response for the ZPA mitogen was one in which test cells responded to a significantly greater degree in the presence of ZPA than with anterior margin. Anchorage Independent Growth Assays were carried out as described in Macintyre et al. (1988). Macrophage assays were carried out on chick bone marrow cells, using the methods described in Wright et al. (1985).

RESULTS

Mitogenesis

ZPAs and ZPA conditioned medium stimulated entry into S-Phase in quiescent fibroblasts to a significantly greater degree than anterior margin tissue (see Bell & McLachlan, 1985). Time lapse video recording revealed that this effect was initiated 15 hours after inital exposure to the ZPA tissue (Fig. 1). Surprisingly, no statistically significant difference could be observed in the intermitotic times of cells exposed to either ZPAs or anterior margins (Fig. 2). The difference in fact lies in the proportion of cells initally present which enter division. Typically, 50% or so of the cell population would divide in the presence of ZPA as opposed to less than 30% in the presence of anterior margin.

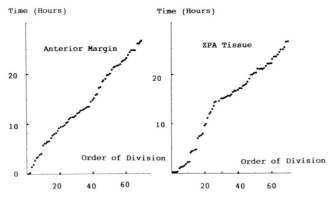

Fig. 1. Cell divisions plotted in order of occurrence against time on the Y axis. The number of divisions per unit time increases at 15 hours for ZPA tissue

The differential effect of the ZPA mitogen was markedly increased by the addition of insulin or $PGF_{2\ alpha}$, but not by FGF, EGF or MSA (Bell, 1986). There was a sharp dose response curve observed in the presence of ZPA tissue, plateauing at the level of 5 pieces per 0.5 ml culture medium, and remaining unchanged up to 30 pieces. In contrast, the effect of anterior margin tissue increased more slowly with increasing number, and 30 pieces of anterior margin failed to reach the plateau value. Although the assays were normally carried out using 3T3 cells, primary chick embryo fibroblast cultures obtained from the limb also responded in the same manner to ZPA and anterior margin tissue. The mitogenic activity of whole limb buds showed a marked temporal pattern, with peak activity between stages 19 and 23 (Smith and McLachlan, 1989).

Nature of the ZPA Mitogen

The mitogenic activity was retained by a 10 kDa molecular weight exclusion filter. The activity was unaffected by co-incubation with Phenyl Sepharose, and decreased but not completely removed by CM and DAEA Sephadex, suggesting that the molecule has weak positive and negative domains. Heparin and Con-A Sepharose removed the differential activity completely.

Cellular Assays

The response of a panel of cell lines to the ZPA mitogen and to various test growth factors suggests that the ZPA mitogen does not behave in the same way as EGF, FGF, insulin and IGFs, or MSA (Bell, 1986). This evidence must be interpreted with some caution, as discussed below.

In Anchorage Independent Growth assays, whole limb buds were observed to secrete a transforming activity (McLachlan et al., 1988). Further study has identified a number of

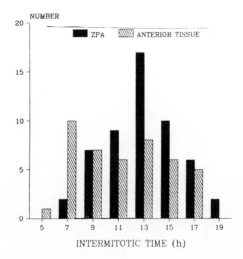

Fig. 2. Distribution of intermitotic times (from 1st division to 2nd division. No significant difference is observed between the populations.

separate components of this activity (Smith & McLachlan, 1990). However, we have not so far observed any difference between ZPA and anterior margin tissue with respect to these components, which include TGF Beta.

Macrophages were not stimulated to form colonies in the presence of medium conditioned by ZPA or anterior margin tissue. However, if colonies were first induced by the presence of 10% FCS, then ZPA conditioned medium significantly increased the number of colonies formed compared to anterior margin tissue (Fig. 3). This demonstrates that the ZPA mitogen is not a Macrophage Colony Stimulating Factor (MCSF), but is a potentiator of MCSF action in some way.

DISCUSSION

These results indicate that the ZPA produces a diffusible mitogenic activity which corresponds to the operational definition of a growth factor (Huang and Huang, 1985). Anterior margin either lacks this activity, or possesses it to a much lesser degree, although it would be rash to speak of a gradient across the limb based on two points. Co-incubation with a variety of separation media give results consistent with the idea that the mitogen is a heparin-binding non-hydrophobic glycoprotein, with weak positive and negative domains. The activity is retained by a 10 kDa molecular weight exclusion filter.

Comparison of the effect of the ZPA mitogen with a number of known growth factors, as assayed on a panel of cell types, suggests that it does not resemble EGF, IGF, MSA, or FGF in a simple way. However, the activity described for convenience here as the "ZPA mitogen" may consist of more than one component, or of a modified version of one of these growth factors.

It can be said with some certainty that the activity is not a transforming growth factor (or factors). No differential activity was observed between the ZPA and anterior margin in assays which detect both TGF Alphas and TGF Betas. NR6 cells were also observed to respond to the ZPA mitogen in the same

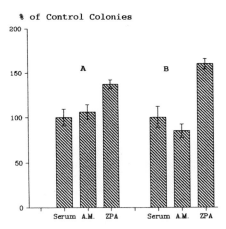

Fig. 3. Colonies as % of serum control in the presence of (A) 0.2 mls CM (B) 0.4 mls CM. Anterior margin tissue is not significantly different from serum.

way as 3T3 cells, although they lack functional EGF receptors (Bell, 1986): this rules out EGF and TGF Alpha by an independent criterion.

FGF might seem a particularly promising candidate, since it binds to heparin, and has been identified in the chick limb bud at stages similar to those reported here. It remains to be seen if any way can be found to reconcile the findings of the cell panel assay with this possibility. PDGF has also been suggested as a possible candidate for the ZPA mitogen (Bell, 1986).

Relationship with Retinoic Acid

There is still uncertainty as to the role of Retinoic Acid (RA) in the limb. As described elsewhere in this volume, it may be the natural morphogen, or it may mimic the morphogen, or it may initiate production of the morphogen, in the sense that it induces a ZPA to form. We have demonstrated that one of the features of the ZPA is that it produces a mitogen, and the relationship of this to RA may prove to be important. In assays of the kind described here, we have not found RA to be mitogenic at any of the concentrations tested. Although RA has been reported to increase cellular growth (Mehta et al. 1989); Ito, this volume) this is measured by increases in confluent cell density over 3-5 days, which may reflect underlying changes in cell communication properties, rather than a direct effect on cell division (see also Bryant, this volume). However, RA implants increase the width of the limb (Lee and Tickle, 1985). If RA is not a direct mitogen, this must reflect an induction of mitogenic activity in the anterior limb cells by RA. This would accord with the notion that RA induces a ZPA to form, and is testable in that it predicts that exposure of anterior margin tissue to RA should initiate detectable levels of ZPA mitogen production *in vitro*.

It also seems that the mitogen is not the morphogen. Apical tissue from the limb also produces a mitogenic activity which as the same dose reponse characteristics as the ZPA mitogen, and is additive in effect in combinations of conditioned media (Bell and McLachlan, 1985). This suggests that the apical mitogen and the ZPA mitogen are the same. Yet apical tissue does not initate reduplications when grafted to the anterior margin, even though it does increase the local labelling index.

It seems likely that when RA binds to its nuclear receptor, it then activates a series of genes which includes the ZPA mitogen. Given that the ZPA mitogen acts over a period which is similar to that over which the ZPA is effective *in vivo*, it seems likely that this induction takes place over a brief time span. Whether RA has this effect as the natural inducer, or as a mimic of the natural inducer, or by first initiating the natural inducer, remains to be established. However, if the identity of the ZPA mitogen could be established, then this might provide a valuable marker at the level of gene activation for the onset of these processes.

Finally, although the assay employed here is that of induction of division in quiescent cells, the ZPA mitogen may operate *in vivo* by maintaining the proportion of cells in division. This would accord with our finding that it does not alter intermitotic times, but rather the proportion of cells in division. This may relate to the observation of Searls and Janners (1971) that when the limb initially grows out it does so as a result of the proportion of dividing cells in the limb bud remaining higher than that of the flank.

ACKNOWLEDGEMENTS

The experiments reported here were carried out over some years with my colleagues Janet Smith and John Macintyre, and my then graduate student, Karen Bell. The time lapse video recordings were carried out in collaboration with John Heath. Paul Rooney assayed conditioned medium for us on chick macrophages. This work was supported in large part by the Medical Research Council.

REFERENCES

Bell, K.M., 1986, The preliminary characterisation of mitogens secreted by embryonic chick wing bud tissues *in vitro*. J. Embryol exp Morph., 93:257.
Bell, K.M. and McLachlan, J.C., 1985, Stimulation of division in mouse 3T3 cells by co-culture with embryonic chick limb tissue. J. Embryol exp Morph., 86:219.
Cooke, J. and Summerbell, D., 1980, Cell cycle and experimental pattern duplication in the chick wing during embryonic development., Nature, 287:697.
Hamburger, V. and Hamilton, H.L., 1951, A series of normal stages in the development of the chick embryo. J. Morph., 88:49.
Honig, L. and Summerbell, D., 1985, Maps of strength of positional signalling activity in the developing chick limb bud, J. Embryol exp Morph., 87:163.
Huang, J.S. and Huang, S.S., 1985, Role of growth factors in oncogenesis: growth factor-proto-oncogene pathways of mitogenesis, in: "Growth Factors in Biology and Medicine", D. Evered, ed., Longmans, London.
Lee, J. and Tickle, C., 1985, Retinoic acid and pattern formation in the developing chick wing: SEM and quantitative studies of the early effects on the apical ectodermal ridge and bud outgrowth, J. Embryol exp Morph., 90:139.
Macintyre, J., McLachlan, J.C., Hume, D., and Smith, J., 1988, A microwell assay for anchorage independent cell growth, Tissue and Cell, 20;331.
McLachlan, J.C., Macintyre, J., Hume, D. and Smith, J., 1988, Direct demonstration of production of transforming growth factor activity by embryonic chick tissue. Experientia, 44:351.
Mehta, P., Bertram, J.S. and Loewenstein, W.R., 1989, The actions of retinoids on cellular growth correlate with their actions on gap junctional communication. J. Cell Biology., 108:1053.
Searls, R.L. and Janners, M., 1971, Initiation of limb bud outgrowth in the embryonic chick. Devl. Biol., 24:198.

Smith, J.C. and Wolpert, L., 1979, Pattern formation along
    the anteroposterior axis of the chick wing: the
    increase in width following a polarising region
    graft and the effect of x-irradiation, J. Embryol
    exp Morph., 63:127.
Smith, J., and McLachlan, J.C., 1989, Developmental pattern
    of growth factor production in chick embryo limb buds.
    J. Anat., 165:159.
Smith, J., and McLachlan, J.C., 1990, Identification of a
    novel growth factor with transforming activity
    secreted by individual chick embryos. Development,
    109:905.
Wright, E.G., Wilson, E.S. and Lorimore, S.A., 1985, A
    cell line producing a colony stimulating factor for
    macrophage culture, IRCS Med. Sci. 13:441.

BIOCHEMICAL AND MOLECULAR ASPECTS OF RETINOID ACTION IN LIMB PATTERN FORMATION

Reinhold Janocha and Gregor Eichele

Harvard Medical School, Department of Cellular and
Molecular Physiology, 25 Shattuck Street
Boston, MA 02115, USA

Retinoids are a class of naturally occurring compounds
consisting of a hydrocarbon ring linked to an unsaturated
hydrocarbon side chain with a polar terminal group. Retinoids
have dramatic effects on differentiation and development. For
example, retinoic acid (RA) alters positional identity in
developing and regenerating systems: in the chick limb bud the
anteroposterior axis is affected by locally applied RA (Tickle *et
al.* 1982, 1985; Summerbell, 1983) and systemic and/or exogenous
treatment with RA changes the rostrocaudal axis of *Xenopus*
(Durston *et al.*, 1989; Sive *et al.*, 1990) and the proximodistal
axis of the regenerating urodele limb (Maden, 1982; Thoms and
Stocum, 1984).

Tickle *et al.* (1982) and Summerbell (1983) discovered that
RA released from a carrier implanted into the anterior chick limb
bud margin will generate digit pattern duplications. Such
duplications can be generated by implanting a piece of posterior
limb bud tissue that is known as the zone of polarizing activity
or ZPA (Saunders and Gasseling, 1968). It has been proposed that
ZPA releases a morphogen that specifies the limb pattern along
the anteroposterior limb axis (Wolpert, 1969). Because RA mimics
ZPA function, Tickle *et al.* (1982) have suggested that RA might
be the hypothetical morphogen. This view is reflected in the
specificity of RA action (pattern duplications are induced in a
dose-, time-, stage-, and position-dependent fashion) and in the
observation that pattern duplications occur at physiological
concentrations of RA (Tickle *et al.*, 1985; Thaller and Eichele,
1987, 1990). Furthermore, RA is present in the limb bud and it
forms a gradient across the limb bud with a high point at the

Developmental Patterning of the Vertebrate Limb
Edited by J.R. Hinchliffe *et al.*, Plenum Press, New York, 1991

posterior margin coincident with the ZPA (Thaller and Eichele, 1987). Limb buds also express cellular retinoic acid binding protein (CRABP; Maden *et al.*, 1988; Perez-Castro *et al.*, 1989) and several retinoic acid receptors (Dollé *et al.*, 1989). These observations are consistent with RA being a gradient-forming morphogen that establishes positional values in the limb system. However, there are at present no experiments that rule out a more indirect mode of action of RA. For example RA could induce a new ZPA, which then releases the "true" morphogenetic signal (see Summerbell and Harvey, 1983).

Our understanding of the mechanism of action of RA was greatly advanced by the discovery of RA receptors (Giguère *et al.*, 1987; Petkovich *et al.*, 1987; Brand *et al.*, 1988; Zelent *et al.*, 1989; Mangelsdorf *et al.*, 1990). These receptors are nuclear proteins and are closely related to steroid- and thyroid hormone receptors (Evans, 1988). It is well established that these nuclear receptor proteins are ligand-dependent transcription factors that regulate the expression of specific response genes. Therefore, the developmental effects of RA can be formulated and analyzed in terms of a *signal transduction process*. The signal is retinoic acid or possibly a gradient of RA, the sensors are the RA-receptors, and the output is represented by the products of the genes targeted by retinoic acid receptors. This way of looking at the problem suggests that the complex problem of pattern formation along the anteroposterior limb axis can be simplified to a tractable biochemical strategy that poses a series of specific questions. For example, which retinoids are present in the limb bud? How are they distributed in space and time and how are they synthesized and degraded? Which receptors are expressed in the limb bud and which genes are regulated directly or indirectly by retinoic acid and its receptors?

Endogenous Retinoids and Retinoid Metabolism

To address the question concerning the presence of endogenous retinoids, HPLC fractionated organic solvent extracts of whole chick embryos were screened for morphogenetically active retinoids using the digit pattern duplication assay (Thaller and Eichele, 1990). The approach using whole embryos makes sense because a putative signalling molecule ought to occur and operate not only in limb buds but also in other regions of the embryo. The assay of the HPLC fractions revealed two broad activity peaks. By means of sequential HPLC fractionations, chemical derivatization, and gas-chromatography/mass spectroscopy the two activities were identified as RA and 3,4-didehydroretinoic acid (ddRA). The structure of ddRA (see Fig. 1) is very similar to that of RA, except for a second double bond in the ionylidine ring forcing the ring into a more planar conformation, a structural change that could selectively affect ligand affinity to RARs and to CRABPs. The concentrations of endogenous RA and ddRA is highest in the early limb bud (Hamburger-Hamilton stages 20 to 22) when the pattern is thought to be specified. Interestingly, the absolute level of ddRA is several times higher than that of RA (ddRA: 150nM, RA: 30nM). The dose-response curves of RA and ddRA are nearly identical, that is the two compounds are essentially equipotent. In addition to RA and ddRA, limb buds also contain retinol, 3,4-didehydroretinol (ddROH), retinal and 3,4-didehydroretinal (Thaller and Eichele, 1990).

How are these various endogenous retinoids connected with each other? A good strategy to find out is to locally release each retinoid in radioactive form from a bead implanted into the limb bud (Thaller and Eichele, 1988, 1990). Such *in vivo* conversion studies revealed that endogenous retinoids can be organized into a network consisting of two parallel pathways (Fig. 1). In agreement with the classical work by Dowling and Wald (1960), retinol is first converted to retinal, which, in turn, is oxidized to retinoic acid. The endogenous concentration of retinol is about 600 nM, considerably higher than that of RA. This is consistent with retinol being the precursor molecule. In contrast, limb buds contain little retinal(15 nM). When ^{3}[H]-retinal is locally applied, it is either reduced to retinol or oxidized to RA. ^{3}[H]-RA, however, cannot be reduced back to retinal but is metabolized to 4-hydroxy-RA and 4-oxo-RA (R.J. and C. Thaller, unpublished observation). There is evidence from other studies that the first step leading from retinol to retinal

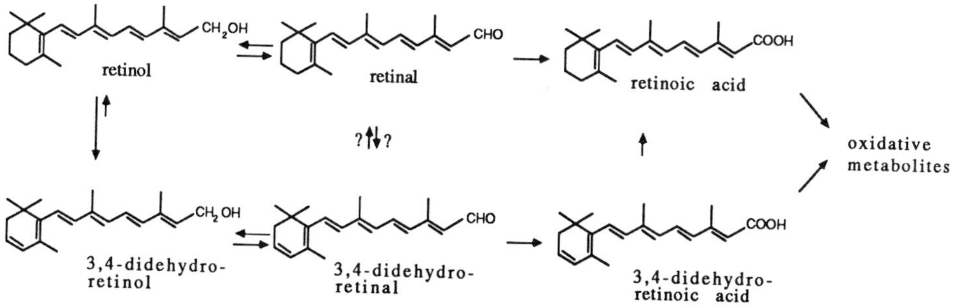

Fig. 1. Reaction pathways of endogenously retinoids in the chick limb bud.

is rate-determining (Napoli and Race, 1987). In addition to its oxidation to retinal and RA, retinol is also converted to ddROH which is then further oxidized to ddRA again through an aldehyde intermediate (lower pathway in Fig. 1). Using ^{3}[H]-retinol and ^{3}[H]-ddROH it was shown that the conversion of ROH to ddROH proceeds mainly in the direction of a dehydrogenation (Thaller and Eichele, 1990). This dehydrogenation reaction is also the main link between the two pathways. It is important to realize that the various interconversion reactions depicted in Fig. 1 take place locally in the limb bud.

Since RA and ddRA are essentially equipotent, could it be that ddRA is not active by itself but needs to be first converted to RA? This does not seem to be the case. When radiolabeled ddRA was applied to wing buds, less than 5% of it reacts to RA. Such a minimal conversion of ddRA to RA can hardly account for the observed equipotency, and therefore ddRA appears to be a morphogenetic signalling molecule on its own right. It is not clear at the moment why there are two active forms of retinoic acid and why there are two pathways. The organization of several enzymatic reactions into a network could provide a mechanism to developmentally control the concentrations of RA and ddRA.

ddROH is the major metabolite of retinol in the chick limb bud. Other tissues that can also generate ddROH from retinol include the human skin (Vahlquist, 1980; Törmä and Vahlquist, 1985), mouse liver (Törmä and Vahlquist, 1988), ocular tissues of tadpoles (Tsin et al., 1985) and the neural tube of chick embryos (Wagner et al., 1990). Thus didehydroretinoids appear to have a wide distribution in vertebrate systems.

Enzymes Involved in Retinoid Metabolism

The enzymes involved in RA synthesis can be classified as retinol-retinal dehydrogenases which require NAD$^+$ and/or NADP as a cofactor (Nicotra and Livrea, 1982; Julia et al., 1986; Napoli, 1986). Classical ethanol/acetaldehyde dehydrogenases are inhibited by 4-methylpyrazole. In contrast, several partially purified enzymatic activities competent to oxidize retinol to retinal and to retinoic acid cannot be blocked by 4-methylpyrazole, but are inhibited by disulfiram, a reagent that generally modifies protein sulfhydryl groups through oxidation and internal crosslinking (Napoli, 1986; Napoli and Race, 1987; Siegenthaler et al., 1990). In addition to cytosolic dehydrogenases, a retinal synthesizing activity was detected in microsomal preparations (Leo et al., 1987).

There are several reports addressing the degradation of retinoic acid (Roberts et al. 1980; Leo et al., 1984; Leo et al., 1989). Generally, RA is oxidized at the ring at the 4 position leading to 4-hydroxy- and then to 4-oxo-RA. In the chick limb bud where RA is relatively short-lived, RA is also degraded through 4-hydroxy-RA and 4-oxo-RA (R.J and C. Thaller, unpublished observation). At present, the metabolic fate of ddRA is unknown. Are these metabolites of biological significance? Interestingly, 4-oxo-RA is quite active in the pattern duplication assay (about 1/10 of RA; G. Eichele, unpublished observation), and if it turned out to be endogenously present at reasonably high concentrations, it might play a role.

Work from Sporn's laboratory suggests that the enzymes involved in RA degradation belong to the group of P-450 dependent monooxygenases. P-450 dependent monooxygenases catalyze numerous oxidation reactions involving a wide variety of substrates (steroids, fatty acids, eicosanoids, carcinogens, pesticides; see Gonzales, 1989, for a review). These enzymes are NADPH-dependent and are localized in microsomes and mitochondria.

RA and other retinoids are hydrophobic and would partition into membranes. To prevent this, cells contain specific high-affinity binding proteins known as CRABP and CRBP (see Blomhoff et al., 1990, for a review). Therefore, most retinoid molecules in a cell are presumably bound to their appropriate binding protein. In other words, CRABP and CRBP act as a buffer to control the concentration of free ligand. One additional function of CRABP and CRBP could be that they help to present the ligand to the appropriate converting enzymes. A precedence for this view comes from studies of retinol esterases present in rat liver microsomes (Ong et al., 1987; Yost et al., 1988).

Do Retinoids Have a Role in the Developing Central Nervous System?

Using the chick limb bud as an assay system, regions of the chick embryo other than the ZPA were tested for their ability to re-specify the pattern along the anteroposterior axis of the wing. Earlier work by Hornbruch and Wolpert (1986) showed that Hensen's node and the notochord have polarizing activity. More recently, Wagner et al. (1990) have reported that floor plate (the ventralmost region of the developing neural tube) but not neurectoderm possesses ZPA-like activity. This is particularly interesting since the floor plate seems to have a "polarizing" effect within the neural tube. One manifestation of this is the capacity of the floor plate to attract commissural axons towards the ventral midline of the neural tube, presumably by releasing a chemoattractant (Placzeck et al., 1990 and references therein). Given that the floor plate imitates the ZPA, is there a relation between the two organizing tissues? One obvious question is whether floor plate contains RA. The small size of the floor plate precludes a direct measurement of endogenous RA levels. In an attempt to overcome this problem Wagner et al. (1990) examined whether floor plate can synthesis RA from radiolabelled precursor retinol. Therefore, floor plate tissue was dissected out of neural tube and incubated in vitro in medium containing $^3[H]$-retinol. The radioactive products were extracted and fractionated sequentially on two HPLC columns. These analyses clearly showed that floor plate produces RA, ddROH and ddRA from retinol. To examine whether floor plate is metabolically more active than non-floor plate neuroectoderm, the production of the major metabolite ddROH in these two tissues was compared. It turned out that per µg protein, the floor plate is 3 to 4 times more active in synthesizing ddROH than non-floor plate neuroectoderm. Similarly, floor plate tissue is somewhat more active in producing RA and ddRA. But all these activities are only moderately enriched in the floor plate and not restricted to it. This is in some disagreement with the transplantation experiments that showed restriction of morphogenetic activity to the floor plate. One explanation is that limb bud cells respond to RA and ddRA only above a certain threshold concentration and that the level generated by non-floor plate neuroectoderm is below this critical threshold. An alternative explanation is that the polarizing effect of the floor-plate in the limb bud is not directly mediated by RA, but by a signal that is downstream of RA such as a gene product targeted by RA.

In summary, research from a number of laboratories over the past few years has considerably advanced our understanding of the mechanism of action of retinoids. This has gsignificantly helped to conceptualize and also clarify the role of retinoids in morphogenesis. While many experiments point to a intriguing role of retinoids in development, more work is needed to determin whether retinoic acid is really a crucial morphogen that determines positional values in the limb and in other embryonic fields.

Acknowledgments

The work from the authors laboratory was supported by a grant from NIH (HD 20209).

References

Brand, N.J. Petkovich, M., Krust., A., Chambon, P., Marchio, A. Tiollais,P., and Dejean, A., 1988, Identification of a second human retinoic acid receptor, Nature 332: 850.

Blomhoff, R., Green, M., Berg, T., and Norum, K., 1990, Transport and storage of vitamin A, Science, 250:399.

Dollé, P. Ruberte, E. Kastner, P., Petkovich, M, Stoner, C.M., Gudas, L.J., and Chambon, 1989, Differential expression of genes encoding alpha, beta and gamma retinoic acid receptors and CRABP in the developing limbs of the mouse. Nature, 342:767.

Dowling, J.E., and Wald, G., 1960, The biological function of vitamin A acid, Proc. natl. Acad. Sci., 46:587.

Durston, A.J., Timmermans, J.P.M., Hage, W.J., Hendriks, H.F.J., de Vries, N.J., Heideveld, M., and Nieuwkoop, P.D., 1989, Retinoic acid causes an anteroposterior transformation in the developing central nervous system, Nature,340:140.

Evans, R., 1988, The steroid and thyroid hormone receptor super-family, Science, 240:889.

Gonzales, F., 1989, The molecular biology of cytochrome P450s, Pharm. Rev.,40:244.

Guiguère, V., Ong, E.S. Segui, P., and Evans, R.M., 1987. Indetification of a receptor for the morphogen retinoic acid, Nature 330: 624.

Hornbruch, A., and Wolpert, L., 1986, Positional signalling by Hensen's node when grafted to the chick limb bud, J. Embryol. Exp. Morphol., 94:257.

Julia, P., Farrés, J., and Parés, X, 1986, Ocular alcohol dehydrogenase in the rat: regional distribution and kinetics of the ADH-1 isoenzyme with retinol and retinal, Exp. Eye Res., 42:305.

Leo, M.A., Iida, S., and Lieber, C., 1984, Retinoic acid metabolism by a system reconstituted with cytochrome P-450, Arch. Biochem. Biophys., 234:305.

Leo, M.A., Kim, C.-I., and Lieber, C., 1987, NAD$^+$-dependent retinol dehydrogenase in liver microsomes, Arch. Biochem. Biophys. 259:241.

Leo, M.A., Lasker, J., Raucy, J., Kim, C.-I., Black, M., and Lieber, C., 1989, Metabolism of retinol and retinoic acid by human liver cytochrome P450IIC8, Arch. Biochem. Biophys., 269:305.

Maden, M., 1982, Vitamin A and pattern formation in the regenerating limb, Nature, 295:672.

Maden, M., Ong, D.E., Summerbell, D., and Chityl, F., 1988, Spatial distribution of cellular protein binding to retinoic acid in the chick limb bud. Nature 335: 733.

Mangelsdorf D.J. D.J., Ong, E.S., Dyck, J.A. and Evans R.M., 1990, Nuclear receptor that defines a novel retinoic acid response pathway. Nature 345:224.

Napoli, J.L., 1986, Retinol metabolism in LLC-PK$_1$ cells, J. Biol. Chem., 261:13592.

Napoli, J.L., and Race, K.R., 1987, The biosynthesis of retinoic acid from retinol by rat tissues in vitro, Arch. Biochem. Biophys., 255:95.

Nicotra, C., and Livrea, 1982, Retinol dehydrogenase from bovine retinal rod outer segments. Kinetic mechanism of the solubilized enzyme, J. Biol. Chem., 257:11836.

Ong, D., Kakkad, B., and MacDonald,P.N., 1987, Acyl-CoA-independent esterification of retinol bound to cellular retinol-binding protein (type II) by microsomes from rat small intestine, J. Biol. Chem., 262:2729.

Perez-Castro, A.V., Toth-Rogler, L.E., Wei, L., Nguyen-Huu, M.C., 1989, Spatial and temporal pattern of expression of cellular retinoic acid binding protein and the cellular retinol binding protein during mouse embryogenesis. Proc. natl. Acad. Sci USA 86: 8813.

Petkovich, M. Brand, N.J., Krust, A., and Chambon, P., 1987, A human retinoic acid receptor that belongs to the family of nuclear receptors. Nature 330: 444.

Placzeck, M., Tessier-Lavigne, M., Jessell, T. and Dodd, J., 1990, Orientation of commissural axons *in vitro* in response to a floor plate-derived chemoattractant. Development, 110:19.

Roberts, A.B., Lamb, L.C., and Sporn, M., 1980, Metabolism of all-trans-retinoic acid in hamster liver microsomes : oxidation of 4-hydroxy- to 4-keto-retinoic acid, Arch. Biochem. Biophys., 374:383.

Saunders, J.W., and Gasseling, M.T., 1968, Ectodermal-mesenchymal interactions in the origin of limb symmetry,in: "Epithelial-Mesenchymal Interactions", Fleischmajer, R., and Billingham, R.E., eds. p. 78. Williams and Wilkins, Baltimore.

Siegenthaler, G., Saurat, J.-H., and Ponec, M., 1990, Retinol and retinal metabolism, Biochem. J., 268:371.

Sive, H., Draper, B., Harland, R., and Weintraub, H., 1990, Identification of a retinoic acid-sensitive period during primary axis formation in *Xenopus laevis*, Genes & Development, 4:932

Summerbell, D., 1983, The effect of local application of retinoic acid to the anterior margin of the developing chick limb, J. Embryol. Exp. Morphol., 78:269.

Summerbell, D., and Harvey, F., 1983, Vitamin A and the control of pattern in developing limbs. Prog. Clin. Biol. Res., 110A: 109.

Thaller, C., and Eichele, G., 1987, Identification and spatial distribution of retinoids in the developing chick limb, Nature, 327:625.

Thaller, C., and Eichele, G., 1988, Characterization of retinoid metabolism in the chick limb bud, Development, 103:473.

Thaller, C., and Eichele, G., 1990, Isolation of 3,4-didehydroretinoic acid, a novel morphogenetic signal in the chick wing bud, Nature, 345:815.

Thoms, S.D., and Stocum, D.L., 1984, Retinoic acid-induced pattern duplication in regenerating urodele limbs, Dev. Biol., 103:319.

Tickle, C., Alberts, B.M., Wolpert, L., and Lee, J., 1982, Local application of retinoic acid to the limb bud mimics the action of the polarising region, Nature, 296:564.

Tickle, C., Lee, J., and Eichele, G., 1985, A quantitative analysis of the effect of all-trans-retinoic acid on the pattern of chick wing development, Dev. Biol., 109:82.

Törmä, H., and Vahlquist, A., 1985, Biosynthesis of 3-dehydroretinol (Vitamin A$_2$) from all-trans-retinol (vitamin A$_1$) in human epidermis, J. Invest. Dermatol., 85:498.

Törmä, H., and Vahlquist, A., 1988, Identification of 3-dehydroretinol (vitamin A$_2$) in mouse liver, Biochim. Biophys. Act., 961:177.

Tsin, A.T., Alvarez, R., Fong, S.-L., and Bridges, C., 1985, Conversion of retinol to 3,4-didehydroretinol in the tadpole, Comp. Biochem. Physiol., 81B:415.

Vahlquist, A., 1980, The identification of dehydroretinol (vitamin A$_2$) in human skin, Experientia, 36:317.

Wagner, M., Thaller, C., Jessell, T., and Eichele, G., 1990, Polarizing activity and retinoid synthesis in the floor plate of the neural tube, Nature, 345:819.

Wolpert, L., 1969, Positional information and the spatial pattern of cellular differentiation, J. theor. Biol. 25: 1.

Yost, R., Harrison, E., and Ross, A.C., 1988, Esterification by rat liver microsomes of retinol bound to cellular retinol-binding protein, J. Biol. Chem., 263:18693.

Zelent, A., Krust, A., Petkovich, M., Kastner P., and Chambon, P., 1989, Cloning of murine alpha and beta retinoic acid receptors and a novel receptor gamma predominantly expressed in skin. Nature 339: 714.

POSITION-DEPENDENT PROPERTIES OF LIMB CELLS

S.V. Bryant, T. Hayamizu, N. Wanek, and D.M. Gardiner

Developmental Biology Center
University of California, Irvine
Irvine, California 92717, USA

An understanding of the way in which cells acquire and utilize positional information in their interactions with one another to generate new pattern, awaits the identification of the molecules that specify positional identity. The vertebrate limb has long been used as an experimental model for the study of pattern formation, and there is currently a considerable body of knowledge about limb development and regeneration at the level of tissues and cells. This information has allowed for the development of a conceptual framework within which to study limbs and from which functional assays for genes involved in pattern formation can be fashioned.

CELL INTERACTIONS IN LIMB PATTERN FORMATION

Almost all the information we have about pattern formation in limbs is a result of the fact that to varying degrees, all vertebrate limbs have the ability to regenerate missing parts and to form supernumerary pattern elements. This regulative ability is most complete in urodele amphibians (Bryant et al. 1987). It is more limited in higher vertebrates due to the restricted location of the outgrowth-permitting apical ectoderm (Saunders 1948), and to the fact that only cells within a few hundred micrometers of the apical ectodermal ridge are actively engaged in pattern formation (Summerbell et al. 1973). Investigators have been able to exploit regulative abilities to test many postulated cause and effect relationships in the development of the limb pattern.

A major conclusion from studies of pattern regulation is that cells in different positions within the developing limb bud are non-equivalent (Lewis and Wolpert 1976) in terms of their positional properties. This non-equivalence is most dramatically demonstrated when cells from different positions are brought into contact by grafting. In the most regulative limbs, grafts of anterior next to posterior, dorsal next to ventral, or proximal next to distal result in the stimulation of growth and the insertion of new pattern elements between the confronted cells (Bryant et al. 1987). The new pattern restores local continuity and eliminates the positional disparities created by grafting. In the case of grafts that confront anterior and posterior or dorsal and ventral the new pattern forms a supernumerary limb. These responses indicate that limb cells in different positions have different positional identities. The relevance of positional identity is that it is required for the cell interactions that lead to the generation or regeneration of limb pattern.

POSITIONAL DIFFERENCES IN LIMBS

Despite the clear demonstration of non-equivalence, very few differences between anterior and posterior cells have so far been documented. At the cellular level, anterior and posterior chick and mouse (but not amphibian (Muneoka and Bryant 1984a)) limb cells behave differently when they interact at a confrontation: anterior cells divide more than posterior ones so that supernumerary structures are composed largely of cells of anterior origin(Honig 1983a; Javois and Iten 1986). Similar differences in growth characteristics of anterior and posterior limb cells have been observed in vitro (Shi and Muneoka 1990; Gardiner et al. 1990). In preliminary studies, we have found that both anterior and posterior cells increase the number of EGF receptors in response to retinoic acid, but that anterior cells show a consistently larger response than posterior cells (Brylski, Oberg and Bryant, unpublished data).

At the level of molecules, a few positional differences have been reported. A monoclonal antibody, AV1, identifies a cell-surface glycoprotein synthesized by anterior-ventral cells in chick limbs (Ohsugi et al. 1988), but little is known about the significance of this molecule for pattern formation. Retinoic acid, which has major effects on limb patterning (see below), is more abundant in posterior than in anterior chick limb tissues (Thaller and Eichele 1987). Since retinol, the precursor to retinoic acid, is uniformly distributed, the asymmetry in retinoic acid distribution presumably arises as a result of positional differences between anterior and posterior cells in their ability to metabolize, stabilize or degrade retinoids. An additional metabolite of retinol, 3,4-didehydroretinoic acid, has recently been reported to be present in chick limbs (Thaller and Eichele 1990). Didehydroretinoic acid is equivalent to retinoic acid in pattern-forming activity and is present in greater abundance; its distribution in limbs is unknown at present. Cellular retinoic acid binding protein (CRABP) is more abundant in anterior and distal chick limb bud cells (Maden et al. 1988). Conflicting data have been reported regarding its distribution in mouse limb buds (Dollé et al. 1989; Perez-Castro et al. 1989). None of the nuclear receptors for retinoic acid (RARα, β and γ) are differentially distributed in undifferentiated limbs (Dollé et al. 1989). The nature of the underlying differences between anterior and posterior limb cells that these results imply is at present unknown.

There is considerable interest in the roles of homeobox genes in vertebrate development (Kessel and Gruss 1990). At the descriptive level there is evidence that homeobox genes are involved in the establishment of pattern in the primary body axis (Duboule and Dollé 1989; Graham et al. 1989). Experimental evidence from transgenic animals (Balling et al. 1989; Wolgemuth et al. 1989), or from injections of mRNA (Ruiz i Altaba and Melton 1989; Wright et al. 1989), or of antibodies to homeobox proteins (Wright et al. 1989), indicates that homeobox gene products are necessary for correct spatial patterning of the primary body axis. The evidence for the involvement of homeobox genes in positional identity in limbs is intriguing, but is so far limited to descriptive studies. XlHbox-1 protein (equivalent to murine Hox-3.3) has been localized to proximal and anterior regions of both Xenopus and mouse forelimb bud mesoderm (Oliver et al. 1988).

GHox-2.1 (equivalent to murine Hox 2.1) has been shown by in situ hybridization to be expressed in anterior and proximal regions of the chick limb bud (Wedden et al. 1989). Several members of the Hox-4 cluster (previously Hox-5) have been studied by in situ hybridization(Dollé et al. 1989) and by protein localization (Oliver et al. 1989), and appear to be differentially expressed in distal and posterior parts of the mouse bud. Other homeobox genes expressed in limbs do not show differential expression along the anterior-posterior axis.

Although most of the experimental data regarding the mechanisms of limb pattern formation have come from studies with chicks and amphibians, in recent years we have expanded our ability to work experimentally on the developing mouse limb. This is of importance because of the considerably more sophisticated molecular genetics that is available, thus eventually allowing for direst tests of gene function. With the availability of exo utero techniques for manipulations during later stages of development (Muneoka et al. 1986; Muneoka et al. 1990), we will be able to study gene expression not only during normal develop;ment, but also in response to grafts confronting anterior and posterior tissues, to amputation and to the local release of retinoic acid from implanted beads. To date we have used these techniques to develop a limb bud staging system to ensure experimental reproducibility (Wanek et al. 1989), have carried out a fate mapping study of the hind limb bud (Muneoka et al. 1989), and have demonstrated regulative abilites in response to both amputation and positional confrontations between anterior and posterior tissue by grafting (Wanek et al. 1989a). In the following sections we discuss preliminary results of work in progress that is designed to extend our ability to investigate the molecular biology of pattern formation in the developing mouse limb bud.

ASSAY FOR POSTERIOR POSITIONAL INFORMATION IN LIMB BUD CELLS

As discussed above, a characteristic of all vertebrate limbs is their ability to undergo pattern regulation in response to confrontations between cells from different positions. It is this postion-specific property of limb cells that operationally defines positional information. This ability is not restricted to interactions within species; equivalent results are obtained when grafts are made heterospecifically (Fallon and Crosby 1977; Sessions et al. 1989)as well as between developing and regenerating tissues (Muneoka and Bryant 1984b). Such results have led to the conclusion that the basic features of the mechanism of outgrowth and patterning are equivalent among all the vertebrates and during development and regeneration. Thus the differences in the details of the development of particular types of limbs point out the ways in which variations in the basic mechanism can exist and yet not affect the conserved design of the final structure.

We have taken advantage of this commonality of mechanism to assay for positional identity in mouse limb cells at those early stages of outgrowth that are not yet accessible by direct surgical techniques. To do this we have expanded upon the original observation of Tickle and colleagues (Tickle et al. 1976) that grafts of posterior mouse limb bud tissue stimulate the formation of supernumerary digits when grafted heterospecifically into an anterior site in a stage 20-22 chick wing bud. As a result we now have a functional assay for an essential positional property of posterior cells, i.e. their ability to signal, or trigger, a regulative response at a positional confrontation with anterior cells (Wanek and Bryant 1990).

Embryonic Swiss Webster mice were collected on days E11-E13 (plug day=E1). Limb buds were staged according to Wanek et al. (1989b). Embryonic chick hosts (stage 20-21) were prepared by standard procedures. Wedges of tissue from stage 3 to 8 donor mouse limb buds were removed with tungsten knives, marked with carbon to maintain dorsal-ventral orientation and inserted into slits made in the anterior wing margin of host chick embryos. Grafted limbs were collected 6 days after grafting and processed for whole mount skeletal analysis. The limbs were analysed for the presence of supernumerary digits.

We found that posterior limb bud cells from stages earlier than stage 7 are positive for posterior identity (signalling ability). All posterior grafts from stages 3 through 6 induced the formation of one or more extra digits. Posterior signalling ability was absent by stage 8, and was intermediate at stage 7. None of the control grafts (anterior wedges from stage 4 hosts transplanted into anterior sites in stage 20-21 chick limbs) stimulated the formation of extra structures (Wanek and Bryant 1990). A similar decline in signalling ability in chick posterior limb cells has been reported previously: high level of signalling through stage 25 (equivalent to stage 6 in the mouse) and declining until absent at stage 30 (equivalent to stage 8/9 in the mouse) (Honig and Summerbell 1985).

STABILITY OF POSITIONAL INFORMATION OF LIMB CELLS *IN VITRO*

Being able to work with limb cells in vitro, and still have them resemble their counterparts in developing limbs in terms of positional properties, would be of great utility in studying the molecular biology of limb pattern formation. A previous study indicated a steep decline in positional properties of chick limb cells during the first day of culture. This report was preliminary and lacked detailed information regarding the conditions under which the cells were grown and assayed (Honig 1983b).

We are presently investigating the retention of posterior positional information by chick and mouse limb cells in vitro as assayed by the ability to stimulate supernumerary digits after grafting to an anterior site in a host chick limb bud (Hayamizu et al. 1990a). Cells from the posterior third of chick (stage 20-21) and mouse (stage 3) limb buds are dissociated and plated either at a monolayer density (1.8×10^4 cells/cm^2) or at micromass density (4×10^5 cells/10μl), using previously established methods (Ahrens et al. 1977; Shi and Muneoka 1990; Gardiner et al. 1990). After various times in culture, cells are harvested and implanted into the anterior of host chick wing buds. Posterior chick cells cultured at 37°C for 18hrs induce the formation of extra digits in all cases. Cultured anterior cells do not induce extra digits. After 24hrs of culture, 80% of the grafts induce digits, and after 36hrs, 60%. At 48hrs none of the grafts of cultured cells induce extra digits (Hayamizu et al. 1990a). For mouse cells, 75% of the posterior cultures show evidence of posterior identity at 0hrs and 6hrs of culture. At 18hrs only 33% and at 24hrs only 11% showed evidence of posterior identity. As in chick cells, mouse cells did not express posterior positional properties after 48hrs in culture (Hayamizu et al. 1990a). These results indicate that there is a window of time during which mouse and chick limb cells can be manipulated in vitro while still maintaining posterior postional information. Although we do not at present have a comparable assay for anterior identity, anterior chick or quail cells grafted into posterior locations have been reported to induce the position-specific formation of extra structures (Carlson 1984a,b; Iten and Murphy 1980). Studies are in progress to test whether mouse anterior cells evoke a response in chicks.

EFFECTS OF RETINOIC ACID ON POSITIONAL PROPERTIES OF LIMB CELLS

Chick: *in vivo*

When retinoic acid-loaded beads are implanted into the anterior of a stage 20 chick wing bud, supernumerary digits are formed adjacent to the bead (Tickle et al. 1982). This result mimics that obtained when posterior edge (ZPA) cells are similarly grafted (Tickle et al. 1975) and has led to the hypothesis that the new digits are specified by different concentrations of retinoic acid at different distances from the bead (Thaller and Eichele 1987). An alternative interpretation that is consistent with the data from other vertebrates, is that retinoic acid sets one of the boundaries of an axis. Hence, in amphibians, retinoic acid changes cells from anterior to posterior (Stocum and Thoms 1984), dorsal to ventral (Ludolf et al. 1990) and distal to proximal (Maden 1982).

To test this idea we implanted retinoic acid-loaded micro-carrier beads (Dowex AG1-X2 ion exchange resin beads) into the anterior of stage 20 chick wing buds (Eichele et al. 1984). The loading regime and dose of retinoic acid (1.0 mg/ml) was such that bead implants yielded results equivalent to ZPA-grafts to the same location. Beads were removed after varying periods of time and embryos were incubated for 90 min, during which time exogenous retinoic acid was depleted (Eichele et al. 1985). A wedge of tissue adjacent to the site of bead implantation was then removed and assayed for posterior positional identity by grafting to an anterior site in a host chick wing bud. We found that cells adjacent to the bead acquired posterior positional information after as little as 15hrs of retinoic acid exposure (Wanek et al. 1990). This is the same as the time required for retinoic acid beads (Eichele et al. 1985) and ZPA grafts (Smith 1980) to induce extra digits. We also transplanted retinoic acid-exposed anterior quail cells into chicks to determine the contribution from the respecified anterior cells (changed into posterior cells) to the supernumerary digits. The contribution pattern was identical to that of control quail ZPA grafts; quail cells only contributed to the digit closest to the graft. The remaining digits were induced in host tissue (Wanek et al. 1990).

These data are consistent with the idea that retinoic acid changes anterior cells into posterior edge cells. Subsequent interactions between anterior cells and the newly respecified posterior cells adjacent to them will lead to super-numerary digit formation. The data do not support the idea that a gradient of exogenous retinoic acid respecifies the digit pattern since the entire wedge of tissue consists of posterior edge only, rather than the anlage for several of the digits, as expected from the gradient idea.

This result raises another issue; if retinoic acid were the limb morphogen produced by the ZPA, it follows from our results that anterior cells exposed to exogenous retinoic acid (released from a bead) are induced to produce retinoic acid, since they now behave as ZPA cells. Therefore one is led to conclude that all limb cells can produce retinoic acid, and that cells exposed to retinoic acid respond by producing retinoic acid, i.e. retinoic acid production is regulated via a positive feedback loop. While positive feedback responses occur in development (Meinhardt 1989), by themselves they are hard to reconcile with the establishment of a gradient. This is because anterior cells exposed to low levels of retinoic acid would respond by increasing the amount of retinoic acid they produce, thus eliminating the gradient. One possible solution to this dilemma is to consider retinoic acid as the activator in a Gierer-Meinhardt system (Gierer and Meinhardt 1972), but in this case retinoic acid would establish the edge of the field and a second molecule, the inhibitor, would form a gradient across the bud that would provide the cells with positional information.

Mouse: *In vivo*

We have tested whether retinoic acid induces digit duplication in mice by implanting retinoic acid-loaded beads into the anterior of stage 4-8 limb buds using exo utero surgical procedures (Muneoka et al. 1986; Muneoka et al. 1990). At all doses tested (soaking solutions of 10-0.01mg/ml), retinoic acid causes reduced limb patterns (Wanek, unpublished data). In our hands, the effective soaking dose for chick duplications is 1mg/ml. Higher doses result in loss of limb structure. We plan to test lower doses (< 0.01mg/ml) of retinoic acid in the bead soaking solution in case mouse limbs are simply more sensitive to retinoic acid than are chicks.

Chick and mouse: *In vitro*

After finding that posterior positional information in both chick and mouse cells was stable in vitro for limited periods of time, we tested whether retinoic acid can respecify limb cells in vitro. Cultures of anterior limb cells were grown at micromass density (Ahrens et al. 1977) in the presence of 5×10^{-8} M retinoic acid (equivalent to the concentration of retinoic acid in the posterior quarter of a chick limb bud (Thaller and Eichele 1987)) for 24hrs, then grafted into the anterior of a host wing bud to assay for posterior identity. Untreated, cultured anterior cells never induced extra digits, whereas retinoic acid-treated anterior chick cells induced extra digits in nearly 60% of the cases. Preliminary results with mouse anterior cells, have given us two positive cases so far (Hayamizu et al. 1990b).

EFFECTS OF RETINOIC ACID ON GROWTH OF MOUSE LIMB CELLS

The duplicated limb pattern resulting from the localized application of retinoic acid in chick limbs (Tickle et al. 1982) involves growth to form supernumerary digits. We have tested the possibility that retinoic acid may act directly to stimulate growth by examining the effect of retinoic acid on growth of cultured mouse limb bud cells. We have found that retinoic acid significantly inhibits rather than stimulates proliferation (Gardiner et al. 1990).

Stage 3 limb buds were collected, dissected into anterior and posterior halves, dissociated and replicate plated in 96 well plates. Six hrs after plating, retinoic acid (10^{-6} to 10^{-10} M) was added to the cultures. Cells were harvested and counted after a total of 72 hrs of culture.

We found that retinoic acid significantly inhibits both anterior and posterior cell proliferation over a wide range of concentrations (10^{-6} to 10^{-9} M), with maximum inhibition of growth (50% of controls) between 10^{-7} and 10^{-8} M. These doses correspond to the measured concentration of retinoic acid in chick and mouse limb buds (Satre and Kochhar 1989; Thaller and Eichele 1987). Our results for the mouse are contradictory to those reported for the chick (Ide and Aono 1988).

The finding that retinoic acid inhibits growth is consistent with the effects of retinoic acid on several different cell types in vitro (Roberts and Sporn 1984), and with the observation that retinoic acid inhibits blastemal growth in amphibians during the period of retinoic acid exposure (Pietsch 1987). Our findings are also consistent with the idea that retinoic acid alters positional identity. When retinoic acid is uniformly applied, as it is in vitro, the result is uniformity of positional identity. Since limb growth is dependent on interactions between cells with different positional identities (Bryant et al. 1987), application of retinoic acid is expected to decrease the diversity of positional identity in the population and therefore to decrease overall growth rate.

SUMMARY

The studies we report here are aimed at developing a combination of functional in vivo and in vitro approaches to the study of pattern formation that will facilitate future molecular analyses. To that end, we have utilized a heterospecific assay for posterior positional information in mouse limb cells and are working on a equivalent assay for anterior positional information. We have used this assay to obtain data indicating an ontogenetic loss of posterior positional information, thus predicting that genes that are causally related to the specification of positional information in limbs will show a corresponding change in their patterns of expression over the same time period. In collaboration with Dr. Ken Muneoka and colleagues at Tulane University, have developed standardized techniques for the primary culture of mouse limb bud cells that allow for the expression of positional differences in growth rates between anterior and posterior cells. In collaboration with the Muneoka lab, we have obtained results indicating that there is a window of time during which mouse and chick limb cells can be manipulated genetically or biochemically in vitro, while still maintaining their posterior identity. Finally, we have demonstrated that limb cells can be respecified by treatment with retinoic acid in vitro . This is an important result because it provides us with the means to study changes in the expression of genes involved in pattern formation as anterior cells are changed into posterior cells under controlled experimental conditions in vitro.

REFERENCES

Ahrens, P. B., M. Solursh and R. S. Reiter. (1977). Stage-related capacity for limb chondrogenesis in cell culture. Devel. Biol. **69**: 69-82.

Balling, R., G. Mutter, P. Gruss and M. Kessel. (1989). Craniofacial abnormalities induced by ectopic expression of the homeobox gene *Hox-1.1* in transgenic mice. Cell. **58**: 337-347.

Bryant, S. V., D. M. Gardiner and K. Muneoka. (1987). Limb development and regeneration. Amer. Zool. **27**: 675-696.

Carlson, B. M. (1984a). The formation of supernumerary structures after grafting anterior quail wing bud mesoderm of various ages into chick wing buds. Dev. Biol. **101**: 97-105.

Carlson, B. M. (1984b). The preservation of the ability of cultured quail wing bud mesoderm to elicit a position-related differentiative response. Dev. Biol. **101**: 106-115.

Dollé, P., J.-C. Izpisua-Benmonte, H. Falkenstein, A. Renucci and D. Duboule. (1989). Coordinate expression of the murine Hox-5 complex homoeobox-containing genes during limb pattern formation. Nature. **342**: 767-772.

Dollé, P., E. Ruberte, P. Kastner, M. Petkovich, C. M. Stoner, L. J. Gudas and P. Chambon. (1989). Differential expression of genes encoding (alpha), ß and (gamma) retinoic acid receptors and CRABP in the developing limbs of the mouse. Nature. **342**: 702-705.

Duboule, D. and P. Dollé. (1989). The structural and functional organization of the murine HOX gene family resembles that of *Drosophila* homeotic genes. EMBO J. **8**: 1497-1505.

Eichele, G., C. Tickle and B. M. Alberts. (1984). Microcontrolled release of biologically active compounds in chick embryos: Beads of 200-μm diameter for the local release of retinoids. Anal. Biochem. **142**: 542-555.

Eichele, G., C. Tickle and B. M. Alberts. (1985). Studies on the mechanism of retinoid-induced pattern duplications in the early chick limb bud: temporal and spatial aspects. J. Cell Biol. **101**: 1913-1920.

Fallon, J. F. and G. M. Crosby. (1977). Polarising zone activity in limb buds of amniotes. Vertebrate Limb and Somite Morphogenesis. Cambridge, Cambridge University Press.

Gardiner, D.M., C. Gaudier and S.V. Bryant. (1990). Retinoic acid inhibits proliferation of mouse limb bud cells *in vitro*. (in preparation).

Gierer, A. and H. Meinhardt. (1972). A theory of biological pattern formation. Kybernetik. **12**: 30-39.

Graham, A., N. Papalopulu and R. Krumlauf. (1989). The murine and *Drosophila* homeobox gene complexes have common features of organization and expression. Cell. **57**: 367-378.

Hayamizu, T. F., C. Trevino, C. Shi, K. Muneoka and S. V. Bryant. (1990a). *In vitro* stability of positional signalling activity in dissociated cells from developing chick and mouse limbs buds. (in preparation).

Hayamizu, T.F., D.M. Gardiner and S.V. Bryant. (1990b). Retinoic acid can respecify limb cells *in vitro*. (in preparation).

Honig, L. S. (1983a). Does anterior (non-polarizing region) tissue signal in the developing chick limb? Dev. Biol. **97**: 424-432.

Honig, L. S. (1983b). Polarizing activity of the avian limb examined on a cellular basis. Limb Development and Regeneration. New York, Alan R. Liss, Inc.

Honig, L. S. and D. Summerbell. (1985). Maps of strength of positional signalling activity in the developing chick wing bud. J. Embryol. Exp. Morph. **87**: 163-174.

Ide, H. and H. Aono. (1988). Retinoic acid promotes proliferation and chondrogenesis in the distal mesodermal cells of chick limb bud. Dev. Biol. **130**: 767-773.

Iten, L. E. and D. J. Murphy. (1980). Pattern regulation in the embryonic chick limb: supernumerary limb formation with anterior (non-ZPA) limb bud tissue. Dev. Biol. **75**: 373-385.

Javois, L. C. and L. E. Iten. (1986). The handedness and origin of supernumerary imb structures following 180° rotation of the chick wing bud on its stump. J. Embryol. Exp. Morph. **91**: 135-152.

Kessel, M. and P. Gruss. (1990). Murine developmental control genes. Science. **249:** 374-379.

Lewis, J. H. and L. Wolpert. (1976). The principle of non-equivalence in development. J. Theor. Biol. **62:** 479-490.

Ludolf, D. C., J. A. Cameron and D. L. Stocum. (1990). The effect of retinoic acid on positional memory in the dorsoventral axis of regenerating axolol limbs. Dev. Biol. **140:** 41-52.

Maden, M. (1982). Vitamin A and pattern formation in the regenerating limb. Nature. **295:** 672-675.

Maden, M., D. E. Ong, D. Summerbell and F. Chytil. (1988). Spatial distribution of cellular protein binding to retinoic acid in the chick limb bud. Nature. **335:** 733-735.

Meinhardt, H. (1989). Models for positional signalling with application to the dorsoventral patterning of insects and segregation into different cell types. Development. Supplement: 169-180.

Muneoka, K. and S. V. Bryant. (1984a). Cellular contribution to supernumerary limbs in the axolotl, *Ambystoma mexicanum*. Dev. Biol. **105:** 166-178.

Muneoka, K. and S. V. Bryant. (1984b). Cellular contribution to supernumerary limbs resulting from the interaction between developing and regenerating tissues in the axolotl. Dev. Biol. **105:** 179-187.

Muneoka, K., N. Wanek and S. V. Bryant. (1986). Mouse embryos develop normally exo utero. J. Exper. Zool. **239:** 289-293.

Muneoka, K., N. Wanek and S. V. Bryant. (1989). Mammalian limb bud development: In situ fate maps of early hindlimb buds. J. Exper. Zool. **249:** 50-54.

Muneoka, K., N. Wanek, C. Trevino and S. V. Bryant. (1990). *Exo utero* surgery. Post-Implantation Mammalian Embryo. Oxford, IRL Press.

Ohsugi, K., H. Ide and T. Momoi. (1988). Temporal and spatial expression of a position specific antigen, AV-1, in chick limb buds. Dev. Biol. **130:** 454-463.

Oliver, G., N. Sidell, W. Fiske, C. Heinzmann, T. Mohandas, R. S. Sparkes and E. M. DeRobertis. (1989). Complementary homeo protein gradients in developing limb buds. Genes and Development. **3:** 641-650.

Oliver, G., C. V. E. Wright, J. Hardwicke and E. M. De Robertis. (1988). A gradient of homeodomain protein in developing forelimbs of *Xenopus* and mouse embryos. Cell. **55:** 1017-1024.

Perez-Castro, A. V., L. E. Toth-Rogler, L. Wei and M. C. Nguyen-Huu. (1989). Spatial and temporal pattern of expression of the cellular retinoic acid-binding protein and the cellular retinol-binding protein during mouse embryogenesis. Proc. Natl. Acad. Sci. **86:** 8813-8817.

Pietsch, P. (1987). The effects of retinoic acid on mitosis during tail and limb regeneration in the axolotl larva, *Ambystoma mexicanum*. Roux's Arch. Dev. Biol. **196:** 169-175.

Roberts, A. B. and M. B. Sporn. (1984). Cellular biology and biochemistry of the retinoids. The Retinoids. London, Academic Press, Inc.

Ruiz i Altaba, A. and D. A. Melton. (1989). Involvement of the *Xenopus* homeobox gene Xhox3 in pattern formation along the anterior-posterior axis. Cell. **57:** 317-326.

Satre, M. A. and D. M. Kochhar. (1989). Elevations in the endogenous levels of the putative morphogen retinoic acid in embryonic mouse limb-buds associated with limb dysmorphogenesis. Dev. Biol. **133:** 529-536.

Saunders, J. W., Jr. (1948). The proximo-distal sequence of origin of the parts of the chick wing and the role of ectoderm. J. Exper. Zool. **108:** 363-403.

Sessions, S. K., D. M. Gardiner and S. V. Bryant. (1989). Compatible limb patterning mechanisms in urodeles and anurans. Dev. Biol. **131**: 294-301.

Shi, C. and K. Muneoka. (1990). Position-dependent growth stimulation of mouse limb bud cells *in vitro*. Dev. Biol. (submitted for publication):

Smith, J. C. (1980). The time required for positional signalling in the chick wing bud. J. Embryol. Exp. Morph. **60**: 321-328.

Stocum, D. L. and S. D. Thoms. (1984). Retinoic-acid-induced pattern completion in regenerating double anterior limbs of urodeles. J. Exper. Zool. **232**: 207-215.

Summerbell, D., J. H. Lewis and L. Wolpert. (1973). Positional information in chick limb morphogenesis. Nature. **244**: 492-496.

Thaller, C. and G. Eichele. (1987). Identification and spatial distribution of retinoids in the developing chick limb bud. Nature. **327**: 625-628.

Thaller, C. and G. Eichele. (1990). Isolation of 3,4-didehydroretinoic acid, a novel morphogenetic signal in the chick wing bud. Nature. **345**: 815-819.

Tickle, C., B. Alberts, L. Wolpert and J. Lee. (1982). Local application of retinoic acid to the limb bond [sic] mimics the action of the polarizing region. Nature. **296**: 564-566.

Tickle, C, G. Shellswell, A. Crawley and L. Wolpert. (1976). Positional signalling by mouse limb polarising region in the chick wing bud. Nature. **259**: 396-397.

Tickle, C., D. Summerbell and L. . Wolpert. (1975). Positional signalling and specification of digits in chick limb morphogenesis. Nature. **254**: 199-202.

Wanek, N. and S. V. Bryant. (1990). Characterization of positional signaling activity in the mouse limb bud using a heterospecific assay. (in preparation):

Wanek, N., K. Muneoka and S. V. Bryant. (1989a). Evidence for regulation following amputation and tissue grafting in the developing mouse limb. J. Exper. Zool. **249**: 55-61.

Wanek, N., K. Muneoka, G. Holler-Dinsmore, R. Burton and S. V. Bryant. (1989b). A staging system for mouse limb development. J. Exper. Zool. **249**: 41-49.

Wanek, N., K. Muneoka and S. V. Bryant. (1990). Retinoic acid treated anterior chick wing bud cells behave as posterior cells in a functional assay. (in preparation):

Wedden, S. E., K. Pang and G. Eichele. (1989). Expression pattern of homeobox-containing genes during chick embryogenesis. Development. **105**: 639-650.

Wolgemuth, D. J., R. R. Behringer, M. P. Mostoller, R. L. Brinster and R. D. Palmiter. (1989). Transgenic mice overexpressing the mouse homoeobox-containing gene *Hox-1.4* exhibit abnormal gut development. Nature. **337**: 464-467.

Wright, C. V. E., K. W. Y. Cho, J. Hardwicke, R. H. Collins and E. M. De Robertis. (1989). Interference with function of a homeobox gene in *Xenopus* embryos produces malformations of the anterior spinal cord. Cell. **59**: 81-93.

RETINOIC ACID AND LIMB PATTERNING AND MORPHOGENESIS

C. Tickle

Department of Anatomy & Developmental Biology
University College & Middlesex School of Medicine
Windeyer Building, Cleveland St.
London, UK

INTRODUCTION

The formation of a little finger at one edge of the hand and a thumb at the other is a prime example of patterning in embryonic development. Digits are good markers for position along the antero-posterior (a-p) axis of a limb. The chick wing in which three digits develop has provided an excellent experimental model for analyzing the mechanisms involved in producing the pattern across the a-p axis (Saunders & Gasseling, 1968; Tickle et al., 1975). The character of a digit appears to depend on distance from the polarizing region, a small group of mesenchyme cells at the posterior edge of the limb bud. When a polarizing region is grafted to the anterior margin of a chick wing bud, a mirror-image symmetrical pattern of digits, 432234, develops. Anterior cells now form posterior structures in response to a signal from the polarizing region cells. Local application of all-trans-retinoic acid, a vitamin A derivative, to the chick wing bud can reproduce the effects of grafts of polarizing region tissue (Tickle et al., 1982; Summerbell, 1984). This finding has opened up a new avenue into the cellular and molecular basis of limb patterning and morphogenesis.

DOES RETINOIC ACID ACT DIRECTLY AS A POSITIONAL SIGNAL OR IS IT PART OF A MORE COMPEX SIGNALLING SYSTEM ?

The close similarities between the pattern changes brought about by local application of retinoic acid and grafts of polarizing region tissue suggested that retinoic acid could be the signal produced by the polarizing region that specifies the character of cells across the a-p axis of the limb (Tickle et al., 1985, Eichele et al., 1985). Thaller & Eichele (1987) showed that retinoic acid occurs endogeneously in chick wing buds and is enriched in the posterior parts of the bud where the polarizing region is located. When retinol is applied to the posterior margin in the intact bud, it is converted into retinoic acid (Thaller & Eichele, 1988). These findings are consistent with a model in which the polarizing region generates a retinoic acid gradient across the a-p axis of the bud. Cells at different positions across the a-p axis would be exposed to differing concentrations of retinoic acid and this would provide a positional signal.

A good example of cells responding to different concentrations of retinoic acid is furnished by experiments on a teratocarcinoma cell line, P19. P19 cells differentiate into cardiac muscle, skeletal muscle or nerve according to the concentration of retinoic acid applied (Edwards & McBurney, 1983). In the limb, retinoic acid does not affect the type of cells that differentiate but alters the pattern of cell differentiation.

Developmental Patterning of the Vertebrate Limb
Edited by J.R. Hinchliffe *et al.*, Plenum Press, New York, 1991

Do cells in the limb respond to the local concentration of retinoic acid? There is no direct evidence that this is the case. However, a gradient of retinoic acid is more effective in producing pattern changes than the same amount of retinoic acid distributed evenly (Eichele et al., 1985). An alternative possibility is that retinoic acid acts indirectly and other signals are involved. This is one interpretation of the recent finding that cell-cell communication between polarizing region and responding cells appears to play a significant role in the production of pattern changes (Allen et al., 1990). Retinoic acid could induce a new polarizing region that produces a positional signal that passes via gap junctions. However another possibility is that the interaction between polarizing region and anterior cells could be necessary for retinoic acid production by polarizing region cells. If this was the case, retinoic acid application would merely bypass this interaction.

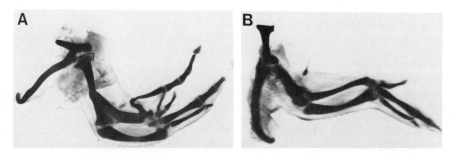

Fig.1 Beads soaked in retinoic acid were implanted to the anterior margin of leg buds (stage 20 embryos). Tissue adjacent to the bead was subsequently cut out and grafted to anterior margin of wing buds (stage 20 embryos). (A) Wing with additional digits. The most anterior digit is a toe, but there is also an additional wing digit 2. Leg tissue taken 24h after implantation of bead soaked in 1 mg/ml retinoic acid. (B) Wing with normal digit pattern and no additional digits. Leg tissue taken 8h after implantation of a bead soaked in 10 mg/ml retinoic acid.

Preliminary experiments suggest that tissue taken from chick limb buds after 22-24h treatment with retinoic acid can act as a polarizing region when grafted to a second bud and additional digits develop (see also Summerbell & Harvey, 1983). Typically, some of the additional digits develop from the grafted tissue but there are also additional digits of host origin (Fig. 1A). This has been shown by implanting beads soaked in retinoic acid to leg buds, for example, and assaying tissue adjacent to the bead for polarizing region activity in wing buds. It is unlikely that the results can be explained by carry-over of retinoic acid because when the tissue is taken after a 8h treatment, no additional digits form in the assay (Fig. 1B). The concentration of applied retinoic acid in the bud tissue reaches a steady state and then gradually declines as the bud grows (Eichele et al., 1985). It is not clear whether the apparent acquisition of polarizing region activity by cells in the bud after exposure to retinoic acid is due to the production of retinoid signals by the new polarizing region or whether a second signal is involved.

RETINOIC ACID AND EXPRESSION OF A HOMEOBOX GENE

The local application of retinoic acid to the anterior margin of chick wing buds leads to a mirror-image pattern of digits. After an 18h exposure to the retinoid, anterior cells are reprogrammed irreversibly to form posterior structures (Eichele et al., 1985). This provides an opportunity to investigate experimentally the cellular and molecular basis of pattern specification.

Position of cells in the limb could be encoded by homeobox genes. In collaboration with Drs. G. Oliver and E. De Robertis, we investigated the role of XlHbox 1 in chick wing patterning (Oliver et al., 1990). XlHbox1 is expressed at the anterior margin and proximally in wing bud mesenchyme. This can be demonstrated by using antibodies that recognize the highly conserved homeoprotein (Oliver et al., 1988a,b). If XlHbox 1 codes for an anterior position in the wing, then one would predict that when retinoic acid is applied to produce a mirror image pattern of digits that XlHbox 1 expression would be switched off in anterior cells. However, the reverse is found. Following retinoic acid application, XlHbox 1 is still expressed at the anterior margin and mesenchyme cells in more posterior and distal regions of the wing bud also express the gene. This result is surprising but it has been found that expansion of XLHbox 1 expression is correlated with shoulder girdle abnormalities (Fig. 2). It is tempting to speculate that expression of this gene is somehow connected with the development of the shoulder. According to fate maps (Saunders, 1948), the normal domain of XlHbox 1 expression appears to map to the region of the bud mesenchyme that gives rise to the girdle.

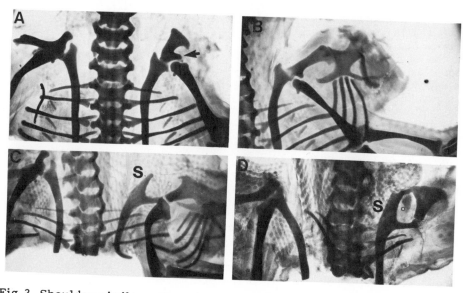

Fig 2. Shoulder girdle pattern in wings following either polarizing region grafts or application of retinoic acid. (A) Polarizing region grafted to anterior margin of right wing bud (stage 18 embryo). Shoulder girdle has a short coracoid and the coracoid process (arrowed) is reduced. Compare with girdle on untreated left side. Digit pattern (not illustrated) 432234. (B) Bead soaked in 0.1 mg/ml retinoic acid implanted at anterior margin of right wing bud (stage 18). Shoulder girdle with very reduced coracoid process. Digit pattern 43234. (C) & (D) Beads soaked in 10 mg/ml retinoic acid implanted at anterior margin of right wing buds (stage 20). Both scapula (S) and coracoid are very misshapen in these girdles. Digit pattern in both cases was 4334. Wing skeleton has been removed from specimen shown in (D).

The development of the shoulder girdle is affected by both retinoic acid application and grafts of polarizing region tissue. The abnormalities of the girdle occur in conjunction with both duplicated and truncated wings. When polarizing region tissue or retinoic acid is applied posteriorly, the wings have a normal pattern and the shoulder is not affected. Changes in shoulder girdle pattern are detected when wing buds are manipulated between stages 16 and 20 (Oliver et al., 1990). At stage 20, high concentrations of retinoic acid are necessary to affect girdle pattern (Fig. 2). Re-examination of the wing patterns following grafting polarizing region tissue to the wing bud forming region of the flank in stage 16 embryos also reveals effects on shoulder pattern that appear similar to those produced by high doses of retinoic acid treatment at stage 20 (Wolpert & Hornbruch, 1987).

XlHbox 1 expression occurs only in anterior mesenchyme cells of the wing bud. When retinoic acid is applied to the apex of the bud, cells in the entire anterior half of the bud express the gene. There is a fairly sharp boundary and cells in the posterior half of the bud do not switch on XlHbox 1. It may be somewhat fanciful but this sharp boundary between anterior and posterior wing mesenchyme is reminiscent of the differences found between anterior and posterior halves of somites (Keynes & Stern, 1988). The behaviour of grafted wing tissue also shows that expression of XlHbox 1 is confined to anterior cells and furthermore is cell autonomous. When anterior cells are grafted posteriorly, the XlHbox 1 gene product persists for at least 18h whereas when posterior cells are grafted to anterior position, XlHbox 1 remains switched off. Other homeobox genes are also expressed in anterior mesenchyme cells of chick wing buds. Wedden et al (1989) isolated and sequenced a gene that they have called Ghox 2.1. In situ hybridization revealed that transcripts of this gene are more abundant in anterior parts of the bud. It is also interesting that the expression pattern of RAR-β, one of the nuclear receptors for retinoic acid is also similar to that of Xl Hbox 1(Smith & Eichele, 1990; also see Rowe, Richman, Ochanda & Brickell, this volume). Furthermore, when retinoic acid is applied to produce duplicated wing patterns, the domain of RAR-β expression, like the domain of Xl Hbox 1, expands into more distal and posterior regions of the bud. It would be interesting to compare the features of XLHbox 1 and RAR-β expression in more detail.

RETINOIC ACID AND CO-ORDINATION OF PATTERNING AND MORPHOGENESIS
The width of a limb bud is related to the number of digits that develop. The key factor is the length of the apical ridge which increases in broad buds and decreases in narrow buds (Lee & Tickle, 1985). Experiments using retinoic acid to manipulate the length of the ridge support the hypothesis that bud shape is not controlled directly by the patterning signal. Instead, positional specification of the mesenchyme could control a local interaction between mesenchyme and epithelium that determines the morphology of the ridge; the thickened ridge could then control locally cell proliferation in the bud mesenchyme (Fig. 3).

Recombination experiments between tissues of retinoid-treated and normal wing buds suggest that an irreversible effect of retinoic acid on the apical ridge can be ruled out (Tickle et al., 1989). When ectoderm taken from wing buds in which the apical ridge has been obliterated by retinoic acid is recombined with normal mesenchyme, the apical ridge is re-established in the recombination and the recombined tissues give rise to digits. Any effects of retinoic acid on the apical ridge are reversible. In contrast, when treated mesenchyme is recombined with normal ectoderm, the development of the recombination shows that irreversible changes have taken place in the mesenchyme. For example, recombinations containing mesenchyme from buds that would develop additional digits, give rise to duplicated digit patterns (Oliver et al., 1990).

The spatial and temporal changes in cell proliferation in buds treated with retinoic acid appear to rule out the possibility that retinoic acid controls directly growth of the mesenchyme (Wilde et al., in preparation). The major changes in cell proliferation occur later than the time at which irreversible re-programming of the mesenchyme has taken place. Furthermore, in buds treated in retinoic acid to produce truncations, a decrease in proliferation occurs after the apical ridge has flattened.

The scheme suggested in Fig. 3 predicts that at least two local signalling molecules are involved in the interplay between mesenchyme and the apical ectodermal ridge. At present, the molecular identity of these signals is unknown. However, recent studies on the distribution of growth factors in vertebrate limbs raise interesting possibilities. For example, Ralphs et al. (1990) showed that insulin-like growth factors (IGFs) can be detected immunohistochemically at the tip of early chick wing buds.

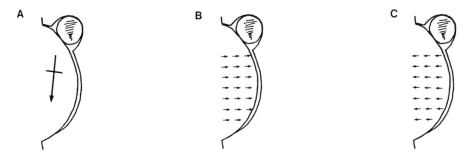

Fig. 3. Scheme for co-ordination of patterning and morphogenesis derived from the results of experiments in which retinoic acid is applied locally to wing buds. (A) retinoic acid diffuses into the bud and leads either directly or indirectly to irreversible respecification of the position of mesenchyme cells in the bud. (B) mesenchyme cells as a result of their positional specification produce a continuous signal that locally controls the morphology of the apical ridge. (C) cells of the thickened apical ridge produce a continuous signal that locally controls cell proliferation.

CONCLUSIONS

Retinoic acid is clearly an important molecule involved in development of vertebrate limbs. Retinoic acid may directly specify position of cells across the antero-posterior axis or could be part of a more complex signalling system. The response of limb bud cells to retinoic acid can be analyzed at both the cellular and molecular level. Expression of XlHbox 1, a homeobox gene expressed in limb anterior mesenchyme can be affected by retinoic acid and may be involved in specification of the pattern of the shoulder girdle. The morphogenetic response to retinoic acid involves changes in bud shape. The shape of the bud appears to be determined by a series of interactions between the mesenchyme and the apical ectodermal ridge. The molecular and cellular analysis of the mechanism of retinoic acid action should lead to a fuller understanding of the process of pattern formation in the vertebrate limb.

ACKNOWLEDGEMENTS

This work is supported by the MRC. I thank Anne Crawley for preparing the figures.

REFERENCES

Allen, F., Tickle, C. & Warner, A., 1990, The role of gap junctions in patterning of the chick limb bud, Development, 108:623.

Edwards, M.K.S. & McBurney, M.W., 1983, The concentration of retinoic acid determines the differentiated cell types formed by a teratocarcinoma cell line, Devl Biol., 98:187.

Eichele, G., Tickle, C. & Alberts, B.M., 1985, Studies on the mechanism of retinoid-induced pattern duplications in the early chick limb bud: temporal and spatial aspects, J. Cell Biol., 101:1913.

Keynes, R.J. & Stern, C.D., 1988, Mechanisms of vertebrate segmentation, Development, 103:413.

Lee, J. & Tickle, C., 1985, Retinoic acid and pattern formation in the developing chick wing: SEM and quantitative studies of early effects on the apical ectodermal ridge and bud outgrowth, J.Embryol. exp. Morph., 90:139.

Oliver, G., Wright, C.V.E., Hardwicke, J. & De Robertis, E.M., 1988a, Differential antero-posterior expression of two proteins encoded by a homeobox gene in Xenopus and mouse embryos, EMBO J., 7:3199.

Oliver,G., Wright, C.V.E., Hardwicke, J. & De Robertis, E.M., 1988b, A gradient of homeodomain protein in developing fore-limbs of Xenopus and mouse embryos, Cell, 55:1017.

Oliver, G., De Robertis, E.M., Wolpert, L. & Tickle, C., 1990, Expression of a homeobox gene in the chick wing bud following application of retinoic acid and grafts of polarizing region tissue, EMBO J., 9:3093.

Ralphs, J.R., Wylie, L. & Hill, D.J., 1990, Distribution of insulin-like growth factor peptides in the developing chick embryo, Development, 109:51.

Saunders, J.W., 1948, The proximo-distal sequence of the origin of the parts of the chick wing and the role of the ectoderm, J.Exp. Zool., 108:363.

Saunders, J.W. & Gasseling, M.T., 1968, Ectodermal-mesenchymal interactions in the origin of limb symmetry, in: "Epithelial-mesenchymal interactions, " R.Fleischmajer & R.E. Billingham, ed., Williams & Wilkins, Baltimore.

Smith, S.M. & Eichele, G., 1990, Development in press.

Summerbell, D., 1984, The effect of local application of retinoic acid to the anterior margin of the developing chick limb bud, J. Embryol. exp. Morph., 78:269.

Summerbell, D. & Harvey, F., 1983, Vitamin A and the control of pattern in developing limbs, in: "Limb development and regeneration" Prt A., J.F. Fallon & A.I. Caplan, ed., Alan R.Liss, New York.

Thaller, C. & Eichele, G., 1987, Identification and spatial distribution of retinoids in the developing chick limb bud, Nature, Lond., 327:625.

Thaller, C. & Eichele, G., 1988, Characterization of retinoid metabolism in the developing chick limb bud, Development, 103:473.

Tickle, C., Summerbell, D. & Wolpert, L., 1975, Positional signalling and specification of digits in chick limb morphogenesis, Nature, Lond., 254:199.

Tickle, C., Alberts, B., Wolpert, L. & Lee, J., 1982, Local application of retinoic acid to the limb bud mimics the action of the polarizing region, Nature, Lond., 296:564.

Tickle, C., Lee, J. & Eichele, G., 1985, A quantitative analysis of the effect of all-trans-retinoic acid on the pattern of chick wing development, Devl Biol., 109:82.

Tickle, C., Crawley, A. & Farrar, J., 1989, Retinoic acid application to chick wing buds leads to a dose-dependent reorganization of the apical ectodermal ridge that is mediated by the mesenchyme, Development, 106:691.

Wedden, S.E., Pang, K. & Eichele, G., 1989, Expression pattern of homeobox-containing genes during chick embryogenesis, Development, 105:639.

Wolpert, L. & Hornbruch, A., 1987, Positional signalling and the development of the humerus in the chick limb bud, Development, 100:333.

DOES RETINOIC ACID ORGANISE A LIMB OR INDUCE A ZPA?

Dennis Summerbell and Nick Waterson

The National Institute for Medical Research
The Ridgeway, Mill Hill
London NW7 1AA UK

The vertebrate limb bud develops as an autonomous embryonic field. The pattern of differentiation within the bud is at least in part controlled by discrete organising regions. While the interactions involved are probably complex the system can be simply modelled as three independent linear orthogonal axes. For some time the fashionable, or to use the correct molecular biology term, the sexy study, has been the antero–posterior (AP) axis and its discrete organisor, the Zone of Polarising Activity (ZPA).

When cells from the posterior margin of a limb bud are grafted to the anterior margin they cause production of supernumerary digits (Saunders and Gasseling, 1968). The type of digit is related to the distance of the responding tissue from the grafted ZPA

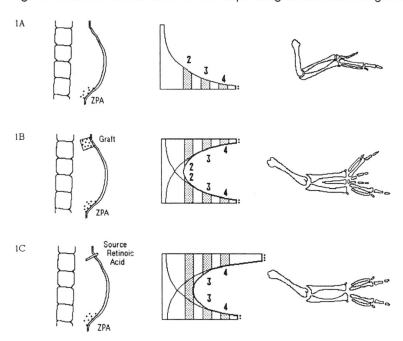

Figure 1   The morphogen gradient theory. The ZPA is the source of a morphogen, Retinoic Acid, that diffuses across the limb forming a monotonic gradient. The identity of digits is specified by the local concentration of the morphogen.

(Tickle, Summerbell and Wolpert, 1975). Local application of all trans Retinoic Acid (RAt) to the anterior margin mimics the effect of a ZPA graft (Summerbell, 1984; Summerbell and Harvey, 1983; Tickle et al, 1982). Retinoic acid is present in normal limbs and is relatively abundant in the posterior (ZPA containing) region (Thaller and Eichele, 1987). These and numerous other observations have led to the working hypothesis that RAt is a natural morphogen controlling pattern formation across the AP axis (See Figure 1).

However it is equally possible that RAt is not the product of the ZPA but rather that RAt acts to induce the formation of a ZPA from normal limb cells. Tissue adjacent to an RAt implant developes ZPA-like properties within 18 to 24 hours (Summerbell and Harvey, 1983). In this paper we examine this observation in more detail. The subject is one of much current interest and is also dealt with in this volume in the contributions by Cheryll Tickle and Susan Bryant.

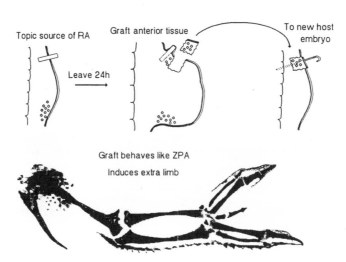

Figure 2 The basic assay. A source of RA is grafted into the primary host. After various time intervals the tissue adjacent to the implant is removed and grafted to a secondary host. It takes about 24 hours for the graft to acquire full ZPA–like activity.

INDUCTION OF ZPA–LIKE ACTIVITY IN TISSUE ADJACENT TO A RAt IMPLANT

The basic experiment involves grafting a source of RAt into the anterior margin of the limb bud. There are two significant differences between our protocols and those reported elsewhere in this volume. Most studies have used positively charged ion exchange beads (AG 1-X2) which have the property of binding retinoic acid. We have instead used small pieces of newsprint (Richmond and Twickenham Times). The 120um thick newsprint is first cut into 1mm squares. Each square is priefly submerged in a 10mM solution of RAt in Dimethylsulphoxide. The square is drained on blotting paper then cut into 9 equal squares of about 330um per side. In our hands we find the amount of RAt incorporated into the newsprint (2.3-4.4ng per implant) is less variable than for beads (2.2-9.6ng per implant). The second difference involves the micro-surgery. We insert the implant into a slit perpendicular to the main body axis (See Figure 2), others implant beads under the apical ectodermal ridge (AER). The latter procedure significantly distorts the structure of the limb bud making studies of the geometry difficult during early stages.. However it has the clear advantage of being operationally a more sensitive bioassay.

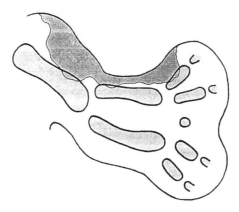

Figure 3 Composite reconstruction. Computer assisted serial section reconstruction is used to estimate the contribution of quail cells to the
supernumerary digits.   Each serial section has been positioned on a
drawing of the intact limb and the extent of the quail tissue indicated.
Clearly this over–emphasises the overall quail contribution.

THE INDUCED–ZPA DOES NOT SELF DIFFERENTIATE INTO EXTRA DIGITS

We next tested that the induced ZPA indeed organised host tissues to form
supernumerary digits. The basic experiment was repeated using quail as the
primary host. The tissue adjacent to the RAt implant was then grafted into a
secondary chick host. Hosts were fixed, sectionned and stained after 24, 48 and
72 hours. Computer assisted serial section reconstruction demonstrated that in all
cases the graft organised host tissue to form the supernumerary structures. Figure
3 shows the maximum contribution of cells from the graft observed at 72 hours.

Effect of implant on first host following early removal after x hours ——————

Effect of implant on second host following depletion for x hours ⋯⋯⋯⋯

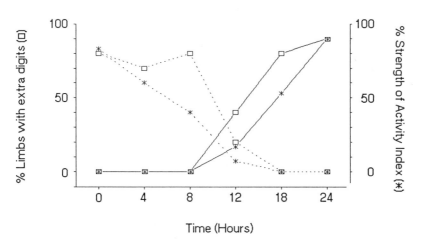

Figure 4 Time taken for the implant to lose its ability to induce supernumerary
digits and for the adjacent tissue to acquire ZPA–like activity.

THE IMPLANT LOSES ITS INDUCTIVE ACTIVITY AFTER 12 HOURS

Following implantation of a source of RAt we assay the tissue adjacent to the graft for ZPA–like activity (Figure 4). 12 hours post implantation the assay is negative. The recipient limbs form few supernumerary digits. 18 hours post implantation most limbs have at least one extra digit. After 24 hours over half of the grafts showed activity comparable to a normal ZPA graft. This time scale is broadly consistent with the time scale for induction by a ZPA graft. During this early period the RAt implant gradually loses its ability to induce. If the implant is removed and grafted into a second host limb it retains some activity for several hours but after 12 hours it has essentially lost the ability to produce supernumerary digits. Thus the implant has lost its ability to induce before the adjacent tissue has gained the ability. This suggests that the phenomenon is not due to passive transfer of RAt from implant to graft. We next tested this possibility directly.

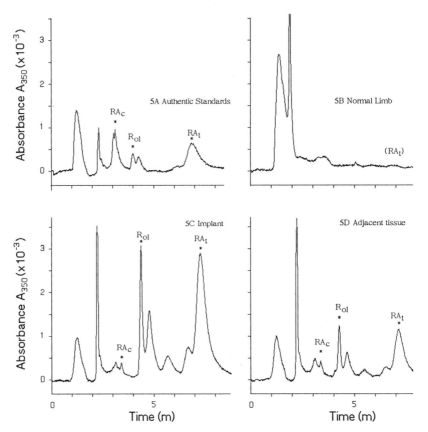

Figure 5 Normal phase HPLC of limb extracts, for details see text.

AFTER 24 HOURS THERE IS LESS RAt IN THE GRAFT THAN IN THE IMPLANT

We performed a standard implantation. After a given time we removed the implant, and that portion of the bud that would be used as a secondary "ZPA" graft. We extracted for retinoids then made direct measurement of concentration using reverse (RP) and normal phase (NP) High Performance Liquid Chromatography (HPLC). Figure 5 shows typical NP results from 8 pooled embryos at 24 hours together with controls. Figure 5A shows a set of mixed internal standards with Retinol (Rol), cis Retinoic Acid (RAc) and RAt indicated. Figure 5B shows an extract from 8 whole normal limb buds. Figure 5C shows an extract from 8 implants. 5D shows an extract from 8 putative induced "ZPAs". It is clear that the "ZPA" contains significantly less RAt than is now left in the implant after 24 hours in the host. Yet the former will succesfully induce supernumerary digits in a secondary host while the latter will not. The absolute concentration of RAt is therefore not the only factor determining ability to induce. Note that the concentration of RAt in the implant is very much higher than in a normal ZPA (compare with Figure 5B). Each induced ZPA contains about 300pg of RA (about 100-fold higher than in a normal ZPA as estimated by Thaller and Eichele).

DISCUSSION

Our data favours the hypothesis that Retinoic Acid induces the formation of a ZPA by the cells adjacent to the implant. This raises the intriguing possibility that a primary action of RAt is the enhancement of synthesis of RAt.

REFERENCES

Saunders, J.W. and Gasseling, M.P. (1968). Ectodermal-mesenchymal interaction In the origin of limb symmetry. In Epithelial- mesenchymal interactions (Ed. Fleischmajer), pp. 78-97. Williams and Wilkins, Baltimore.

Summerbell, D. (1983). The effect of local application of retinoic acid to the anterior margin of the developing chick limb. J. Embryol. exp. Morph. **78**, 269-289.

Summerbell, D. and Harvey, F. (1983). Vitamin A and the control of pattern in developing limbs. Prog. Clin. Biol. Res. **110**, 109-118.

Thaller, C. and Eichele, G. (1987). Identification and spatial distribution of retinoids in the developing chick limb bud. Nature, **327**, 625-628.

Tickle, C., Alberts, B., Wolpert, L. and Lee, J. (1982). Local application of retinoic acid to the limb bond mimics the action of the polarising region. Nature **296**, 564-565.

Tickle, C., Summerbell, D. and Wolpert, L. (1975). Positional signalling and specification of digits in chick limb morphogenesis. Nature, Lond., 254, l99-202.

RETINOIC ACID EFFECTS ON EXPERIMENTAL CHICK WING BUDS:

PATTERNS OF CELL DEATH AND SKELETOGENESIS

Sharon C. Frost and J. Richard Hinchliffe

Dept of Biological Sciences
University College of Wales, Aberystwyth, Dyfed, Wales, UK

INTRODUCTION

Tickle, Lee and Eichele (1985) have shown that retinoic acid (RA), a derivative of vitamin A, is capable of mimicking the polarising action of the ZPA. If a latex bead is soaked in RA and implanted into the anterior of a Hamburger-Hamilton stage 19-22 wing bud, duplication of digits occurs, as when a ZPA is grafted to a similar position. Eichele (1989) has recently shown that an RA-soaked bead (RA-bead) can cause digits to form in posterior half (and hence ZPA) deleted limb buds. These limb buds would otherwise produce no digits (Hinchliffe and Gumpel-Pinot, 1981) whereas a bead in the anterior half of the limb leads to a digit pattern that is the reverse of normal, ie 432 instead of 234. However, this does not occur with 100% success. The percentage of total reversals (432) is low; 9% in Eichele's work.

The low success rate in producing a fully-reversed skeleton is hard to explain. It suggests that the ZPA may have more than one effect within the limb; maintaining cells in addition to signalling position. Todt and Fallon (1987) have proposed a two-factor ZPA control system. If the ZPA, (or a posterior region more extensive than the ZPA) also has a maintenance role, the relatively poor development of the deleted limb buds could be explained by the lack of maintenance factor, since the RA implanted beads may act only as a signal source.

If the posterior half of the limb bud is amputated massive cell death (inhibiting normal skeletal development) occurs in the distal mesenchyme of the remaining anterior half (Wilson and Hinchliffe, 1987). This is caused by removal of the ZPA, since it can be reversed by a ZPA graft. These results are consistent with the maintenance role theory. To examine this theory, the pattern of cell death in RA bead-implanted anterior half limbs was analysed. Was cell death inhibited in the half limbs in which a reversal of the digital pattern subsequently occurred?

EXPERIMENTS AND RESULTS

Experiments were carried out *in ovo*, on the anterior halves of stage 19-21 wing buds. Cell death patterns at 24 or 48 hours post-operation were monitored by Neutral Red staining. Skeletal pattern was examined at 9 days of development (6 days post-operation) by methylene blue staining. Latex beads of 200 mm diameter were soaked in RA at a concentration of 0.05 or (in

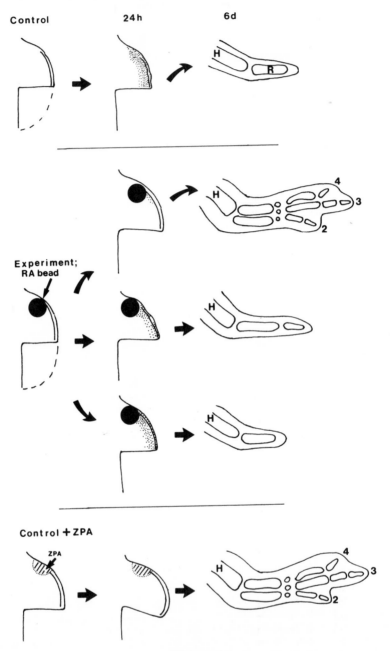

Control 24h 6d

H R

4
3
2

H

Experiment;
RA bead

H

H

Control + ZPA

ZPA

4
3
2

H

Figure 1

Diagram showing effects of RA-bead implantation (see text). Control
(posterior half removal) results in substantial cell death and poor skeletal
development. Experimental RA-bead implantation results in a range of
effects from complete cell death inhibition, (assumed to result in full
digital skeletal development) to a failure in inhibition. Control ZPA
implants always inhibit the cell death and usually result in full digital
skeletal development. (Stippling represents cell death).

most cases) 0.1 mg/ml in DMSO. These concentrations are known to be the most effective in producing digital duplications (Tickle et al., 1985).

Experiments

These were performed as follows:
1. Control: isolated anterior halves.

2. Experimental implantations: RA-beads placed into anterior halves under the AER, at its anterior end.

3. Control implantations: i) RA-beads implanted into normal wing buds.
                          ii) ZPA grafts in anterior halves.

Results

1. Control: - Extensive distal mesenchymal cell death occurred and the AER regressed, with later poor skeletal development, as reported by Wilson and Hinchliffe, 1987.

2. Experimental implantations: - There was a variable pattern of cell death after 24 or 48 hours. Some anterior halves (52%, 51/98) showed little inhibition of cell death, with regressing apical ridges. Others, (48%, 47/98) showed partial inhibition of cell death, and had patchy or sometimes complete survival of the AER.

Skeletal development varied correspondingly, with the following structures obtained:

a. humerus-radius (short) in 63% (93/148) as in control posterior half amputations.
b. humerus-radius-digital element in 20% (30/148). (Digit lacking any specific character).
c. humerus-radius-plus complete set of reversed digits (bead432) in 17% (25/148).

The skeletal development results reported here correspond to those of Eichele (1989), which showed a range from bead432 (9%) bead32 or bead43 (4.5%) to humerus-radius only (46%). We propose that those limb buds with inhibition of cell death and some AER survival develop a totally (bead432) or partially complete set of reversed digits.

3. Control implantations: - Duplications resulted from RA- beads implanted pre-axially into normal wing buds. Full duplication (bead432234 or 43234) was obtained in almost all cases (12/15).

ZPA grafts in anterior halves always resulted in cell death inhibition (see also Wilson and Hinchliffe, 1987), and subsequently in a reversal in digit pattern (g432). The ZPA grafted half limbs had a noticeably more normal appearance 24 hours after the operation than the RA-bead implanted half limbs, with a larger volume of tissue and normal cell death patterns.

DISCUSSION

In these experiments an RA-soaked bead substitutes successfully for the ZPA in only a relatively low percentage of cases. The results obtained show a low percentage of cell death inhibition corresponding to the low number of skeletal reversals seen both here and in Eichele's work.

It is difficult to explain the variability of the results, but the low success rate is consistent with Todt and Fallon's two-factor theory. The ZPA (or posterior area larger than the ZPA) may have a maintenance role in addition to position specification, and implanting RA-soaked beads will not fulfill the complete function of a ZPA graft. The ZPA may itself produce growth-promoting substances or have a growth-promoting effect on adjacent tissue. This interpretation would explain the high success rate in obtaining reversed skeletal development with a ZPA graft to anterior half limbs, and in obtaining duplication with either an RA-bead or a ZPA grafted pre-axially into a normal wing, which still has a functioning ZPA.

This theory is supported by the work of McLachlan (these proceedings), which shows that in *in vitro* cell culture the ZPA mitogen, as opposed to the morphogen, is found not to be retinoic acid. Retinoic acid will not cause an increase in the proportion of cells entering division, whereas an as yet unidentified diffusible mitogen produced by the ZPA will do so.

Further evidence is provided by Fallon, Dvorak and Simandl (these proceedings). In studies on *talpid*[2] mutants, removal of the posterior half of a wingbud does not lead to massive cell death in the remaining anterior half, in contrast to the wild type experiment described here. This would suggest that the *talpid* mutant, which has a weak ZPA, also has maintenance activity widespread throughout the limb. Removal of the ZPA does not lead to necrosis in the remaining mesenchyme, and a maintenance factor must therefore be acting on the tissue. This supports the two-factor theory previously mentioned.

REFERENCES

EICHELE, G., (1989). Retinoic acid induces a pattern of digits in anterior half wing buds that lack the zone of polarising acitvity. Development. 107: 863-867.
FALLON, J. F., DVORAK, I., SIMANDL, B. K. (1991). Insights into limb development and pattern formation from studies of the limbless and talpid[2] chick mutants. (These proceedings)
HINCHLIFFE, J. R., GUMPEL-PINOT, M. (1981). Control of maintenance and antero-posterior skeletal differentiation of the anterior mesenchyme of the chick wing bud by its posterior margin (the ZPA). J. Embryol exp. Morphol. 62: 63-83.
McLACHLAN, J., (1991). Growth factors produced by the polarising zone, and their relationship to retinoic acid. (These proceedings)
TICKLE, C., LEE, J., EICHELE, G. (1985). A quantitative analysis of the effect of all-trans retinoic acid on the pattern of chick wing development. Devl Biol. 109: 82-95.
TODT, W. I, FALLON, J. F. (1987). Posterior apical ectodermal ridge removal in the chick wing bud triggers a series of events resulting in defective anterior pattern formation. Development. 101: 505-515.
WILSON, D. J., HINCHLIFFE, J. R. (1987). The effect of removing the zone of polarising activity on the anterior half of the chick wing bud. Development. 99: 99-108.

A COMPARISON OF REACTION DIFFUSION AND MECHANOCHEMICAL

MODELS FOR LIMB DEVELOPMENT

Philip Maini

Centre for Mathematical Biology
Mathematical Institute
24-29 St. Giles'
Oxford OX1 3LB

Several theoretical models have been proposed to attempt to elucidate the underlying mechanisms involved in the spatial patterning of skeletal elements in the limb. Here, I briefly compare two such models - reaction diffusion (RD) and mechano-chemical (MC) - and highlight their properties and predictions.

Although similar mathematically, these models are based on fundamentally different biological approaches. RD theory asserts that a chemical prepattern is first set up in the limb by a system of reacting and diffusing chemicals (Turing, 1952). Cells then respond to this prepattern by differentiating wherever the concentration of one of the chemicals lies above a threshold value (Wolpert, 1969, 1981). Thus, the spatial pattern of chemical concentration is reflected by the spatial pattern of cell differentiation.

The MC theory (Oster et.al., 1983, Oster et.al., 1985) proposes that the mechanical and chemical interactions of cells with their external environment - the extracellular matrix - leads to a spatial pattern in cell density. The cells in the aggregates thus formed differentiate. Thus, the spatial pattern of cell differentiation overlies the spatial pattern of cell density.

The spatial patterns predicted by both RD and MC theory, from linear analysis, are essentially the eigenfunctions of the Laplacian operator (Maini and Solursh, 1990). The more complex tensor form of the MC model may lead to a richer set of structures than those exhibited by RD models. This possibility has yet to be analysed. Thus, at the moment, one cannot distinguish between these models on a purely mathematical basis.

The similarity of the mathematical formulation of the above hypotheses does, however, enable one to make some general predictions on the properties of the skeletal pattern in the limb independent of the biological basis of the model. For

Developmental Patterning of the Vertebrate Limb
Edited by J.R. Hinchliffe *et al.*, Plenum Press, New York, 1991

example, both models suggest that there are only a limited number of ways in which elements may arise and bifurcate. This has lead to a more precise formulation of the idea of developmental constraints on vertebrate limb evolution (Oster et.al., 1988, Oster and Murray, 1989).

Both models predict that the complexity and form of spatial pattern is dependent on scale, geometry and boundary conditions. For example, they predict that decreasing limb bud width will result in a loss of elements. This agrees with experimental observation (Alberch and Gale, 1983).

Recently it has been shown that if two stage 19 anterior chick limb halves are combined, the recombinant forms two humeral elements (Wolpert and Hornbruch, 1990). As the recombinant has the same size as a normal limb, both the above models would predict a single humerus. The observation suggests that the anterior stage 19 limb half contains cells that have already differentiated yet, at this stage, there is no visible aggregation of cells. This is inconsistent with the MC approach but may be consistent with the RD approach assuming that the prepattern is laid down and stabilised before stage 19. However, another interpretation of the result is that during normal development a prepattern may divide the early limb bud into domains in which cells are either competent or incompetent to differentiate into cartilage. Such a prepattern may be set up by a gradient in homeobox gene expression or in retinoid distribution (Maini and Solursh, 1990). The recombination experiment would then be seen as combining two such cell populations and, in effect, doubling the domain width. The eventual developmental fate of cells could then be determined by either of the above models.

This scenario suggests that pattern formation is a hierarchal process wherein each mechanism provides the initial condition for subsequent processes. This possibility warrants further experimental and theoretical investigation.

ACKNOWLEDGEMENT   This work was supported in part by the National Science Foundation under Grant No. DMS 8901388.

REFERENCES

P. Alberch and E. Gale, Size dependency during the development of the amphibian foot. Colchicine induced digital loss and reduction, J. Embryol. exp. Morphol. 76, 177-197 (1983).

P.K. Maini and M. Solursh, Cellular mechanisms of pattern formation in the developing limb, Int. Rev. Cytology (to appear).

G.F. Oster, J.D. Murray, and A.K. Harris, Mechanical aspects of mesenchymal morphogenesis, J. Embryol. exp. Morphol. 78, 83-125 (1983).

G.F. Oster, J.D. Murray, and P.K. Maini, A model for chondrogenic condensations in the developing limb: the role of extracellular matrix and cell tractions, J. Embryol. exp. Morphol. 89, 93-112 (1985).

G.F. Oster, N. Shubin, J.D. Murray, and P. Alberch, Evolution and morphogenetic rules. The shape of the vertebrate limb in ontogeny and phylogeny, Evolution 45, 862-884 (1988).

G.F. Oster and J.D. Murray, Pattern formation models and developmental constraints, in J.P. Trinkaus Anniversary Volume 1989 (in press).

A.M. Turing, The chemical basis of morphogenesis, Phil. Trans. Roy. Soc. Lond. B237, 37-72 (1952).

L. Wolpert, Positional information and the spatial pattern of cellular differentiation, J. Theor. Biol. 25, 1-47 (1969).

L. Wolpert, Positional information and pattern formation, Phil. Trans. Roy. Soc. Lond. B295, 441-450 (1981).

L. Wolpert and A. Hornbruch, Double anterior chick limb buds and models for cartilage rudiment specification, Development 109, 961-966 (1990).

THE PATHWAY OF POLARIZING ACTIVITY FROM HENSEN'S NODE TO THE

WING BUD IN THE CHICK EMBRYO

Amata Hornbruch

Department of Anatomy and Developmental Biology
University College and Middlesex School of Medicine
London W1P 6DB U.K.

INTRODUCTION

The skeletal elements of the normal wing of a chick embryo are humerus, radius and ulna, the wrist and in antero-posterior order the digits *2,3,4* . The traditional experiment to demonstrate that cells from the posterior edge of a stage 20 limb bud transmit a signal which lays down the antero-posterior axis of the digits is to transplant a cube about 100 μm^3 of these cells to the anterior edge of another similarly staged limb bud. The result is widely known as the mirror-image reduplication with a digit combination of *432234* (Fig.1a & 1b) along the proximo-distal axis of the wing. (Saunders and Gasseling, 1968). This signal was interpreted in terms of a diffusible morphogen produced by the cells at the posterior part of the developing bud (Tickle et al., 1975). Increasingly retinoic acid and many of its analogues which produce similar reduplications in the 10 day old wing when applied to the early limb bud (Summerbell, 1983; Tickle et al., 1985) are thought to be the signal itself or at least are instrumental in its transmission or reception (Summerbell and Maden, 1990).

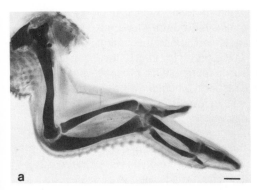
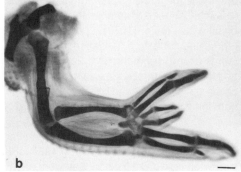

Fig.1 Wholemounts of 10 day old embryonic chick wings stained with alcian green; a) normal wing; b) reduplicated wing after a ZPA graft. Scale bar = 1 mm.

Developmental Patterning of the Vertebrate Limb
Edited by J.R. Hinchliffe *et al.*, Plenum Press, New York, 1991

The natural distribution of retinoic acid throughout the limb bud appears to be graded with the highest concentration at the posterior part of the bud (Thaller and Eichele, 1987). This may suggest that retinoic acid could be the endogenous morphogen (Brockes, 1989).

The importance of the functions and the distribution of retinoic acid and its derivatives within the limb bud will be discussed extensively elsewhere in this volume.

Maps showing the distribution of polarizing activity of the wing bud stages have been available by MacCabe et al (1973), and more recently by Honig and Summerbell (1985).

I am reporting here the spatial and temporal distribution of polarizing activity from its first appearance in Hensen's node at stage 4 spreading wide through the flank tissue to concentrate at the posterior wing field at stage 16.

It is important to point out that investigations of this kind cannot be done with technology on the molecular level at the present time. The very small amounts of tissue available and the low concentration of polarizing activity can only be measured with our very sensitive bioassay.

METHODS

Fertile chicken eggs from a local breeder were incubated on fixed shelves at $38 +/- 1^{o}C$ in a humidified atmosphere from 18h to 4 days. All embryos were staged according to the criteria of Hamburger and Hamilton (1951). Host embryos were windowed on the third or fourth day of incubation and used when they had reached stage 20.

Donor embryos were used from stage 4 to stage 16. The eggs were windowed in the usual way. The blastoderm was cut out with scissors washed in PBS (Oxoid) and pinned out dorsal side up in Hank's BSS (Gibco) in a small petri dish coated with Sylgard 180 (Dow Corning Corporation), this is a transparent rubbery polymer. The graft tissue was dissected from the flank parallel to each somite starting from a cranial position and proceeding caudally. Right and left hand flank tissue was used from each embryo from the same antero-posterior positions.

The size of the graft varied from 100 - 200 μm in antero- posterior and 150 - 300 μm in medio-lateral direction due to the increasing width of the somites in older embryos. To determine the exact location of the flank tissue in relation to the somites in embryos with less than 25 pairs of somites the unsegmented somitic mesoderm was divided into 10 equal parts . Meier (1979) reported that the unsegmented somitic mesoderm between the regressing node and the last segmented somite is pre-patterned into 10 to 12 somito-meres prior to segmentation. Each of these parts is to represent the prospective somites and they were named as somite position and the respective number which each piece will obtain during segmentation . The nomenclature somite level is used for graft tissue taken from the flank opposite already segmented somites.

The dissected graft tissue was placed under the previously loosened apical ectodermal ridge (see Hornbruch and Wolpert 1986 for detailed method). Embryos were further incubated until day 10 when they were sacrificed and both wings were fixed in 5% TCA, stained in alcian green, dehydrated in graded alcohols and cleared in methyl salicylate (Summerbell and Wolpert 1973). These wholemounts were examined for irregularities in the digit pattern.

STRENGTH OF POLARIZING ACTIVITY INDEX

The index of polarizing activity is an assessment of the strength of the signal in percent (Honig et al., 1981). All grafts for each somite level were asigned points for every additional digit; 0 points for a normal wing, 1 point for digit *2*, 2 points for digit *3*, and 3 points for digit *4*. The sum of points divided by the highest score possible (number of wings in the set multiplied by 3) expresses the activity in percent.

RESULTS

Hensen's node from embryos stage 4 to 9 when grafted to the wing bud gave extra digits *2* and *3* in 80% of all cases. There was never an extra digit *4* in any of the sets. The index of polarizing activity lay around 50% for each of these stages except for stage 8 it was 29% (Table 1). Hensen's node from stage 10 and older stages did not give rise to extra digits. The head process from stage 5 had very low activity with only 5% but tissue from the same area at stage 6 showed 30% activity and 48% for stage 9. The primitive streak from immediately posterior to the node was tested for stages 9 and 10 and showed an index of 16% and 25% polarizing activity respectively. No digits *4* were monitored in this tissue either. Single somites 1 to 6 for stages 7 and 8 were found to have no polarizing activity (Table 1).

Lateral plate mesoderm form the level of Hensen's node was tested for stage 8 (4 to 6 pairs of somites) and resulted in only 5 wings with an extra digit *2* from 88 grafts. Stage 9 embryos (7 to 9 pairs of somites) gave 11 wings with an extra digit *2* and 2 wings with an extra digit *3* from the same positions and a similar number of grafts.

Tissue from the flank between the somite positions 12 and 25 was taken from embryos stage 10 (10 to 12 pairs of somites) to stage 16 (28 to 30 pairs of somites). This contains the flank regions anterior and posterior to and including the wing field which lies between somites 15 and 20. Polarizing activity for stage 10 hardly rose above 10% with no graft giving rise to an extra digit *4*. Stage 11 (13 to 15 pairs of somites) showed a marked increase in activity for the anterior wing field opposite somites 15 to 18 to around 25%. Several grafts gave wings with extra digits *4* for this stage. Stage 12 (16 to 18 pairs of somites) displayed the highest activity at somite level 18 with 33%, this is the middle of the wing field. The anterior wing field has lost most of the earlier activity.

Table 1   Grafts of Hensen's node and adjacent tissues for stages 4 to 10

Stage	4	5	6	7	8	9	10	
Node	48	52	42	40	29	50	0	% Index
Head process	-	5	30	-	-	48	0	
Prim.streak	-	-	-	-	-	16	25	
Somites 1-6	-	-	-	0	0	-	-	

Prim. streak; primitive streak i.e. tissue posterior to the node, -; no grafts were performed for those tissues.

Table 2   % Strength of polarizing activity index for somite position 12 to 25 for stages 10 to 16.

Somite position	Stage 10	Stage 11	Stage 12	Stage 13	Stage 14	Stage 15	Stage 16
12	3						
13	3						
14	5	13					
15	11	24	8				
16	6	25	6				
17	11	16	8	13	20		
18	9	28	33	27	16	22	
19	6	20	19	35	19	28	33
20	7	10	17	16	48	78	93
21	7	10	2	18	53	77	80
22	6	9	3	18	44	47	73
23				8	18	42	52
24				6		9	43
25				toe			27

Stages 13 (19 to 21 pairs of somites) to 15 (25 to 27 pairs of somites) showed further increases in polarizing activity and a transfer of the peak to the posterior part of the wing field (Table 2). Stage 16 is the last of the pre-wing bud stages. Polarizing activity is over 90% for somite level 20 and can be considered as "full strength". There is also relatively high activity in the flank tissue posterior to the wing bud, 27% at somite level 25.

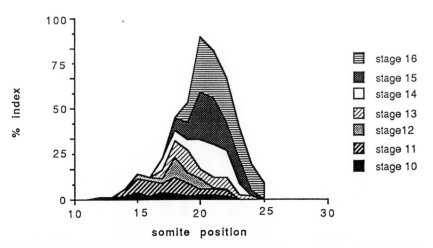

Fig.2 Graph showing the strength of polarizing activity in percent. Data for stages 10 - 16 are superimposed. The wing field lies between somite 15 and somite 20.

DISCUSSION

I have investigated the pathway of the spatial and temporal distribution of polarizing activity in the early embryo from Hensen's node to the pre-limb bud stage 16.

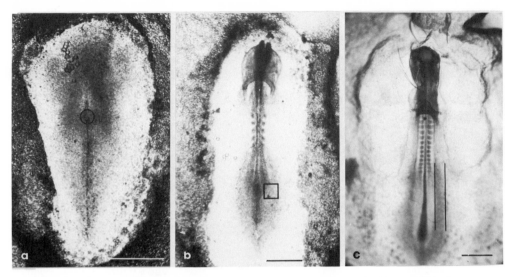

Fig.3 Ventral view of chick embryos in New culture; a) stage 5, O, Hensen's node;
b) stage 8, 5 pairs of somites, □,lateral plate mesoderm tested at the level of
the node; c) stage 11, 14 pairs of somites, flank tissue tested is indicated by the
vertical lines. Scale bar = 1 mm.

The average strength of polarizing activity for tissues from Hensen's node stage 4 to 9 was 50%. Only extra digits *2* and *3* occured. None of the grafts gave an extra digit *4* . The presents of polarizing activity in Hensen's node suggests that the signal from the node which is also capabale of organising additional rows of somites when grafted to the early blastoderm (Hornbruch et al., 1979) is similar to the signal from the ZPA. However a graft of ZPA tissue to a stage 4 or 5 blastoderm will not result in any kind of extra structure. Though embryos may be conservative in the number and the type of signals they use developmentally earlier signals may be easier to read or interpret for developmentally older tissue than the reverse. There was moderate activity in the head process of stages 6 and 9 as well as in the primitive streak posterior to the node for stages 9 and 10. The lateral plate mesoderm at the level of Hensen's node showed hardly any activity for stages 8 and 9 although activity was still present in the node.

It is important to point out here that the wing field lies between somites 15 and 20. The flank tissue lateral to the unsegmented somitic mesoderm for stage 10 had a fairly uniform distribution of activity, never higher than 11% for most somite positions. Stage 11 showed activity around 25% in the anterior wing field for somite positions 15 to 18 but only around 10% for the posterior part of the wing field. By stage 12 there was a definitive peak of activity at the middle of the wing field at somite 18. It is of interest to note that the first elevation of the apical ectodermal ridge also appears at the level of somite 18 some 24 to 30 hours later at stage 18 (Todt and Fallon, 1984). There was a distinct shift of the peak of activity from the middle of the wing field for stages 11 and 12 to the posterior edge of the wing field at somite level 20 for stages 15 and 16.

The development of the ZPA in the limb shows a well defined spatial and temporal pattern.However, it has not been possible to demonstrate a continuum of active tissue between the node and the limbs. This suggests that polarizing activity is not clonally derived from a single population of cells originating in the node but appears *de novo* in lateral flank mesenchyme.

ACKNOWLEDGMENTS

I am indebted to Lewis Wolpert for his patience while I completed this work. My special thanks go to the computer experts Dennis Summerbell for trying to educate me in computer science and to Primus Mullis for assistence with the graph.

REFERENCES

Brockes, J.P., 1989, Retinoids, homeobox genes, and limb morphogenesis, Neuron, 2:1285.

Hamburger, V. and Hamilton, H.L., 1951, A series of normal stages in the development of the chick embryo. J. Morph., 66: 49.

Honig, L.S., Smith, J.C., Hornbruch, A. and Wolpert,L., 1981, Effects of biochemical inhibitors on positional signalling in the chick limb bud. J. Embryol. exp. Morph., 62: 203.

Honig, L.S. and Summerbell, D., 1985, Maps of strength of positional signalling activity in the developing chick wing bud, J. Embryol. exp. Morph., 87: 163.

Hornbruch, A., Summerbell, D. and Wolpert, L., 1979, Somite formation in the early chick embryo following grafts of Hensen's node, J. Embryol. exp. Morph., 51: 51.

Hornbruch, A. and Wolpert, L.,1986, Positional signalling by Hensen's node when grafted to the chick limb bud, J. Embryol. exp. Morph., 94: 257.

MacCabe, A.B., Gasseling, M. and Saunders, J.W.,1973, Spaciotemporal distribution of mechanisms that control outgrowth and antero-posterior polarisation of the limb bud in the chick embryo, Mech. Ageing Develop., 2: 1

Meier, S.,1979, Development of the chick embryo mesoblast. Formation the embryonic axis and establishment of the metameric pattern, Dev. Biol., 73:25.

Saunders, J.W. and Gasseling, M.,1968, Ectodermal-mesenchymal interaction in the origin of limb symmetry, in: "Epithelial-msenchymal interaction", R. Fleischmayer and R.E. Billingham, ed., Williams and Wilkins, Baltimore.

Summerbell, D.,1983, The effect of local application of retinoic to the anterior margin of the developing chick limb, J. Embryol. exp. Morph.,78:269.

Summerbell, D. and Wolpert, L. (1973) Precision od development in the chick limb morphogenesis, Nature, Lond., 244:228.

Summerbell, D. and Maden, M.,1990, Retinoic acid, a developmental signalling molecule, Trends Neurosci., 13:142.

Thaller, C. and Eichele, G.,1987, Identification and spatial distribution of retinoids in the developing chick limb bud.Nature, 327:625.

Tickle, C., Lee, J. and Eichele, G.,1985, A quantitative analysis of the effect of all-trans-retinoic acid on the pattern of the chick wing development, Dev. Biol. 109:82.

Tickle, C., Summerbell, D. and Wolpert, L.,1975, Positional signalling and the specification of digits in chick limb morphogenesis, Nature. Lond.,254:199.

Todt, W.L. and Fallon, J.Fallon, 1984, Development of the apical ectodermal ridge in the chick wing bud,J. Embryol. exp. Morph. ,80:21.

EXTRACELLULAR MATRIX IN EARLY LIMB DEVELOPMENT

Michael Solursh

Department of Biology
University of Iowa
Iowa City, Iowa 52242 USA

INTRODUCTION

The aim of this chapter is to provide an overview of the role of extracellular matrix (ECM) in limb patterning and to help place some of the other chapters on related topics in perspective with other aspects of pattern formation in the developing limb. A very tentative scheme is illustrated in Fig. 1. Retinoids or some comparable signalling mechanism is assumed to modulate the spatial expression of various DNA binding proteins. These in turn might alter the expression of genes that act more distally to regulate growth, morphogenesis, and differentiation. These later factors include cell adhesion molecules (CAM's), substrate attached molecules (SAM's), extracellular matrix components (ECM), growth factors and their receptors, and proteases.

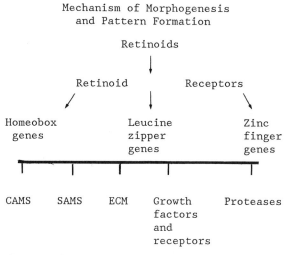

Fig. 1. Hypothetical scheme to indicate relationships between an integrative signalling mechanism, changes in expression of transcription factors and the expression of gene products which are involved directly in morphogenesis.

The discussion here will focus on a number of ECM components that have already been studied in the early limb bud. Some of these are listed in Table 1.

Table 1. Some ECM molecules in the early limb bud

collagen I
collagen IV
fibronectin
hyaluronan
laminin
basement membrane heparan sulfate proteoglycan
cell surface heparan sulfate proteoglycan
large mesenchymal chondroitin sulfate proteoglycan (PG-M)

The available data are largely descriptive in terms of the temporal and spatial expression of various known ECM components or their mRNA's with little functional information. Nevertheless, it is clear that the ECM is an early indicator of pattern formation.

THE DISTRIBUTION OF ECM COMPONENTS

The distribution of a number of ECM components has been considered in a recent review (Solursh, 1990). One ECM molecule which exhibits early localization, prior to cytodifferentiation, is the large mesenchymal proteoglycan, PG-M (Shinomura et al., 1990). Based on immunohistochemistry of the chick wing bud, it is not detected until after Hamburger and Hamilton stage 17, when it is found in the distal limb mesenchyme of the progress zone. With further development, it is found in the forming pre-chondrogenic core, as well as in the sub-ectodermal mesenchyme and the mesenchyme at the distal tip of the wing bud. The fact that PG-M can interact with several other ECM molecules, including fibronectin, hyaluronan, and collagen I, makes it a good candidate for having a regulatory role in mediating changes in the ECM (see Shinomura and Kimata, 1990).

The distribution of PG-M differs considerably from fibrillar ECM components, as well as from that of basement membrane components. Molecules such as collagen I and fibronectin accumulate relatively uniformly throughout the mesenchyme, but then become particularly abundant in the sub-ectodermal basement membrane and associated mesenchyme (Dessau et al., 1980; Shinomura et al., 1990). Subsequently, both of these components become more localized in the forming pre-chondrogenic regions. Fibronectin is also localized in association with the vascular system (Kosher et al., 1982), which becomes patterned very early in development. Subsequently, after stage 22, when the dorsal and ventral pre-muscle masses begin to form (Schramm and Solursh, 1990), fibronectin appears relatively reduced in the myogenic areas (Chiquet et al., 1981). Basement membrane components, such as collagen IV and basement membrane heparan sulfate proteoglycan, are not only detected in the sub-ectodermal basement membrane, but are widely distributed throughout the mesoderm (Solursh and Jensen, 1988). Only later with cytodifferentiation do they become restricted to the muscle areas. Laminin is more restricted in distribution, being localized to the sub-ectodermal basement membrane and appearing by stage 23 in the pre-muscle areas. While this is prior to overt myogenesis, some muscle pre-mRNA's are first detected at this stage (Swiderski and Solursh, 1990). In light of the dramatic effects of laminin on myoblast proliferation (von der Mark and Kuhl, 1985) and migration (Ocalan et al., 1988) _in vitro_, it is interesting to consider that this early localization of laminin to the pre-myogenic areas is involved in onset of myogenesis. While myoblasts synthesize laminin _in vitro_ (Kuhl et al., 1982), the source of laminin _in situ_ is not known.

Hyaluronan is a major ECM constituent of embryonic mesenchyme. Because of difficulties in fixing it _in situ_, localization studies have been difficult to carry out. By adding tannic acid to the primary fixative leaching of hyaluronan can be reduced (Singley and Solursh, 1980). Based on this procedure, hyaluronan can be shown to accumulate progressively with development, particularly in the periphery of the limb bud in the vicinity of the ectoderm. In addition, hyaluronan is drastically reduced in the myogenic regions at the time of onset of overt myogenesis at stage 26 (Singley and Solursh, 1981). The possible roles of hyaluronan and its receptor in chondrogenesis and myogenesis are discussed in the chapter by Toole.

ECM RECEPTORS

The interactions between cells and the ECM are mediated by cell surface receptors. While these receptors are under intensive investigation (see Albelda and Buck, 1990, for review), their role in limb pattern formation has still received little attention.

One ECM receptor, which is expressed in the developing limb bud at very early stages of development, is the cell surface heparan sulfate proteoglycan, syndecan (Solursh et al., 1990). This integral membrane proteoglycan contains both heparan sulfate and chondroitin sulfate chains and links the cytoskeleton to interstitial ECM components, including collagen and fibronectin. Based on immunolocalization, it can be detected at a high level on both ectodermal and limb mesenchymal cells at the earliest stages in the appearance of the limb bud in the mouse. As limb bud outgrowth continues, antigen is lost in the apical ectodermal ridge and on differentiating muscle and cartilage cells. The high affinity of syndecan for ECM components and its distribution in the early limb bud are consistent with a role in maintaining the morphologic integrity of the limb bud during the period of rapid outgrowth.

The integrins represent another large family of proteins that function as both cell-substratum and cell-cell adhesion receptors. Information concerning the regulation of integrin expression during limb patterning is still incomplete. There are indications that the $beta_1$ integrin complex increases as mesenchymal cells leave the proximity of the apical ectodermal ridge, based on results obtained by immunoprecipitation of metabolically labeled cell surface antigens from proximal and distal limb cells (Swalla el al., 1991). In addition, cells from the developmentally more mature proximal region of the limb bud are more susceptible to agents that promote cell flattening. Such developmentally regulated changes in the expression of ECM receptors might have important functions in morphogenesis and differentiation.

FUNCTIONS OF ECM

It is difficult to assess the functions of ECM components in limb pattern formation without mutational analysis. As considered below, such analysis is now becoming feasible. Based on tissue culture approaches, cell-ECM interactions have been implicated in pattern formation affecting chondrogenesis and myogenesis. Some such studies from the author's laboratory are briefly considered here.

Chondrogenesis

It is now well established that limb mesenchyme will differentiate into chondrocytes under conditions where there is little cell-ECM interaction (See Zanetti and Solursh, 1989, for review). Even single mesenchyme cells can become cartilage cells under such favorable conditions. Recent studies suggest that the extent of polymerization of beta actin into microfilaments is inversely related to the onset of chondrogenic differentiation (Daniels et al., 1989). It is possible that whether or not a mesenchymal cell becomes a chondrocyte can be regulated through effects of ECM on cytoskeletal organization.

In regard to pattern formation, it is noteworthy that a diffusible factor derived from the limb ectoderm inhibits chondrogenesis in vitro by limb mesenchymal cells through an actin cytoskeletal mediated mechanism (see Zanetti and Solursh, 1989). The antichondrogenic action of ectodermal conditioned medium involves cell spreading and can be alleviated by agents, such as cytochalasin D, that induce cell rounding (Zanetti et al., 1990). Based on such in vitro studies, we have postulated that the ectoderm might function in vivo through a similar diffusible factor to regulate the spatial patterning of cartilage and soft connective tissue formation in the limb (Solursh et al, 1981). As considered in the chapter by Hurle, there is now evidence such a mechanism might operate in situ.

Myogenesis

The skeletal musculature is derived from somitic cells which migrate into the somatopleure (see chapter by Christ). These cells become organized into dorsal and ventral pre-muscle masses prior to cytodifferentiation. Recent studies (Schramm and Solursh, 1990) indicate that myogenic cells are initially distributed throughout the proximal region of the limb, but gradually become depleted in the central pre-chondrogenic core concurrently with early chondrogenic events, including the appearance of PG-M. A similar sequence is observed in high density cell cultures, where it might be possible to carry out more mechanistic studies, including an examination of the role of ECM molecules produced in the pre-chondrogenic core in stimulating the redistribution of migrating myogenic cells.

FUTURE DIRECTIONS

This brief overview emphasizes the potential importance of the extracellular matrix in pattern formation during early development. Some specific ECM components are considered in more detail in several of the chapters included in this volume. It is clear that a great deal of descriptive information is still needed concerning additional, presently unknown, ECM molecules, as well as characterization of their structure and receptors. Most promising is the use of molecular methods to either block or modify the expression of specific ECM molecules. In this way their contribution to pattern formation can be assessed more directly than is possible by descriptive studies. Then it will be possible to begin to focus on the regulatory mechanisms which control the expression of the genes for specific ECM molecules and relate the presence of particular response elements in ECM genes to the actions of specific DNA binding proteins, as indicated in Fig. 1.

REFERENCES

Albelda, S.M., and Buck, C.A., 1990, Integrins and other cell adhesion molecules, FASEB J., 4:2868-2880.

Chiquet, M., Eppenberger, H.M., and Turner, D.C., 1981, Muscle morphogenesis: evidence for an organizing function of exogenous fibronectin, Dev. Biol., 88:220-235.

Daniels, K.J., Hay, E.D., and Solursh, M., 1989, Differential accumulation of beta and gamma actin during initiation of chondrogenesis, J. Cell. Biol., 109:41a.

Dessau, W., von der Mark, H., von der Mark, K., and Fischer, S., 1980, Changes in the patterns of collagens and fibronectin during limb-bud chondrogenesis, J. Embryol. Exp. Morphol., 57:51-60.

Kosher, R.A., Kordasey, H.W., and Ledger, P.W., 1982, Temporal and spatial distribution of fibronectin during development of the embryonic chick limb bud, Cell Differ., 11:217-228.

Kuhl, U., Timpl, R., and von der Mark, K., 1982, Synthesis of type IV collagen and laminin in cultures of skeletal muscle cells and their assembly on the surface of myotubes, Dev. Biol., 93:344-354.

Ocalan, M., Goodman, S.L., Kuhle, A., Hauschka, S.D., and von der Mark, K., 1988, Laminin alters cell shape and stimulates motility and proliferation of murine skeletal myoblasts, Dev. Biol., 125:158-167.

Schramm, C., and Solursh, M., 1990, The formation of premuscle masses during chick wing bud development, Anat. Embryol., 182:235-247.

Shinomura, T., and Kimata, K., 1990, Precartilage condensation during skeletal pattern formation, Develop. Growth & Differ., 32:243-248.

Shinomura, T., Jensen, K.L., Yamagata, M., Kimata, K., and Solursh, M., 1990, The distribution of mesenchyme proteoglycan (PG-M) during wing bud outgrowth, Anat. Embryol., 181:227-233.

Singley, C.T., and Solursh, M., 1980, The use of tannic acid for the ultrastructural visualization of hyaluronic acid, Histochemistry, 65:93-102.

Singley, C.T., and Solursh, M., 1981, The spatial distribution of hyaluronic acid and mesenchymal condensation in the embryonic chick wing, Dev. Biol., 84:102-120.

Solursh, M., 1990, The role of extracellular matrix molecules in early limb development, Seminars in Dev. Biol., 1:45-53.

Solursh, M., and Jensen, K.L., The accumulation of basement membrane components during the onset of chondrogenesis and myogenesis in the chick wing bud, Development, 104:41-49.

Solursh, M., Reiter, R.S., Jensen, K.L., Kato, M., and Bernfield, M., 1990, Transient expression of a cell surface heparan sulfate proteoglycan (syndecan) during limb development, Dev. Biol., 140:83-92.

Solursh, M., Singley, C.T., Reiter, R.S., 1981, The influence of epithelia on cartilage and loose connective tissue formation by limb mesenchyme cultures, Dev. Biol., 86:471-482.

Swalla, B.J., Jungles, S., and Solursh, M., 1991, Differences in integrin expression by proximal and distal chick wing bud cell subpopulations, Dev. Biol., (submitted).

Swiderski, R.E., and Solursh, M., 1990, Precocious appearance of cardiac troponin T pre-mRNAs during early avian embryonic skeletal muscle development in ovo, Dev. Biol., 140:73-82.

von der Mark, K., and Kuhl, U., 1985, Laminin and its receptor, Biochim. biophys. Acta, 823:147-160.

Zanetti, N.C., and Solursh, M., 1989, Effect of cell shape on cartilage differentiation, in: "Cell Shape: Determinants, Regulation, and Regulatory Role," W.D. Stein, and F. Bonner, eds., Academic Press, Inc., New York.

Zanetti, N.C., Dress, V.M., and Solursh, M., 1990, Comparison between ectoderm-conditioned medium and fibronectin in their effects on chondrogenesis by limb bud mesenchymal cells, Dev. Biol., 139:383-395.

MOLECULAR HETEROGENEITY OF CHONDROITIN SULPHATE IN THE DEVELOPING CHICK LIMB

Charles W. Archer, Marian Fernandez-Teran*,
Fiona Craig, and Micheal Bayliss[+]

Dept. of Orthopaedic Surgery, Institute of
Orthopaedics, University College and Middlesex
School of Medicine, Stanmore, Middx. HA7 4LP
*Dept. of Anatomy and Cell Biology, University of
Cantabria, Santander, Spain
+Division of Biochemistry, Kennedy Institute
of Rheumatology, Bute Gardens, Hammersmith,
London W6

INTRODUCTION

Glycosaminoglycans (GAGs) are matrix components which have
a ubiquitous distribution throughout the vertebrate body.
Normally they are covalently linked to core proteins thus
constituting proteoglycans. Although their distribution is
largely considered to be extracellular, intracellular and
transmembrane locations have also been reported (Rouslahti,
1988). In the developing limb, GAGs have been implicated in the
maintenance of distal 'progress zone' mesenchyme (Kosher and
Savage, 1981), epithelial/mesenchymal interactions (Hall,
1983a,b), facilitating neural crest migration (Morriss-Kay and
Tuckett, 1989) and constituting the prechondrogenic
proteoglycan, PG-M (Kimata et al., 1986). During the
differentiation of connective tissues, spatial and temporal
variations in GAGs have been reported in skeletal elements
(Craig et al., 1987a; Sorrell and Caterson, 1989), joint-
associated soft tissues (Craig et al., 1987b; Sorrell and
Caterson, 1989) and muscles (Caplan, this volume).

It is becoming increasingly apparent that considerable
molecular heterogeneity exists within the GAG chains both in the
degree and position of sulphation within the repeating
disaccharide sub-units (Uchiyama et al., 1987). Whilst the

significance of this heterogeneity remains unclear, several antibodies are now available that can be used to detect molecular variations in GAG chains, particularly keratan sulphates and chondroitin sulphates (Caterson, et al.,1987; Sorrell and Caterson, 1989; Caterson, et al., 1990). Furthermore, whilst the epitopes recognised by these antibodies are widely expressed during development, some are suppressed to a large extent after skeletal maturity. However, during pathological changes which involve renewed growth, they are again expressed (Caterson et al., 1990). In this chapter, we document the temporal and spatial distributions of a variety of chondroitin suphate epitopes in the developing chick limb.

MATERIALS AND METHODS

Staged embryos: Fertilised White Leghorn X Sykes Tinted eggs from a local breeder were incubated at 38 °C +/- 1°C in a humidified atmosphere. Embryos between stages 20 - 29 (Hamburger and Hamilton, 1951) were fixed in 10% formal-saline, dehydrated through a series of alcohols, and wax-embedded. Serial 7μm sections were cut and mounted on glass slides.

Immunocytochemical labelling: Sections were dewaxed in xylene (x3, 5 min. each), re-hydrated through graded ethanols to 0.02M phosphate buffered saline (PBS). Sections were either incubated directly in primary antibody (1:100 dilution) or after bacterial chondroitinase treatment (0.1 u/ml for 30 min at 37°C) depending on the antibody used and epitope required for detection (see below). In this work we will describe three antibodies that detect four separate epitopes:

> 3B3 - with chondroitinase pre-treatment recognises chondroitin-6-sulphate.

> 3B3 - without chondroitinase pre-treatment recognises a native epitope at the non-reducing termini of selected chondroitin sulphate chains.

> 2B6 - with chondroitinase pre-treatment recognises chondroitin-4-sulphate/dermatan sulphate, henceforth referred to as C-4-S

7D4 - without chondroitinase pre-treatment
recognises an oversulphated region near the
linkage region within the glycosaminoglycan
chain.

Primary antibody binding was visualised either by
fluorescently-conjugated secondary antibody (rabbit anti-mouse
IgG) or peroxidase anti-peroxidase (PAP) followed by silver
enhancement (Amersham Int.). Sections were photographed on a
Zeiss photomicroscope III.

RESULTS

Chondroitin-4-sulphate/dermantan sulphate and chondroitin-6-sulphate

There were distinct differences in the pattern of distribution
of chondroitins 4 - and 6-sulphates (C-4-S, C-6-S) as
recognised by antibodies 3B3 and 2B6 (after chondroitinase
pretreatment) which varied with embryonic stage. At stage 20,
there was light labelling of C-6-S throughout the limb
mesenchyme and also within the ectoderm including the apical
ectodermal ridge (AER)(fig. 1). In contrast, C-4-S
immunoreactive material was concentrated beneath the limb
ectoderm (in close association with the basement membrane),
but was absent beneath the ectoderm of the trunk (fig. 2). In
addition, strong positive labelling was found around the
notochord and mesenephric ducts which contrasted that found
for C-6-S. The pattern of C-4-S changed little from stages 20
to 24, except that the sub-ectodermal labelling became less
pronounced. C-6-S labelling, however, showed variation in its
distribution. By stage 22, the ectoderm was no longer
immunoreactive and the mesenchyme directly beneath the AER was
also negative (fig. 3). However, it was seen that dorsal
proximal mesenchyme labelled more intensely than ventral
mesenchyme (fig. 4).
    At the onset of chondrogenic differentiation, there were
again some differences in the distribution of the two
epitopes. Using 3B3 for C-6-S, the chondrogenic anlagen
labelled evenly throughout the limb and incorporated the

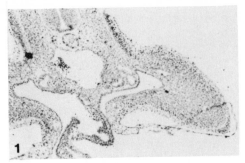
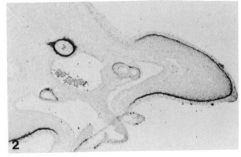

Figure 1. Stage 20 embryo labelled with 3B3 for chondroitin-6-
    sulphate. There is light labelling throughout the
    mesenchyme and includes the ectoderm.

Figure 2. Stage 20 embryo labelled with 2B6 for
    chondroitin-4-/dermatan sulphate. Most label is
    concentrated at the basement membrane of the limb
    ectoderm, but is absent beneath the ectoderm of the
    trunk.

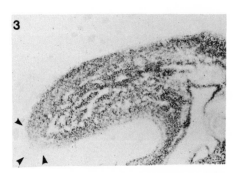
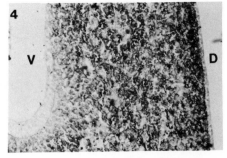

Figure 3. Stage 22 wing bud in longitudinal section stained
    for chondroitin-6-sulphate. most of the mesenchyme
    remains positively stained interspersed by negative
    'pockets'. In addition, the distal-most mesenchyme
    (progress zone) and ectoderm is also negative. Arrowheads
    indicate the distal margins of the limb.

Figure 4. Higher power micrograph of proximal limb shown in
    figure 3. Note the greater intensity of staining in the
    dorsal aspect of the limb. D = Dorsal surface.V = ventral
    surface.

Figure 5. Stage 25 embryo labelled for chondroitin-6-
    sulphate. Again extensive labelling is seen throughout
    the mesenchyme. The differentiating cartilage elements
    are strongly labelled. Note also that the distal
    mesenchyme is again positive.

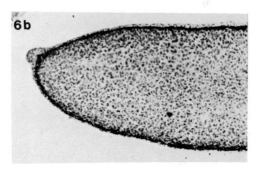

Figure 6a. Stage 25 embryo labelled for chondroitin-
    4/dermatan-sulphate. Intense labelling is seen beneath
    the ectoderm and within the cartilage anlagen.
    Note in both figures the strong staining seen around and
    within the notochord.
Figure 6b. Distal mesenchyme of limb shown in figure 6a. Note
    the strong sub-ectodermal labelling which incorporates
    the notch beneath the AER.

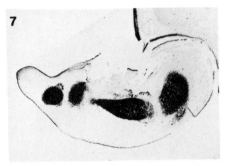 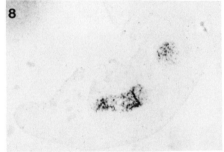

Figure 7. Stage 29 wing bud labelled for chondroitin-
4/dermatan-sulphate. The cartilage elements label
strongly bot do not incorporate the developing joint
regions. The sub-ectoderm continues to label positively
at this late stage.

Figure 8. Stage 29 wing bud labelled for novel native epitope
recognised by antibody 3B3 (without chondroitinase
pretreatment). Only the hypertrophic and flattened cell
zones label.

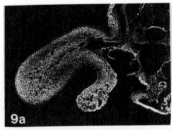 

Figure 9. a and b. Stage 20 and 23 wing buds respectively
labelled with antibody 7D4. Note extensive labelling
throughout the mesenchyme. The ectoderm remains
negative.

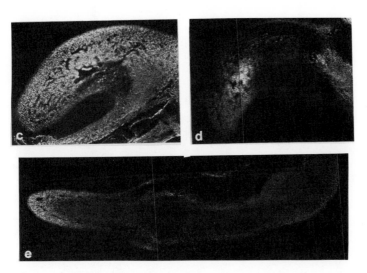

Figure 9. c and d. At stages 24 and 25, the ventral (stage
    24; c) and dorsal (stage 25; d) premyogenic muscle
    masses do not label positively. The chondrogenic
    core and distal mesenchyme show strong
    immunoreactivity. e. At stage 28, labelling for 7D4
    is restricted to the distal mesenchyme, sub-
    ectodermal mesenchyme and perichondrium. Faint
    immunoreactivity is also seen in the cartilage.

prechondrogenic condensations of the distal mesenchyme (fig.
5). Distal mesenchyme also labelled lightly. In contrast,
proximal chondrifying areas of the limb labelled intensely for
C-4-S as did the sub-ectodermal/basement membrane areas (fig.
6a). Again, only the limb sub-ectoderm was immunoreactive
which like pre-chondrogenic stages included the AER (fig. 6b).
Unlike earlier stages, however, the notochord and surrounding
mesenchyme was strongly positive for both C-4 and-6 sulphates
(figs 5 and 6a). At later stages (29) the pattern of labelling
remained essentially the same in that C-4-S was restricted to
the cartilage elements and basement membrane (fig. 7).
Similarly, C-6-S also labelled the entire cartilage elements
and soft tissue elements associated with the developing
musculature (not shown).

In the absence of chondroitinase pretreatment, 3B3
recognises novel epitopes on the non-reducing termini of
selected GAG chains. Up to stage 27, binding to limb
structures was not found. However, at stage 27 and in
subsequent stages, labelling was found exclusively in the
flattened and hypertrophic cell zones of the cartilaginous
rudiments (fig.8)(see also Caterson et al., 1990). At still
later stages, some additional staining was found in
association with the developing joint regions.

Antibody 7D4 recognises an over-sulphation pattern near
the linkage region of the GAG chains. At stages 20, 22 and 23
extensive labelling was found throughout the mesenchyme but
absent from the ectoderm (figs. 9a and b). By stage 24, some
areas of the ventral mesenchyme showed a paucity in labelling
(fig. 9c) and by stage stage 25 labelling was restricted to
sub-ectodermal areas, central core and distal mesenchymes
(fig.9d). At later stages, labelling was found in
perichondrial, peri-articular and and distal mesenchyme. Faint
immunoreactivity was also seen throughout the cartilage (fig.
9e).

DISCUSSION

The results describe the spatial and temporal distributions of
chondoitin-4-/dermatan sulphate and chondroitin-6-sulphate in
addition to two novel GAG epitopes on chondroitin sulphate in
the developing chick limb. Similar studies on the bursae of

Fabricius (Sorrell et al., 1988) have also described stage dependent changes in the distributions of the components recognised by these antibodies. Some general patterns emerge. Chondroitin-6-sulphate was widely distributed throughout the pre-chondrogenic mesenchyme but was excluded from the distal mesenchyme beneath the AER by stage 22. After overt chondrogenesis at stage 25, C-6-S remains distributed throughout the entire cartilaginous skeleton, but also in soft connective tissues such as the muscle, muscle/tendon complexes and periarticular structures. In contrast, chondroitin-4/dermatan-sulphate was localised to the limb ectodermal basement membrane (and to a lesser extent the ectoderm) and in precartilaginous condensation mesoderm. At stage 25, strong labelling is found in the differentiating cartilage elements in addition to the ectodermal basement membrane. At later stages 2B6 (C-4-S) labelled metaphyseal and diaphyseal cartilage intensely, but was absent in the developing joint regions. Other authors (Sorrell and Caterson, 1989) have reported a similar distribution in the developing chick knee, and they also correlated this distribution with that of various keratan sulphate epitopes. A similar distribution pattern for keratan sulphate and collagens types I and II have been reported for the chick metatarsalphalangeal joint (Craig et al., 1987a).

The distribution of C-6-S is consistent with this GAG comprising the prechondrogenic PG-M reported by Kimata et al.,(1986) and is distributed throughout the mesenchyme at early stages. In comparison to cartilage proteoglycan, PG-M comprises larger chondroitin chains, larger core protein but overall is undersulphated. Nevertheless, it appears by the distribution of antibody 7D4 that it is likely that these chains have oversulphated regions. The widespread distribution of C-6-S in other tissues within the limb subsequent to skeletal differentiation is also consistent with proteoglycan and GAG species being associated with muscle (see Caplan, this volume) and tendon (Craig et al.,1987b). Whilst the present study is informative in respect to the distribution of ECM components, it tells us little of matrix turnover. For instance what are the mechanisms that give rise to the paucity in C-6-S beneath the AER in stage 22 limbs? To this end Critchlow and Hinchliffe (in this volume) have conducted autoradiographic pulse/chase experiments utlizing $^{35}SO_4$ which give some preliminary insights.

The sulphation pattern recognised by antibody 7D4 was extensively distributed throughout the young limb bud. However, by stages 24 and 25, premyogenic muscle masses were negative for 7D4 although distal mesenchyme remained strongly positive. Further differentiation shows staining restricted to sub-ectodermal and perichondrial regions but again more importantly, to undifferentiated distal mesenchyme.

The significance of these staining patterns remains obscure, but they appear to be associated with growth. For example, in adult human and dog cartilage expression of 3B3 (native epitope) is suppressed, but during pathological changes such as in osteoarthritis (where cartilage proliferation is initiated) 3B3 is re-expressed (Caterson et al., 1990). A similar situation pertains to the native epitope recognised by antibody 7D4. In the developing chick limb, higher mitotic indices are found in the distal mesenchyme where labelling of 7D4 is most prevalent particularly after overt chondrogenesis. Labelling for 3B3 (native epitope), however, is restricted to mature chondrogenic zones which are characterised by low levels of proliferation. Clearly, much more data is required in order to determine the relationships between the expression of these epitopes during normal development and disease, and also between proliferation and differentiation of the respective tissue types.

REFERENCES

Caterson, B., Calabro, T., and Hampton, A. (1987). Monoclonal antibodies as probes for elucidating proteoglycan structure and function, in: Bilogy of Proteoglycans, T. Wright and R. Mecham, eds., pp.1-26. Academic Press, New York.

Caterson, B., Mahmoodian, F., Sorrell, J., Hardingham, T., Bayliss, M., Carney, S., Ratcliffe, A. and Muir, H. (1990). Modulation of native chondroitin sulphate structure in tissue development and in disease. J. Cell Sci.(in press).

Craig, F., Bentley, G. and Archer, C. (1987a). The temporal and spatial pattern of collagens I and II and keratan sulphate in the developing chick metatarsophalangeal joint. _Development_ 99: 383-391.

Craig, F., Ralphs, J., Bentley, G. and Archer C. (1987b). MZ15, a monoclonal antibody recognizing keratan sulphate, stains chick tendon. _Histochem.J._ 19: 651-657.

Hall, B. K.(1983a) Epithelial-mesenchymal interactions in cartilage and bone development, _in_: Epithelial-Mesenchymal Interactions in Development, R. Sawyer and J. Fallon, eds., pp.189-214. Praeger Press, New York.

Hall, B.K. (1983b). Tissue interactions and chondrogenesis, _in_: Cartilage, Vol. 2. Development, Differentiation and Growth, B. K. Hall, ed., pp. 187-222. Academic Press, Orlando.

Hamburger, V. and Hamilton, H. (1951). A series of normal stages in the development of the chick embryo. _J. Morph._ 88: 49-92.

Kimata, K., Oike, Y., Tani, K.,Shinomura, T., Yamagata, M., Uritani, M. and Suzuki, S. (1986). A large chondroitin sulphate proteoglycan (PG-M) synthesised before chondrogenesis in the limb bud of the chick embryo. _J. Biol.Chem._ 261: 13517-13525.

Kosher, R. and Savage, M. (1981). Glycosaminoglycan synthesis by the apical ectodermal ridge of the chick limb bud. _Nature_, 291: 231-234.

Morriss-Kay, G. and Tuckett, F. (1989). Immunohistochemical localisation of chondroitin sulphate proteoglycans and the effects of chondroitinase ABC in 9- to 11-day rat embryos. _Development_, 106: 787-798.

Rouslahti, E. (1988). Structure and biology of proteoglycans. _Ann. Rev. Cell. Biol._ 4: 229-255.

Sorrell, J., Lintala, A., Mahmoodian, F. and Caterson, B.(1988). Epitope-specific changes in chondroitin sulphate/dermatan sulphate proteoglycans as markers in the lymphopoietic and granulopoietic compartments of developing bursae of Fabricius. _J. Immunol._ 140: 4262-4270.

Sorrell and Caterson, B. (1989). Detection of age-related changes in the distributions of keratan sulphates in developing chick limbs: an immunocytochemical study. _Development_, 106: 657-664.

Uchiyama, H., Kikuchi, K., Ogamo, A. and Nagasawa, K. (1987).
Determination of the distribution of constituent
disaccharide units within the chain near the linkage region
of shark cartilage chondroitin sulphate C. Biochim.
Biophys. Acta 926: 239-248.

SPECIAL PROPERTIES OF THE AER/MESENCHYME INTERFACE IN THE

CHICK WING BUD: ANALYSIS OF ECM MOLECULES AND INTEGRIN RECEPTORS

Matthew A. Critchlow[*] and J. Richard Hinchliffe

Department of Biological Sciences
University College of Wales
Aberystwyth, Wales, UK

INTRODUCTION

During chick wing morphogenesis, reciprocal interactions
take place between the apical ectodermal ridge (AER) and the
subjacent mesoderm. The AER (a pseudostratified columnar
epithelium) induces limb outgrowth, while the limb mesoderm
induces the formation of the apical ridge and subsequently
maintains both its specialized morphology and its functional
properties (reviewed by Fallon, et al., 1983). Although the
morphogenetic importance of AER-mesenchymal communication is now
well established, the underlying mechanisms involved have not
been identified. However, studies in other developing systems
have provided good evidence that the extracellular matrix (ECM)
is involved in mediating some epithelial-mesenchymal interactions
(Bernfield et al., 1984). Furthermore, the ECM is known to
control many aspects of cell behaviour in culture, including
motility, morphology and differentiation (Trelstad, 1984).
Despite this, comparatively few studies have explored the
possibility that the ECM also functions in mediating ridge-
mesenchymal interactions (Tomasek and Brier, 1986). The aim of
the present study was to identify specialized features of the ECM
at the ridge/mesenchyme interface which may function in the
maintenance of ridge structure, or the promotion of wing
outgrowth.

Using a range of monoclonal and polyclonal antibodies, we
have mapped the distribution of the following ECM components in
the normal wing bud: collagen type I, fibronectin (FN), tenascin
(TN), chondroitin sulphate proteoglycan (CSPG), collagen type IV,
heparan sulphate proteoglycan (HSPG) and laminin (LM). Of
particular interest were the distribution patterns of these ECM
moelcules along the distal, epithelial-mesenchymal interface. We
have also established the spatial distribution of avian integrin;
a transmembrane glycoprotein complex which functions as a
receptor for FN, LM and possibly vitronectin and collagen IV (see

*Authors current address: Department of Anatomy, St. George's
Hospital Medical School, Cranmer Terrace, London, S.W.17 ORE, UK.

Developmental Patterning of the Vertebrate Limb
Edited by J.R. Hinchliffe *et al.*, Plenum Press, New York, 1991

review by Buck and Horwitz, 1987). Avian integrin, which belongs
to a large family of ECM receptors (Hynes, 1987) also binds to
talin (Horwitz et al., 1986) and is therefore in a position to
mediate interactions between the ECM and the cytoskeleton. The
distribution of each antigen was also examined in experimental
wing buds, in a situation where the AER collapses and distal
outgrowth is inhibited. These changes occur in the preaxial
region of the wing bud following the insertion of an impermeable
barrier in the 'mid-position' at stage 21 (Hinchliffe and
Griffiths, 1984). The barrier is positioned in such a way as to
prevent communication along the antero-posterior axis of the wing
bud. The aim of these experiments was to examine whether ridge
collapse, and the cessation of limb outgrowth, correlate with
changes in the composition of ECM or the expression pattern of
integrin receptors at the wing tip.

Previous ultrastructural studies (Newman et al., 1981) and
our own unpublished observations (Griffiths, Critchlow and
Hinchliffe, in preparation) have shown that the basal lamina
beneath the AER is more diffuse and disorganized than elsewhere
in the limb. The maintenance of this disorganized basal lamina
may require rapid turnover of its constituent ECM molecules.
Consistent with this view, Kulyk and Kosher (1987) have found
relatively high levels of hyaluronidase activity in the distal,
AER-containing limb ectoderm. In an attempt to identify regional
differences in the rate of sulphated glycosaminoglycan (GAG)
turnover in the basement membrane, pulse-chase experiments have
been carried out using [^{35}S]sulphate. Labelled wing buds were
grafted to the wing stumps of unlabelled host embryos (see
Methods). The distribution of labelled material was then
examined, at different time intervals, using autoradiography.

METHODS

Immunohistochemistry

ECM molecules and the avian integrin complex were located in
the chick wing bud using indirect immunofluorescence (Hurle et
al., 1990).

Antibodies. A polyclonal, rabbit anti-human plasma FN
antiserum was purchased from the Behring Institute. A monoclonal
antibody (mAb) which recognizes CSPG, designated as CS-56, was
obtained from Sigma. This antibody binds to the GAG moieties of
native CSPG (Avnur and Geiger, 1984). MAb's which recognize
basal lamina HSPG, designated as No. 33 (Bayne et al., 1984), and
the integrin FN/LM receptor, designated as JG22 (Greve and
Gottlieb, 1982), were purchased from the Developmental Studies
Hybridoma Bank (maintained under NICHD contract NO1-HD-6-2915).
Both antibodies were raised against immunogens of chick origin.
Polyclonal antibodies against collagen I, TN and LM were all
raised in rabbit, using immunogens of chick origin, and were
generous gifts of Drs Charles Archer (Orthpaedic Institute,
Stanmore, London); Phillip Gordon-Weekes (Anatomy Department,
Kings College, London) and Ruth Chiquet-Ehrismann (Friedrich
Miescher Institute, Switzerland). A polyclonal, rabbit anti-
human collagen IV antiserum was kindly provided by Dr Victor
Duance (AFRC, Bristol). The appropriate, FITC-conjugated
secondary antibodies were purchased from Dako Ltd.

The spatial distribution patterns of all the antigens listed
above were examined in control wing buds at stages 19, 21, 23,
25, 27, 29 and 31 (Hamburger and Hamilton, 1951) and in
experimental wing buds at 12 and 24 hr following the insertion of
a tantalum foil barrier in the 'mid-position' (intersomite level
17/18) at stage 21 (see Hinchliffe and Griffiths, 1984). In the
majority of cases, sections were prepared from wax-embedded
tissue which had been fixed in acetic-alcohol (Hurle et al.,
1990), since this produced both excellent histology and bright
fluorescence. However, with mAb's JG22 and 33, staining was
carried-out on frozen sections since little or no fluorescence
was observed with these antibodies on wax sections. All frozen
sections were cut from acetic-alcohol fixed tissue which had been
infiltrated with sucrose and gelatin (as described by Krotoski et
al., 1986). Experimental wing buds were sectioned in a region
anterior to the barrier, and contra-lateral control wing buds in
a corresponding anterior position. In certain cases, sections
were pretreated with chondroitinase ABC or testicular
hyaluronidase (both from Sigma) prior to immunostaining.

Sulphate turnover experiments

Stage 21 and 23 chick embryos were labelled for 2 hr with 80
μCi of [^{35}S]sulphate in ovo (as described by Hinchliffe, 1977).
After labelling, the wing buds were removed and grafted to the
wing stumps of unlabelled host embryos of the same stage. It has
previously been shown that limb buds grafted in this manner
continue outgrowth and subsequently undergo differentiation
(Hampé, 1959). In the present work, grafted wing buds were
harvested at either 4, 6, 8 or 12 hr post-operation and processed
for autoradiography. Control wing buds were fixed and processed
immediately after a 2 hr pulse. In some cases, sections were
treated with chondroitinase ABC prior to the addition of
stripping film. In a separate series of expriments on grafted
wing buds, Indian ink injections were used to map functional
blood vessels. These were observed, at a density similar to that
seen in contral-lateral control wing buds, at 3.5 hr post-
operation. This observation agrees with the findings of Yallup
(1984), who carried out similar grafting operations, and suggests
that the effects of anoxia are likely to be limited following the
grafting procedure.

RESULTS AND DISCUSSION

ECM distribution in the normal wing bud

The results of immunohistochemical analysis reveal that the
basement membrane of the apical ridge differs from that beneath
the non-ridge epithelium in terms of both molecular composition
and the relative abundance of certain ECM molecules. Our
observations confirm the report by Tomasek et al. (1982) that
fibronectin (FN) is particularly abundant in the indentation, or
notch (Todt and Fallon, 1984), at the base of the AER (Fig. 1a).
However, this was only apparent between stages 23 and 27. Prior
to stage 23, FN was uniformly distributed along the distal
epithelial-mesenchymal interface. We have also found enhanced
labelling for collagen type IV at the base of the apical ridge
(Fig. 2a), but this again was only detectable between stages 23
and 27. In contrast to the recent report by Solursh and Jensen
(1988), positive staining for collagen IV was found in all

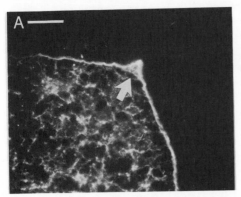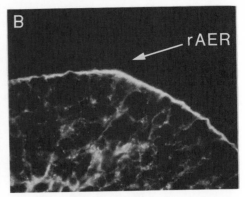

Fig. 1. Immunofluorescent localization of fibronectin in control and experimental wing buds. (A) Stage 25 control. Immuno-reactive material is abundant in the sub-ridge notch (arrow). (B) Experimental wing bud at 24 hr following barrier insertion (stage 25). The notch and its associated accumulation of FN are absent. rAER = regressing apical ridge. Scale bar = 20 μm.

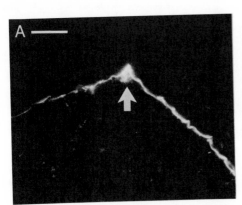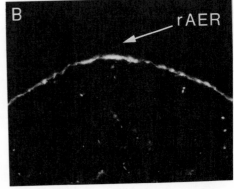

Fig. 2. Immunofluorescent localization of collagen IV in control and experimental wing buds. (A) Stage 25 control. Note the intense sub-ridge staining. (B) Experimental wing bud at 24 hr following barrier insertion (stage 25). Uniform staining is seen along the epithelial-mesenchymal interface. rAER = regressing apical ridge. Scale bar = 20 μm.

regions of the ectodermal basement membrane as early as stage 19 and at all subsequent stages examined. In addition to FN and collagen IV, LM also appeared to be particularly abundant in the sub-ridge notch region (not shown), but this was only observed between stages 25 and 27.

In contrast to the ECM components described above, collagen type I was not detected beneath the ectoderm at the wing tip until stage 21. Furthermore, between stages 21 and 25, collagen I appeared to be far more abundant beneath the non-ridge epithelium than beneath the apical ridge (not shown). CSPGs, as determined using the mAb CS-56, were present along the entire epithelial-mesenchymal interface of the wing bud at stage 19. At

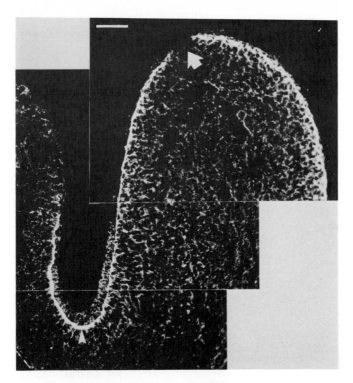

Fig. 3. Immunolocalization of CSPG in a stage 21 wing bud: paraffin section stained with mAb CS-56. Note that immunoreactive material is absent in the sub-ridge basement membrane (arrow). Bar = 50 μm.

stage 21, however, CS-56 immunoreactivity was no longer detected in the sub-ridge basement membrane, although intense labelling was still observed beneath the non-ridge epithelium (Fig. 3). This striking heterogeneity in the distribution of CSPG at the wing tip (as revealed with mAb CS-56) was evident between stages 21-27. By stage 27, CSPG was no longer detected in the basement membrane region at more proximal levels of the wing bud and by stage 31, positive staining was not detected in any region of the subectodermal basement membrane. Basal lamina HSPG, located using mAb No. 33, was found in all regions of the epithelial-mesenchymal interface and at all stages examined. Regional variation in the intensity of anti-HSPG staining was never observed in the basement membrane. TN was not detected in the the wing until stage 27, and although positive staining was seen along the distal tissue interface, distinct differences between ridge and non-ridge basement membranes were not consistently observed.

It is clear from the results described above that CSPG is a particularly good indicator of differences between the ridge and non-ridge basement membranes. Furthermore, the heterogeneous distribution of CSPG at the wing tip coincides with the period in which a heightened, pseudostratified AER is formed. This raises the possibility that CSPGs are involved in maintaining morphological differences between ridge and non-ridge epithelium. Many studies have shown that intact CSPGs inhibit the adhesion of cultured fibroblasts to FN and collagenous substrata (Rich et al., 1981; Brennan et al., 1983; Yamagata et al., 1989). One species of CSPG, isolated from yolk-sac tumour cells and termed serglycin, appears specifically to inhibit binding between integrin receptors and FN (Hautenan et al., 1989). On the basis of these findings it seems possible that the CSPG beneath the non-ridge epithelium reduces the extent of cell adhesion to FN and collagen in the basement membrane. Conversely, the apparent absence of intact CSPGs beneath the ridge may enhance the extent of cell binding to the basement membrane, serving to anchor the slender ridge cell processes more firmly around the notch. This interpretation may also explain the accumulations of FN, collagen IV and LM in the sub-ridge indentation. We have also found particularly strong labelling of the integrin FN/LM-receptor at the base of the central ridge cells (between stages 23-27) (Fig. 4; see below). Interestingly, Akiyama et al. (1989) have suggested that an aggregation of integrin receptors at the cell surface may lead to greater net affinity for its ligands. According to this view, clustering of such receptors at the base of the AER may also lead to enhanced cell adhesion to the basement membrane.

It is important to note, however, that the distribution of CSPGs established using mAb CS-56, differs from that observed with other anti-CSPG mAbs. For instance, Shinomura et al. (1990) have recently reported that a large CSPG, termed PG-M, is localized in the sub-ridge basement membrane during early wing outgrowth. This was determined using mAb MY-174, which binds to the core protein of PG-M (Solursh, personal communication). In addition, Archer et al. (this volume) have shown that mAb 2-B-6, which recognizes chondroitin-4-sulphate and dermatan sulphate moieties after chondroitinase digestion (Caterson et al., 1985), also stains the sub-ridge basement membrane of the early wing bud. MAb CS-56 has been reported to bind to native epitopes on chondroitin-4 and -6-sulphate chains, but not to purified dermatan sulphate (Avnur and Geiger, 1984). Therefore, the absence of CS-56 epitopes in the sub-ridge basement membrane may indicate that dermatan sulphate-containing PGs are the predominant species expressed in this region[1]. Alternatively, the contrasting staining patterns observed with CS-56, MY-174 and 2-B-6, may be a demonstration that the CSPGs of the sub-ridge region are largely degraded, as suggested by turnover experiments (see below). Since CS-56 recognizes epitopes on native chondroitin sulphate chains (Avnur and Geiger, 1984), degradation of the proteoglycan may lead to their elimination.

Distribution of ECM molecules in experimental wing buds

When an impermeable barrier is inserted in the mid-position of a stage 21 wing bud, a series of degenerative changes are observed in the preaxial region, leading to the loss of distal skeletal elements (Hinchliffe and Griffiths, 1984). By 12 hr post-operation (stage 23), the AER has flattened and mesodermal outgrowth has ceased. By 24 hr (stage 25), extensive necrosis is observed in the distal mesoderm and throughout the overlying, flattened AER. In the present study, the distribution patterns of all the ECM molecules listed above have been examined in the

---

1. Although PG-M comprises mainly chondroitin sulphate side chains, a certain number of dermatan sulphate chains are also bound to the core protein of this PG (Kimata et al., 1986).

preaxial region of experimental wing buds after barrier
insertion. The results show that at both 12 and 24 hr post-
operation, FN (Fig. 1b), collagen IV (Fig. 2b), collagen, I, LM
and HSPG are all distributed uniformly along the distal,
epithelial-mesenchymal interface. Furthermore, the sub-ridge
indentation, or notch, was never observed in these experimental
wing buds. At 12 hr post-operation, CS-56 immunoreactivity was
not detected in the sub-ridge basement membrane but by 24 hr,
positive staining with this mAb was seen in all regions of the
epithelial-mesenchymal interface (not shown). Thus by 24 hr, the
basement membrane beneath the regressing ridge appears identical
to that beneath the non-ridge, dorsal and ventral ectoderm.

Newman et al. (1981) have suggested that sub-ridge FN may be
involved in promoting limb outgrowth. It was argued that the
distal mesenchymal cells would to tend to migrate along the FN-
rich basement membrane of the AER. FN is known to promote cell
migration *in vitro* and recent studies indicate that it plays a
vital role in embryonic cell translocation *in vivo* (see review by
Dufour et al., 1988). We have shown here that the cessation
of wing outgrowth, induced by barrier insertion, is accompanied
by the loss of the sub-ridge notch and its associated
accumulation of FN. Although this finding is consistent with
views of Newman et al. (1981), observations on the temporal
distribution of FN argue against the idea that it acts in
directing wing outgrowth. Enhanced deposition of FN beneath the
AER was not apparent until stage 23, by which time considerable
outgrowth has already occurred.

It clear from the results described here that ridge
flattening, induced by barrier insertion, occurs concomitantly
with changes in the underlying basement membrane. This
strengthens the idea that the specialized features of the normal
sub-ridge basement membrane are involved in the maintenance of
AER structure. In particular, our observations on normal and
truncated wing buds suggest that CSPG may be important in
regulating differences between ridge and non-ridge epithelium.

Distribution of integrin in the wing bud

Avian integrin belongs to a large family of heterdimeric
glycoproteins which function as cell surface receptors for ECM
molecules. The family can be divided into distinct subfamilies,
in which the members share a common β-subunit noncovalently bound
to a unique α-subunit. The avian integrin complex contains
several discrete heterodimers, each one comprising of a β_1-
subunit combined separately with a different α-subunit (Hynes et
al., 1989). In the present study, the spatial distribution of
integrin was examined in normal wing buds using mAb JG22 (Greve
and Gottlieb, 1982), which binds to the β_1-subunit (Buck et al.,
1986). At all stages examined, bright staining was seen along
the epithelial-mesenchymal interface, with lower levels present
in the mesoderm. After stage 19, immunoreactive material
appeared particularly concentrated around blood vessels and at
stages 23 and 25, a moderate increase in staining was observed in
the chondrogenic core region of the wing bud (Fig. 4a). No
enhanced labelling was observed in the myogenic regions. In the
non-ridge epithelium, JG22 immunoreactivity was found exclusively
along the basal surfaces of the cells. In contrast, bright
labelling was seen throughout the AER, as early as stage 19, in a
pattern which appeared to delineate the cell surface (Fig. 4a &

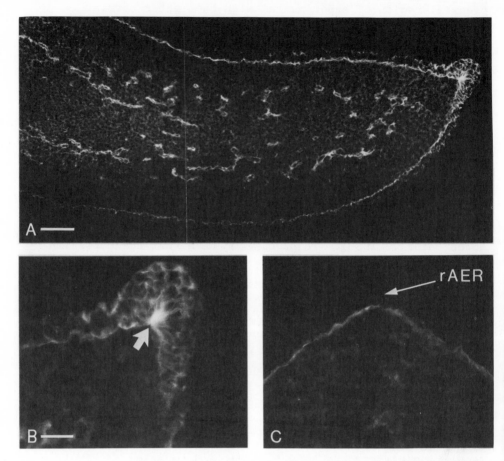

Fig. 4. Immunofluorescent localization of the chick integrin
complex: fixed/frozen sections stained with mAb JG22. (A)
Stage 25 control wing bud. Bar = 50 µm. (B) Detailed view
of the AER from the wing shown in (A). Note the overall
cell surface staining in the AER. Particularly intense
labelling can be seen at the base of the central ridge cells
(arrow). (C) Experimental wing at 24 hr following barrier
insertion (stage 25). Positive staining is absent in the
regressing apical ridge (rAER). The pattern of fluorescence
in the mesoderm, however, appears normal. Scale bar for (B)
and (C) = 20 µm.

4b). This finding is somewhat surprising, since neither FN, LM
or collagen IV were found in the AER, even in sections pretreated
with testicular hyaluronidase or chondroitinase ABC. However,
the expression of integrin receptors along the entire plasma
membrane of the ridge cells is consistent with the idea that they
play a role in cell-cell adhesion.

Immunohistochemical studies have shown that certain β_1-
integrin receptors exhibit an overall cell surface distribution
in the epithelium of human epidermis (Klein et al., 1990),
intestine (Choy et al., 1990), and developing kidney (Korhonen et
al., 1990). In addition, β_1-receptors have been localized to

sites of intercellular contact in cultured human tumour cells and keratinocytes (Kaufmann et al., 1989; Larjava et al., 1990). Recent *in vitro* analysis has also shown that mAbs specific for the β_1-subunit disrupt the formation and maintenance of cell-cell contact sites in cultured keratinocytes (Carter et al., 1990; Larjava et al., 1990). These studies all indicate that certain β_1-integrins play a role in cell-cell adhesion. Chick (β_1) integrin may therefore play a similar role in the AER. It is also interesting to speculate on the role of integrin in maintaining the specialized morphology of the AER. Histological and ultrastructural studies have shown clearly that the columnar cells of the AER are packed far more closely together than the cuboidal cells of non-ridge epithelium (Todt and Fallon, 1984; 1986). The selective expression of integrin complexes along the lateral surfaces of the ridge epithelium may lead to enhanced cell-cell adhesion and therefore closer cell packing. It should be noted, however, that by stage 27, positive labelling of the JG22 epitope(s) was observed only at the basal surface of the AER. The reason for this change is unclear, but it may represent an early stage in the flattening of the AER which occurs during later stages of normal development.

The distribution of integrin receptors was also examined during experimentally induced ridge regression. The latter was again achieved by inserting an impermeable barrier in the 'mid-position' (intersomite level 17/18) at stage 21. By 12 hr post-operation, the ridge cells of the preaxial region had undergone a striking reduction in the intensity of JG22 immunostaining and at 24 hr, the antigen was no longer detectable along the lateral surfaces of the flattened cells (Fig. 4c). However, positive labelling was still observed at the basal cell surface of the experimental ridge at both intervals after the operation. Ultrastructural analysis has shown that by 24 hr following barrier insertion, the characteristic fan-like arrangement of columnar cells disappears in the regressing ridge. Furthermore, a moderate increase in the size of intercellular spaces has been observed (Griffiths, et al., in preparation). These findings are consistent with the idea that integrin receptors are involved in the maintenance of cell-cell contact sites and/or epithelial cell organization in the normal AER.

Sulphate turnover experiments

In the present study, pulse-chase experiments have been carried out, using [^{35}S]sulphate, in an attempt to identify differences in the rate of sulphated GAG turnover in different regions of the wing bud basement membrane. Stage 21 and 23 wing buds were labelled for 2 hr with [^{35}S]sulphate and then grafted to the wing stumps of unlabelled host embryos. In control wing buds, fixed immediately after a 2 hr pulse, the radiolabel was found throughout the basement membrane (Fig. 5a, 6a & 6d). Interestingly, the highest density of silver grains in this region was found beneath the AER. Pretreatment of sections with chondroitinase ABC removed much of the label from the epithelial-mesenchymal interface (Fig. 5b). This indicates that it is incorporated into chondroitin sulphate. Chondroitinase ABC degrades chondroitin-4 and -6-sulphates, as well as dermatan sulphate, but not heparan sulphate (Yamagata et al., 1968). At least some of the label which remains following enzyme treatment is probably incorporated into HSPGs, which were detected in the distal basement membrane using immunohistochemistry.

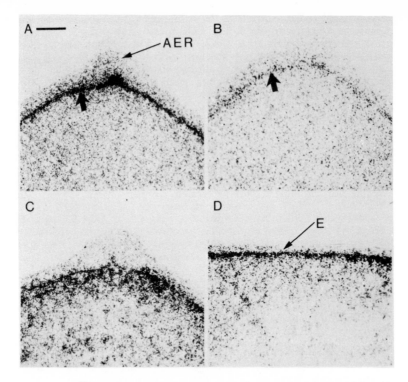

Fig. 5. [^{35}S]sulphate autoradiographs of stage 23 wing buds. (A) Control wing bud labelled for 2 hr. Note the accumulation of label along the epithelial-mesenchymal interface (arrow). (B) Control wing bud labelled for 2 hr and treated with chondroitinase ABC. There is a marked reduction in labelling at the tissue interface (arrow). (C) Labelled wing bud grafted to an unlabelled host for 4 hr. Labelled material is present as a diffuse band beneath the AER. (D) Proximal region of the graft shown in (C). Silver grains are present in a narrow band beneath the ectoderm (E). Scale bar = 30 μm.

In labelled wing buds grafted for 4 hr, silver grains were distributed in a diffuse pattern along the distal, epithelial-mesenchymal interface (Fig. 5c). However, at more proximal levels of the graft, the label remained localized within a narrow band beneath the ectoderm (Fig. 5d). By 6 hr, the label had been removed almost completely from the distal tissue interface (Fig. 6b & 6e), although it remained abundant along the interface at more proximal levels (Fig. 6c & 6e). By 8 hr, the sub-ridge region again lacked any accumulation of radiolabelled sulphate, although in some cases silver grains were still concentrated within the more proximal basement membrane. At 12 hr, the label had been removed from all regions of the epithelial-mesenchymal interface, although low levels were often observed in the chondrogenic regions of the mesenchyme. These observations indicate that sulphated molecules, including CSPG, are degraded more rapidly from the distal tissue interface than they are from the interface at more proximal levels of the wing bud.

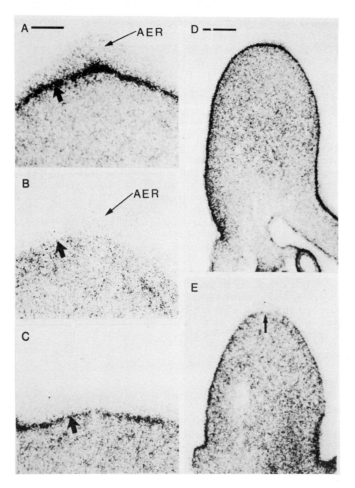

Fig. 6. [^{35}S]sulphate autoradiographs of stage 21 wing
buds. (A) Control wing labelled for 2 hr. (B)
Experimental wing bud labelled for 2 hr and
grafted to an unlabelled host for 6 hr. The label
has been removed almost completely from the
AER/mesenchyme interface (arrow). (C) Proximal
region of the graft shown in (B). Labelled
material is abundant beneath the ectoderm (arrow).
(D) Low power view of a control wing bud. (E) Low
power view of an experimental wing bud labelled
for 2 hr and grafted to a unlabelled host for 6
hr. Arrow indicates sub-ridge region. Scale bar
for (A), (B) and (C) = 30 µm. Scale bar for (D)
and (E) = 100 µm.

This suggestion is consistent with two other observations: (1)
negative staining in the sub-ridge region with the anti-CSPG mAb
CS-56 and (2) the disorganized basal lamina beneath the ridge.
It is also noteworthy that mAb 3-B-3, which recognizes
chondroitin-6-sulphate after chondroitinase digestion (Caterson
et al., 1985), stains much of the subepithelial region in the
wing bud but is negative beneath the ridge (Archer et al., this
volume). These observations are all indicative of rapid CSPG
turnover beneath the ridge. In other systems, mesenchymal cells

have been reported to degrade sulphated GAGs in the epithelial basal lamina (Smith and Bernfield, 1982). If the limb mesoderm serves a similar role, this may represent a mechanism for establishing the heterogeneity in CSPG localization at the wing tip.

Conclusions

The distribution of two molecules, in particular, proved to be useful markers of the AER: absence of CSPG in the sub-ridge basement membrane (established using mAb CS-56) and presence of integrin along both lateral and basal ridge cell surfaces. Autoradiographic pulse-chase experiments suggest that sulphated molecules, including CSPG, undergo rapid turnover beneath the apical ectoderm. This may be related to the absence of CS-56 immunoreactivity beneath the AER. Regarding the morphogenetic significance of these and other observations, it appears more likely that the sub-ridge FN plays a role in control of AER cell profile rather than enhancing distal outgrowth of the mesenchyme (as suggested by Newman et al., 1981). The accumulations of FN, collagen IV and LM, in combination with the aggregation of integrin receptors and the absence of CSPG at the ridge base, may serve to enhance cell adhesion to the basement membrane, anchoring the ridge cells more firmly around the notch. Integrin receptors may also function in elongating the ridge cells, by maintaining large areas of cell-cell contact.

REFERENCES

Akiyama, S.K., Yamada, S.S., Chen, W.-T., and Yamada, K.M., 1989, Analysis of fibronectin receptor function with monoclonal antibodies: Roles in cell adhesion, migration, matrix assembly and cytoskeletal organization, J. Cell Biol., 109:863-875.
Avnur, Z., and Geiger, B., 1984, Immunocytochemical localization of native chondroitin-sulfate in tissues and cultured cells using specific monoclonal antibody, Cell., 38:811-822.
Bayne, E.K., Anderson., M.J., and Fambrough, D.M., 1984, Extracellular matrix organization in developing muscle: correlation with acetylcholine receptor aggregates, J. Cell Biol., 99:1486-1501.
Bernfield, M., Banerjee, S.D., Koda, J.E., and Rapreager, A.C., 1984, Remodelling of the basement membrane as a mechanism of morphogenetic tissue interaction, in: "The Role of Extracellular Matrix in Development," R.L. Trelstad, ed., Alan Liss, New York.
Brennan, M.J., Oldberg, Å., Hayman, E.G., and Ruoslahti, E., 1983, Effect of a proteoglycan produced by rat tumor cells on their adhesion to fibronectin-collagen substrata, Cancer Res., 43:4302-4307.
Buck, C.A., and Horwitz, A.F., 1987, Cell surface receptors for extracellular matrix molecules, Ann. Rev. Cell Biol., 3:179-205.
Buck, C.A., Shea, E., Duggan, K., and Horwitz, A.F., 1986, Integrin (the CSAT antigen): functionality requires oligomeric integrity, J. Cell Biol., 103:2421-2428.
Caterson, B., Christner, J.E., Baker., J.R., and Couchman, J.R., 1985, Production and characterization of monoclonal antibodies directed against connective tissue proteoglycans, Fed. Proc., 44:386-393.

Carter, W.G., Wayner, E.A., Bouchard, T.S., and Kaur, P., 1990, The role of integrins α2β1 and α3β1 in cell-cell and cell-substrate adhesion of human epidermal cells, J. Cell Biol., 110:1387-1401.

Choy, M.-Y., Richman, P.I., Horton, M.A., and MacDonald, T.T., 1990, Expression of the VLA family of integrins in human intestine, J. Pathol., 160:35-40.

Dufour, S., Duband, J.-L., Kornblihtt, A.R., and Thiery, J.-P., 1988, The role of fibronectins in embryonic cell migrations, TIGS, 4:198-203.

Fallon, J.F., Frederick, J.M., Carrington, J.L., Lanser, M.R., and Simandl, B.K., 1983, Epithelial-mesenchymal interactions in chick wing morphogenesis, in: "Epithelial-Mesenchymal Interactions in Development," R.H. Sawyer and J.F. Fallon, eds., Praeger Scientific, New York.

Greve, J., and Gottlieb, D., 1982, Monoclonal antibodies which alter the morphology of cultured chick myogenic cells, J. Cell Biochem., 18:221-230.

Hamburger, V., and Hamilton, H.L., 1951, A series of normal stages in the development of the chick embryo, J. Morph., 88:49-92.

Hampé, A., 1959, Contribution à l'étude du dévelopment et de la régulation des déficiences et des excédents dans la patte de l'embryon de poulet, Arch. d'Anat. Micro. Morph. Exp., 48:345-478.

Hautenan, A., Gailit, J., Mann, D.M., and Ruoslahti, E., 1989, Effects of modifications of the RGD sequence and its context on recognition by the fibronectin receptor, J. Biol. Chem., 264:1437-1442.

Hinchliffe, J.R., 1977, The chondrogenic pattern in chick limb morphogenesis: a problem of development and evolution, in: "Vertebrate Limb and Somite Morphogenesis," D.A. Ede, J.R. Hinchliffe and M. Balls, eds., Cambridge University Press, Cambridge.

Hinchliffe, J.R., and Griffiths, P.J., 1984, Experimental analysis of the control of cell death in chick limb development, in: "Cell Ageing and Cell Death," I. Davies and D.C. Sigee, eds., Cambridge University Press, Cambridge.

Horwitz, A., Duggan, K., Buck., Beckerle, M.C., and Burridge, M.C., 1986, Interaction of plasma membrane fibronectin receptor with talin, a transmembrane linkage, Nature, 320:531-533.

Hurle, J.M., Ros, M.A., Gañan, Y., Macios, D., Critchlow, M.A., and Hinchliffe, J.R., 1990, Experimental analysis of the role of ECM in the patterning of the distal tendons in the developing limb bud, Cell Differ., 30:97-108.

Hynes, R., 1987, Integrins: A family of cell surface receptors, Cell., 48:549-554.

Hynes, R.O., Marcantonio, E.E., Stepp, M.A., Urry, L.A., and Yee, G.H., 1989, Integrin heterodimer complexity in avian and mammalian cells, J. Cell Biol., 109:409-420

Kaufmann, R., Frosch, D., Westphal, C., Weber, L., and Klein, C.E., 1989, Integrin VLA-3 -ultrastructural localization at cell-cell contact sites of human cell cultures, J. Cell Biol., 109:1807-1815.

Kimata, K., Oike, Y., Tani, K., Shinomura, T., Yamagata, M., Uritani, M., and Suzuki, S., 1986, A large chondroitin sulphate proteoglycan (PG-M) synthesized before chondro-genesis in the limb bud of chick embryos, J. Biol. Chem., 261:13517-13525.

Klein, C.E., Steinmayer, T., Mattes, J.M., Kaufmann, R., and

Weber, L., 1990, Integrins of normal human epidermis: differential expression, synthesis and molecular structure, Brit. J. Dermatol., 123:171-178.

Korhonen, M., Ylänne,J., Laitinen, L., and Virtanen, I., 1990, The α_1-α_6 subunits of integrins are characteristically expressed in distinct segments of developing and adult human nephron, J. Cell Biol., 111:1245-1254

Krotoski, D.M., Domingo, C., and Bronner-Fraser, M., 1986, Distribution of a putative cell surface receptor for fibronectin and laminin in the avian embryo, J. Cell Biol., 114:504-518.

Kulyk, W.M., and Kosher, R.A., 1987, Temporal and spatial analysis of hyaluronidase activity during development of the embryonic chick limb bud. Dev. Biol., 120:535-541.

Larjava, H., Peltonen, J., Akiyama, S.K., Yamada, S.S., Gralnick, H.R., Uitto, J., and Yamada, K.M., 1990, Novel function for β_1 integrins in keratinocyte cell-cell interactions, J. Cell Biol., 110:803-815.

Newman, S.A., Frisch, H.L., Perle, M.A., and Tomasek, J.J., 1981, Limb development: Aspects of differentiation pattern formation and morphogenesis, in: "Morphogenesis and Pattern Formation," T.G. Connolly et al., eds., Raven Press, New York.

Rich, A.M., Pearlstein, E., Weissman, G., and Hoffstein, S.T., 1981, Cartilage proteoglycans inhibit fibronectin-mediated adhesion, Nature, 293:224-226.

Shinomura, T., Jensen, K.L, Yamagata, M., Kimata, K., and Solursh, M., 1990, The distribution of mesenchyme proteoglycan (PG-M) during wing bud outgrowth, Anat. Embryol., 181:227-223.

Smith, R.L., and Bernfield, M., 1982, Mesenchyme cells degrade epithelial basal lamina glycosaminoglycan, Dev. Biol., 94:378-390..

Solursh, M., and Jensen, K.L., 1988, The accumulation of basement membrane components during the onset of chondrogenesis and myogenesis in the chick wing bud, Development, 104:41-49.

Todt, W.L., and Fallon, J.F., 1984, Development of the apical ectodermal ridge in the chick wing bud, J. Embryol. Exp. Morph., 80:21-41.

Tomasek, J.J., and Brier, J., 1986, Extracellular matrix maintains apical ectodermal ridge in culture, in: "Progress in Developmental Biology, Part B," H. Slavkin, ed., Alan Liss, New York.

Tomasek, J.J., Mazurkiewicz, J.J., and Newman, S.A., 1982, Non-uniform distribution of fibronectin during avian limb development, Dev. Biol., 90:118-126.

Trelstad, R.L., 1984, "The Role of Extracellular matrix in Development," Alan Liss, New York.

Yallup, B.L., 1984, Analysis of regulation along the anteroposterior axis of the developing chick limb, Ph.D. thesis. University College of Wales, Aberystwyth.

Yamagata, M., Saito, H., Habuchi, O., and Suzuki, S., 1968, Purification and properties of bacterial chondroitinases and chondrosulfatases, J. Biol. Chem., 243:1523-1535.

Yamagata, M., Suzuki, S., Akiyama, S.K., Yamada, K.M., and Kimata, K., 1989, Regulation of cell-substrate adhesion by proteoglycans immobilized on extracellular substrates, J. Biol. Chem., 264:8012-8018.

PHYSICAL FORCES AND PATTERN FORMATION

IN LIMB DEVELOPMENT

Albert K. Harris

Department of Biology
University of North Carolina
Chapel Hill, NC 27599-3280 USA

INTRODUCTION

For those cells of the body whose differentiated cell type is determined while they are at one position, but which then move elsewhere to form their definitive anatomical structures, the geometry of these structures must be explained in terms of whatever physical forces are responsible for moving the cells into their eventual positions. Positional control of gene expression is not enough. The skeletal muscle cells of the limb are a good example: their cells are known to be derived originally from the somites, from which they migrate into the developing limb bud (Chevallier, Kieny and Mauger, 1977). In principle, the forces responsible might include those of growth, adhesion, active cellular locomotion or other processes; but this article is concerned specifically with the ubiquitous locomotory forces called "traction".

Unfortunately, it is difficult to make direct observations of cell movements within developing limb buds; nor are there adequate methods for mapping distributions of physical forces within tissues. Thus, almost all of what is known about the forces of cell locomotion is based on studies in tissue culture and organ culture; only in such artificial, controlled environments has it been possible (so far) to make cellular forces visible in an unambiguous way. Consequently, this article is an attempt to extrapolate from *in vitro* observations to the situation inside the developing limb itself.

Of special relevance to limb morphogenesis is the evidence that traction forces exerted by cells located at one position can pull collagen fibers and other cells toward them over distances of several millimeters or even centimeters. This capability has suggested a novel hypothesis for the morphogenesis of skeletal muscles: foci of strong traction at the future muscle's origin and insertion sites would "reel in" collagen fibers, thus stretching and aligning fibers and muscle cells into the proper anatomical form. Even though this type of mechanism has been shown to work effectively in organ culture (Stopak & Harris, 1982), and although *in vivo* experiments involving the injection of fluorescently labeled collagen have confirmed that comparable patterns of tensile force also exist inside developing limb buds, many questions remain. In particular, how can it be possible for any such simple mechanism to achieve enough specificity to generate the complex yet regular patterns of the musculature?

OBSERVATIONS OF CELL TRACTION

The exertion of traction by tissue cells was first demonstrated unambiguously by the use of extremely thin, elastic substrata made of silicone rubber (Harris, Wild & Stopak, 1980).

Developmental Patterning of the Vertebrate Limb
Edited by J.R. Hinchliffe *et al.*, Plenum Press, New York, 1991

Prior to this, tissue culturists had long noticed that substrata made of clotted blood plasma become progressively distorted in centripetal patterns around explants of most, but not all, cell types. This effect, however, had been interpreted as the result of some kind of shrinkage of these protein substrata themselves, rather than as effects of forces exerted on them by the cells. This supposed shrinkage was thought to be induced by biochemical activities, especially syneresis due to absorption of water by the cells' metabolism. The principal originator of this dehydration theory was the great embryologist Paul Weiss (1955), who reacted vigorously to any suggestion that physical forces might be involved. It is instructive to read his powerful condemnation (Weiss, 1952) of an interesting report by Katzberg (1951) that the percentages of explants becoming connected by tracts of aligned fibrin after a given time varies inversely with the square of the distance between the explants. Even though Katzberg explicitly accepted the dehydration theory, the inverse square proportionality together with Katzberg's choice of the name "attraction fields" for the patterns were too suggestive of physical forces to be permissible in the literature.

The silicone rubber substrata were designed specifically to refute this dehydration theory. They are made from ordinary silicone fluid (polydimethylsiloxane; of 30,000 to 60,000 centipoise viscosity) simply by exposing the fluid surface directly to a Bunsen burner flame for approximately one second (the most complete description of the method is in Harris, 1988). Cells are then plated out directly onto the material produced when exposure to the flame chemically cross-linked the silicone polymer. Because silicone rubber is one of the most inert materials known, the distortions produced in it by crawling tissue cells can only be due to physical forces. In particular, it is impossible that they could result from dehydration, since the rubber is very hydrophobic and is never hydrated.

THE SILICONE RUBBER CULTURE SUBSTRATUM TECHNIQUE

Figure number 1 is a schematic cross section of a fibroblast spreading upon and wrinkling one of these silicone rubber substrata. The rubber layer is an approximately 1 μm skim on the surface of a pool of the liquid polymer many microns thick. This fluid pool lubricates the free lateral movement and wrinkling of the rubber sheet, and its greater refractive index relative to the culture medium is important in making the wrinkles as visible as they are to phase contrast microscopy. Within one or two seconds after the detachment of such a cell, the wrinkles which its traction had formed disappear as the rubber resumes a planar shape. This is important because it means that the continued existence of wrinkles beneath and around a spread cell is evidence that the cell is continuing to exert contractile forces. Time lapse films of these wrinkles, and of lateral movements of marker particles attached to the rubber surface, can give a fairly dramatic indication of the patterns of forces exerted by tissue culture cells.

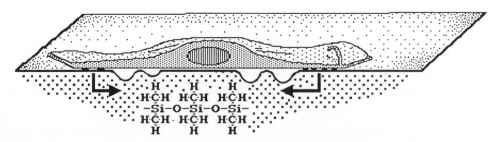

fig. 1

Two classes of wrinkles are observed: compression wrinkles and tension wrinkles. The former lie directly beneath the cell, usually transverse to its predominant direction of spreading or elongation, if there is one. They result from the exertion of forces in the direction perpendicular to their long axes, as can be seen either by observing the lateral movement of the rubber surface itself during their formation or by producing them with the tip of a microneedle pushing sideways on the rubber surface. Tension wrinkles, on the other hand, are indicative of a stretching of the rubber surface in the direction parallel to the wrinkle's axis. The usual pattern is to have compression wrinkles underneath the cell bodies and tension wrinkles radiating out away from their spreading margins.

Time lapse observations of the patterns of wrinkling indicate that the forces are exerted as a centripetal shear force, distributed over relatively broad areas, much as if the bottom of the cell were occupied by something equivalent to a tractor tread or a conveyor belt. I interpret this to mean that the bulk of the force is exerted by the steady pull of the adhesion plaques (or their equivalents) rather than by any cycles of reach, grab and pull, either by filopodia or by lamellipodia along the cell margins. Nevertheless, the direction in which these strong shear forces (traction) are exerted is strongly correlated with the locations of areas of active ruffling movements along the cell margin. The direction of the traction force is consistently *away from* those parts of the cell margins undergoing the most active protrusive activity, such as ruffling. Quite often, both ends of a fibroblast or myoblast will exert traction directed toward the cell center. It is by means of this internal tug of war that cells spread into their typical flattened or elongate shapes. The stress thus produced continues to be exerted for as long as the cells remain spread, and the wrinkles in the rubber spring back within about a second of breakage of adhesions.

Use of the rubber substrata indicates that traction forces are exerted by cells of virtually all differentiated cell types (all of those that are capable of locomotion), but that the strengths of the forces exerted vary greatly according to cell type. Among motile cell types, only nerve growth cones and polymorphonuclear leucocytes (granulocytes) failed to wrinkle rubber substrata on which they crawled; I believe that this is merely because their traction is too weak to be detected in this way, rather than because of any fundamental difference in their propulsion.. Other cell types form a series, with macrophages exerting very weak traction, epithelial cells of a wide variety of types exerting somewhat stronger traction forces, and glial cells and fibroblasts exerting the strongest traction (although tiny individual blood platelets can produce distortions not much smaller than those of whole fibroblasts, and thus seem to be strongest of all).

Based on the numbers of cells needed to produce a given degree of distortion, I would estimate that the strongest fibroblasts exert traction forces as much as 100 fold stronger than macrophages. In addition, when a fibroblast has compressed the rubber beneath itself into many wrinkles, and then detaches from the rubber along one side, one observes that the elastic recoil will pull the cell laterally at speeds several hundred times greater (many μm per second) than the speeds at which these cells actually crawl (less than a μm per minute); since the force of the elastic recoil is the result of the traction force, and can be no stronger than was the traction which produced it, we can again conclude that fibroblast traction is hundreds of times stronger than would be needed if its only function were the displacement of the cells exerting this force.

By using calibrated microneedles to wrinkle the same rubber surfaces on which fibroblasts were being cultured, and comparing the amount of bending of a needle required to distort the rubber to the same degree as did the individual cells, I previously calculated that these fibroblasts were exerting approximately 10^{-3} dyne per μm of of cell margin. However, the fibroblasts observed in those initial studies seem to have been considerably weaker than some of the cells subsequently studied, so that number may be somewhat low. More efforts are needed to make this method truly quantitative.

More attention also needs to be given to the relationship (or distinction) between traction and contraction. Although traction is, presumably, the cumulative result of many coordinated intracellular contractions, it is important to notice that the amplitude of the traction force is not limited to the length of the cells themselves, much less to the lengths of their surface protrusions. Fibroblasts are perfectly capable of conveying or "winching" many millimeters of rubber sheet past their lower surfaces over a period of days. They can likewise "reel in" collagen fibers over such distances, tens or even hundreds of times their own maximum lengths. Thus, traction can serve to displace other materials and other cells toward those cells exerting the traction force, and can do so over distances which are very large relative to the sizes of early embryos. In the particular case of fibroblasts, it seems

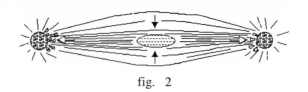

fig. 2

likely that this winching effect, and not the translocation of the fibroblasts themselves, is the real or primary function of the phenomenon we observe as "locomotion" when these cells are cultured on surfaces like glass or plastic. Because these substrata are too rigid to be distorted appreciably by their traction forces, the locomotion is all that one sees. The fibroblast's style of locomotion is certainly better designed for the winching function than for simple locomotion; in particular, the strengths of the forces they are exert are wastefully excessive, relative to their slow speed and erratic wandering.

EFFECTS OF FIBROBLAST TRACTION ON COLLAGEN GELS

Traction can also distort collagen gels, reconstituted from rat tail tendon collagen by methods similar to those of Elsdale & Bard (1972). In fact, when the collagen is laid down as a very thin, flat sheet, the effect of fibroblast traction is to produce wrinkles nearly identical in shape and arrangement to those produced on the rubber substrata. In thicker gels, one still observes patterns of centripetal distortion around fibroblasts, but not wrinkles; however, when a rubber substratum is overlain with a collagen gel, then fibroblasts spreading on the rubber surface (and wrinkling it) will simultaneously produce a pattern of centripetal distortion of the collagen around them. Apparently the spatial pattern of force-exertion is just the same in collagen as it is on rubber; in fact, it also appears that the "tops" of fibroblasts can exert traction equally well as the "bottoms", so long as there is something on top to attach to.

Collagen can be displaced laterally by fibroblast traction, but the most dramatic effect is that of long-range alignment (Harris, Stopak and Wild, 1981). These reconstituted collagen gels possess a truly remarkable sensitivity to long-range alignment by the effects of fibroblast traction. In fact, their behavior in this respect seems to me comparable to the behavior of nematic liquid crystals (like those in digital watch displays): small, local perturbations can give rise to long-range, relatively large-scale alignments of fibers extending long distances beyond the site of the force being imposed. The effect seems virtually autocatalytic.

The kinds of results one observes are diagrammed above in figure 2. Explants of chick hearts or other tissues known to contain strong fibroblasts, when placed on or in collagen gels, will distort these gels centripetally around themselves; at first, the only effect is that nearby collagen fibers are pulled into the explant and compressed in its surface to form a sort of capsule, while fibers further away become reoriented into a radial direction; when there are two or more such explants within a few centimeters of each other, however, the eventual result is the formation of a discrete tract of laterally-compressed collagen fibers running in a straight line from one explant to the other. As diagrammed in the drawing, the lengths of these tracts greatly exceed the maximum distances of outward migration reached by any of the fibroblasts at the time of the formation of the aligned tracts (1-3 days in culture). In fact, these tracts can easily extend across the width of a petri dish and the longest one Stopak and I ever produced was just over 6 centimeters long. We have published several photographs of these, and their general appearance is sketched below in figure 3.

fig. 3

These tracts are similar, but about 10-fold longer, than those seen in Katzberg's studies of fibrin clots; it would be interesting to determine whether the time required for their formation is likewise inversely proportional to the square of distance. A simple mathematical argument can be made as to why such a proportionality might be expected, based on the flux of material converging symmetrically on a point sink being inversely proportional to the square of the distance outward from this sink. However, since the gels were in both cases effectively planar, perhaps one should expect the inward flux of collagen to vary inversely as the first power of distance, rather than the second.

Aligned tracts similar to those formed between one fibroblast explant and another will also form between one explant and a "fixed point," such as a plastic peg attached to the bottom of the petri dish. In fact, when fibroblasts are dispersed through the gel, equivalent tracts will even form between one fixed point and another. Photographs of all these kinds of tracts were published by Stopak and Harris (1982) and Stopak made many additional time lapse films of the stages of distortion of collagen gels under conditions of mechanical stress produced by fibroblast traction. Among other things, these show the progressive distortions of spherical bubble holes into ellipsoids, which provide a kind of map of the 3-dimensional patterns of force distribution in the gels. They also show that, once a tract has formed between two clumps of fibroblasts, the displacement of the fibers is no longer radially toward the closest of these explants, but laterally toward the side of the tract itself. These and other phenomena need to be looked for *in vivo*, perhaps borrowing the methods of Beloussov and his colleagues (1975, 1980).

Other studies with this system led to the observation that fibroblast traction can also cause the spontaneous break up of initially almost homogeneous suspensions of cells in collagen gels; this happens when the margin of the gel is effectively "clamped" in position by entanglement in a ring of glass fiber filter paper. The net effect is analogous to that of reaction-diffusion systems: small initial heterogeneities are magnified by various sorts of mechanical positive feedback; if there are a few more cells at one position than in neighboring areas, then the cells will be pulled from the low density areas into the high density ones (Harris, Stopak and Warner, 1984) These instabilities can even propagate, because the depletion of cells from one area will permit the area beyond that to pull more cells toward itself, first on the side facing the original depletion, and then on the farther side. A mathematical analysis of this category of instabilities (Oster, Murray and Harris, 1983) showed how they could, in principle, generate the patterns of dermal feather germs in bird skin as well as the pattern of cartilage condensations in the development of the vertebrate limb bud.

FIBROBLAST TRACTION AND THE FORMATION OF MUSCLES AND TENDONS

The cartoon drawing below (figure 4) is meant to illustrate graphically the hypothetical model originally proposed by Stopak and Harris (1982) to explain the mechanical formation of skeletal muscles, by means of cell traction. The idea is that especially strong or active traction would be exerted by those fibroblasts located at the "origin" and "insertion" sites of skeletal surfaces, thereby producing the effect of so many little winches located at these sites, reeling in collagen fibers so as to displace, stretch and align these fibers, along with the myoblasts attached to them.

Although we considered that any elongation of the cartilages which occurred during the time of muscle formation could also contribute to the stretching process, it was not our idea that such elongation would be necessary. In fact, in the organ culture studies which

fig. 4

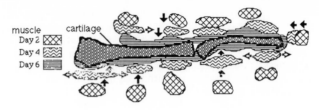

fig. 5

supported at least the practicality of this mechanism, the limb cartilages used actually elongated by only a few percent. On the other hand, Vandenbergh (1982) has proposed that cartilage elongation, by causing increases in distances between muscle origins and insertions, is what causes the alignment of muscle cells and collagen fibers in the formation of skeletal muscles. By using a type of flexible substratum (different from that described here) he directly imposed stretching forces on cultured muscle cells and showed that this results in their alignment and organization into spatial patterns which are histologically very similar to normal muscles. There is no reason to believe that muscle cells would respond any differently to stretching forces which resulted from traction rather than actual elongation, or that the cells would have any way of distinguishing the sources of the forces imposed on them.

Another theory concerning the formation of skeletal muscles has arisen from studies of grasshopper development (Ball, Ho and Goodman, 1985; Jellies, 1990). These authors describe how muscle formation in the grasshopper limb is preceded by the movement of a founder cell initial to the site, and by the elongation of this initial cell at the site where the muscle will subsequently form. The idea is that these initial cells are forming a prepattern for the subsequent accumulation and alignment of the muscle cells. As to the physical causation of either the elongation of the initial cell, or the subsequent accumulation of muscle cells around it, none is specified, but cell traction could easily produce such effects.

Experimental evidence exists both supporting and opposing the hypothesis that muscles and tendons are formed by traction exerted at or near the attachment sites. In support are the results of organ culture experiments in which embryonic cartilages were plated out in collagen gels together with explants of embryonic back muscle (Stopak and Harris, 1982). The result was that traction forces exerted by perichondrial fibroblasts created tracts of aligned collagen in the surrounding collagen gel and that the fragments of muscle tissue were pulled in toward the cartilages and gradually aligned along the sides of the the cartilages, to which they had been connected by means of the tracts of aligned collagen. We concluded that these tracts are equivalent to tendons, that the muscle fragments had been realigned and stretched into facsimiles of actual muscles, and that traction exerted by perichondrial fibroblasts would therefore be entirely capable of causing the morphogenesis of skeletal muscles.

Figure #5 of this paper (shown below) is a diagrammatic summary of figure #15 from Stopak & Harris (1982). Tracings from the 3 photographic montages were digitized with a scanner. The two cartilages (the tarsometatarsus and tibiotarsus of a 10 day chick embryo) are shown as if immobile, having the lengths and position that they had on day 4 of the experiment, even though both of them actually elongated slightly and shifted in relative position (especially between days 4 and 6, due to the contraction of the muscles). The sequential positions and shapes of the 10 explanted masses of back muscle are shown relative to these cartilages. Notice how these masses of muscle cells are progressively pulled inward toward the cartilages (solid arrows) at the same time that they are stretched along the axis of the cartilages (hollow arrows). Considerably more detail is visible in the original published photographs.

In a subsequent series of experiments (Stopak, Wessells and Harris, 1985), isolated type I collagen was covalently labeled with fluorescein and injected into developing chick limb buds. Our hypothesis predicted that this exogenous collagen would be physically remodeled to correspond to and become part of whatever collagenous structure happened to be forming normally at the site of injection. This prediction was confirmed, in that the injected collagen sometimes became compressed and aligned to form parts of tendons, sometimes parts of blood vessel walls, sometimes parts of the dermis, and sometimes other structures, depending on its location. But while this certainly shows that collagen neither has to be secreted into place nor have special self assembly properties corresponding to its eventual structural arrangement, we had no way of proving that the forces responsible for this physical remodeling were necessarily the same as we had observed causing the equivalent rearrangements in tissue culture and organ culture, that is, traction.

Several arguments have been advanced in opposition to the idea that traction is responsible for remodeling collagen inside the embryo, and particularly against the hypothesis that traction is responsible for the morphogenesis of skeletal muscles. One of the most telling is the observed ability of relatively normal-looking muscles to form even in limb buds which have been experimentally altered so that that their distal attachment sites (insertions) are either missing or were never formed. In such cases, Shellswell and Wolpert (1977) reported, the muscles and tendons end "in mid-air". Such a result would indeed be contrary to what the hypothesis would have predicted. On the other hand, in the organ culture experiments, we frequently did often see one-ended tracts extending outward from the sides of cartilages (see figure 6 below), showing that it is not impossible to form muscle and tendon-like alignment patterns with traction only being exerted at one end. A related example was the formation of tracts between fibroblast explants and fixed points in collagen gels, as described above.

fig. 6

Another argument that has been made against this hypothetical mechanism is that it seems to lack sufficient specificity to be able to generate such geometrically complex patterns as one finds in the vertebrate musculature, particularly in the limb. Why, one can ask, wouldn't muscles and/or tendons just form interconnecting all possible pairs of origin and insertion sites? What would prevent abnormal (and often functionless) muscles forming so as to connect the insertion site of muscle A with the insertion site of muscle B? To the former question one can reply that, as with any other possible mechanism, some additional control system would have to be superimposed so as to prevent this randomness. One possibility would be a temporal control system in which the different pairs of origins and insertions exert their traction at sequential times, with connections forming preferentially between those sites which were simultaneously active. To the latter question, why wouldn't muscles sometimes form between the wrong combinations of attachment sites, it so happens that close studies of cadavers have revealed that the human musculature is actually rather variable between individuals and that a particularly frequent form of abnormality is the formation of extra muscles that connect inappropriate pairs of origin and insertion sites. For example, Barlow (1935) describes the incidence of the sternalis muscle, which runs between the attachment sites of the rectus abdominus and the sterno-cleido-mastoid muscles, at the bottom and top of the sternum, respectively, in about 6% of humans.

REFERENCES

Ball, E. E., Ho, R. K. and Goodman, C. S., 1985, Muscle development in the grasshopper embryo I. Muscles, nerves and apodemes in the metathoracic leg, Dev. Biol. 111: 383-398.

Barlow, R.N., 1935, The sternalis muscle in American whites and Negroes, Anat. Rec. 61: 413-426

Beloussov, L. V., 1980, The role of tensile fields and contact cell polarization in the morphogenesis of amphibian axial rudiments, Wilh. Roux' Arch Dev. Biol. 188, 1-7.

Beloussov, L. V., Dorfman, J. G. and Cherdanzev, V. G., 1975, Mechanical stresses and morphological patterns in amphibian embryos, J. Embyol. exp. Morph. 34, 559-174 (1980)

Chevallier, A., Kieny, M. and Mauger, A., 1977, Limb-somite relationship: origin of the limb musculature, J. Embryol. exp. Morph. 41, 245-58

Elsdale, T. and Bard, J., 1972 Collagen substrata for studies on cell behaviour, J. Cell Biol. 41: 298-305.

Harris, A. K., 1988, Fibroblasts and myofibroblasts, Methods in Enzymology 163: 623-642

Harris, A.K., Stopak D. and Wild P., 1981, Fibroblast traction as a mechanism for collagen morphogenesis, Nature 290:249-251.

Harris, A.K., Stopak D. and Warner, P., 1984, Generation of spatially periodic patterns by a mechanical instability: a mechanical alternative to the Turing model, J. Embryol. exp. Morph. 80: 1-20.

Harris A.K., Wild P. and Stopak D., 1980, Silicone rubber substrata: a new wrinkle in the study of cell locomotion, Science. 208:177-179.

Jellies, J., 1990, Muscle assembly in simple systems, Trends in Neurosciences 13: 126-131.

Katzberg, A. A., 1951, Distance as a factor in the development of attraction fields between growing tissues in culture, Science 114: 431-432.

Oster, G. F., Murray, J. D. and Harris, A. K., 1983, Mechanical aspects of mesenchymal morphogenesis, J. Embryol. exp. Morph. 78: 83-125.

Shellswell, G. B. and Wolpert, L., 1977, The pattern of muscle and tendon development in the chick wing, in Vertebrate Limb and Somite Development eds. Ede, D.A., Hinchliffe, J. R. and Balls, M. 3rd. S.E.B. Symposium pp. 71-86.

Stopak, D. and Harris, A. K., 1982, Connective tissue morphogenesis by fibroblast traction, Developmental Biology 90, 383-398.

Stopak, D., Wessells, N. K., and Harris, A. K., 1985, Morphogenetic rearrangement of injected collagen in developing chicken limb buds, Proc. Natl. Acad. Sci. 82. 2804-8.

Vandenbergh, H. H., 1982, Dynamic mechanical orientation of skeletal myotubes *in vitro*, Dev. Biol. 93: 438-443.

Weiss, P., 1955, Nervous system. in "Analysis of Development" (B. H. Willier, P. Weiss and V. Hamburger, eds.) pp. 346-401. Saunders, Philadelphia.

Weiss, P., 1952, Comments and Communications:"Attraction Fields" between growing tissue cultures, Science 115: 293-295.

EXTRACELLULAR MATERIAL ORGANIZATION AND LONG TENDON FORMATION IN THE CHICK LEG AUTOPODIUM. IN VIVO AND IN VITRO STUDY

M.A. Ros[1], J. R. Hinchliffe[2], D. Macias[3], J. M. Hurle[1], and M.A. Critchlow[2]

[1] Dept. Anatomia y Biologia Celular. Facultad de Medicina U. Cantabria. 39011 Santander, Spain
[2] Dept. Biological Sciences. U. College of Wales Aberystwyth, Wales, U.K.
[3] Dpto. Ciencias Morfológicas. U. Extremadura. Spain

1. The limb bud exhibits a pretendinous extracellular matrix pattern: the mesenchyme lamina

During later developmental stages (27-36 HH) the formation of the long autopodial tendons of the chick leg seems to be governed by a precise pattern of extracellular material (ECM). The main element of this pattern is a tenascin-rich sheet termed the "mesenchyme lamina" (ML) which originates at the basement membrane distally and runs proximally through the mesoderm, subjacent and parallel to the basement membrane (Hurlé et al, 1989).Immunohistochemistry showed that the lamina strongly reacted to tenascin and collagen type I but weakly to fibronectin and failed to react to laminin and collagen type IV, clearly differentiating it from the basement membrane.

2. The formation of tendon blastemas takes place in close association to the mesenchymal lamina

The ML arises at the level of the advancing front of chondrogenic skeletal differentiation and becomes progressively displaced concomitantly with the distal progression of the limb development. When the lamina first appears (stage 27) it is related with the distal tip of the premuscular masses. In the course of development the pretendinous mesenchymal cells condense on the inner surface of the lamina (Fig 1) to give rise to the extensor digitorum longus tendon (dorsally) and to the flexor digitorum longus tendon (ventrally). These blastemas form in a precise longitudinal direction along the dorsal and ventral part of each digital ray using the lamina as a template. Lateral to the blastemas the lamina disintegrates and disappears (Fig. 2) .Proximally each blastema joins the corresponding muscular bellies as the individual muscles are developing.

Developmental Patterning of the Vertebrate Limb
Edited by J.R. Hinchliffe *et al.*, Plenum Press, New York, 1991

211

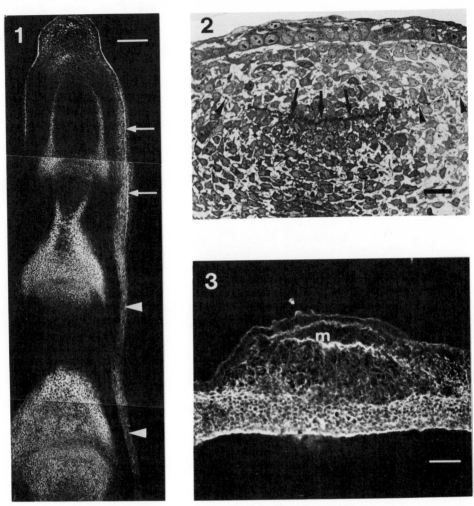

Fig. 1. Tenascin localization in a longitudinal section of stage 34 limb bud. The elongated blastema of the extensor digitorum longus tendon (arrowheads) is shown: distally it runs into the ML (arrows). Bar=200μm

Fig. 2. Transverse semithin section of a distal zone of the third digit of a stage 34 limb bud. The pretendinous blastema can be seen localized mainly on the inner side of the ML (arrows). Note that laterally to the blastema the ML disintegrates (arrowheads). Bar=25μm

Fig. 3. Localization of TN in micromass cultures with ecto-mesenchyme grafts. The picture is of a 3 days micromass culture grafted with a stage 24 explant and cultured for 3 days. (m: mesenchyme lamina like structure). Bar=200μm

3. Experimental modifications of the pretendinous mesenchyme lamina resulted in parallel changes in the pattern of tendon formation

The experimental truncation of the limb through apical ectodermal ridge removal or extradigit formation by local extodermal removal of the third interdigital space resulted in modifications of the ML which led to the development of abnormal flexor and extensor tendons. In truncation the ML becomes detached from the basement membrane, and dorsal and ventral parts link together round the distal end of the terminal phalange. Later dorsal and ventral tendons are linked distally in precisely the same position. In induced extradigits the appearance of tendons is dependent of normal adjacent tendons and always preceded by the appearance o a lamina (Hurlé et al., 1990). These experiments support the interpretation of the ML as a tendon precursor.

4. In vitro analysis

In order to study the factors controlling the development of the ML an in vitro analysis was designed. High-density cultures of mesenchyme from the progress zone of stage 24 limb buds develop into a sheet of cartilage (Ahrens, P.B. et al., 1977) after three-four days of culture. We cultured small fragments of stage 24 distal ectoderm and subjacent mesenchyme on top of these high-density cultures. This represents a day and a half before the lamina was first detected "in vivo". After two days of culture the grafts have developed a distinctive laminar pattern of tenascin positive material in the mesoderm (Fig.3). Organ culture of similar ecto-mesenchyme explants also developed a ML-like structure after one day of culture. Interestingly in the case of organ cultures a cartilage nodule always developed at the bottom of the culture so that the laminar structure located between the cartilage and the ectoderm. The results of our "in vitro" analysis suggests that the association of ecto-mesenchyme with cartilage in areas laking intervening myogenic masses modulates the spatial organization of the ECM

ACKNOWLEDGEMENTS

This work was supported by a grant from the DGICYT (PS87-0095)

REFERENCES

Ahrens. P.B., Solursh, M., Reiter, R.S. (1977). Stage-related capacity for limb chondrogenesis in cell cultures. Dev Biol. 60: 69-82.
Hurlé , J.M., Hinchliffe, J.R., Ros, M.A., Critchlow, M.A. and Genis-Galvez, J.M. (1989). The extracellular matrix architecture relating to myotendinous pattern formation in the distal part of the developing chick limb: an ultrastructural, histochemical and immunocyotchemical analysis. Cell Differ. Develop., 27: 103-120.
Hurle, J. M., Ros, M. A., Gañan, Y., Macias, D., Chritchlow, M., & Hinchliffe, J. R. (1990). Experimental analysis of the role of ECM in the patterning of the distal tendons of the developing limb bud. Cell Differ. Develop.,30: 97-108

HYALURONAN-CELL INTERACTIONS IN LIMB DEVELOPMENT

Bryan Toole*, Shib Banerjee*, Raymond Turner*,
Syeda Munaim*, and Cheryl Knudson+

*Department of Anatomy and Cellular Biology, Tufts University
Health Science Schools, Boston, MA 02111, USA
+Departments of Biochemistry and Pathology, Rush/Presbyterian/
St. Luke's Medical School, Chicago, IL 60612, USA

Hyaluronan is a ubiquitous component of the extracellular matrices in which cells migrate and proliferate during embryonic development (reviewed in Toole, 1981). Its physical and chemical properties contribute to an extracellular milieu which is important both to the structural integrity of embryonic tissues and to the morphogenetic processes that take place within them. One way in which hyaluronan participates in tissue structure arises from its ability to form meshworks that exert osmotic pressure (Comper and Laurent, 1978; Meyer, 1983). The resultant swelling pressure within the tissue can lead to separation of cellular or fibrous structures or deformation of the tissue, possibly facilitating cell and tissue movements (Toole, 1981; Morris-Wiman and Brinkley, 1990). In addition, some embryonic cells exhibit large, hyaluronan-dependent, pericellular coats (Knudson and Toole, 1985) that may influence cell-cell adhesion (Underhill and Toole, 1981; Knudson, 1990a), cell-substratum adhesion (Barnhart et al., 1979), cell proliferation (Brecht et al., 1986), migration (Turley et al., 1985; Schor et al., 1989) or differentiation (Kujawa et al., 1986).

Interaction of hyaluronan with the cell surface is mediated by membrane-bound receptors (reviewed in Toole, 1990). Recent work has demonstrated that the hyaluronan receptor of baby hamster kidney and transformed 3T3 cells is a glycoprotein of molecular weight, ~85kDa (Underhill et al., 1987) and is closely related to the CD44 lymphocyte homing receptor (Aruffo et al., 1990; Lesley et al., 1990; Miyake et al., 1990). We have recently prepared a monoclonal antibody, MAb 4D4, that recognizes chick embryo hyaluronan-binding proteins of molecular weight 93, 91 and 69kDa; these proteins are widely distributed in embryonic tissues, and are probably related to the 85kDa receptor since the antibody blocks binding of hyaluronan to transformed 3T3 cells (Banerjee and Toole, 1990). The antibody also inhibits production of hyaluronan-dependent pericellular coats (Yu et al., 1990), implying that interaction of hyaluronan with the receptor proteins is necessary for coat formation.

MESODERMAL HYALURONAN-DEPENDENT PERICELLULAR COATS

Mesodermal cells in the early chick embryo limb bud are separated by extensive hyaluronan-rich spaces (Singley and Solursh, 1981) and these cells

Developmental Patterning of the Vertebrate Limb
Edited by J.R. Hinchliffe *et al.*, Plenum Press, New York, 1991

215

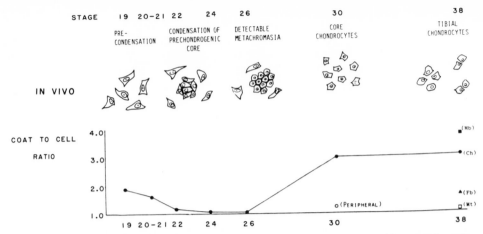

Figure 1. **Comparison of pericellular coats in vitro with cell organization in vivo**. The sizes of pericellular coats are represented as mean coat-to-cell ratios (ratios of coat perimeter to cell perimeter) obtained with cells cultured from limbs at the stages indicated. Prior to condensation, the mesodermal cells in vivo are widely separated by hyaluronan-rich matrix and in vitro they exhibit large hyaluronan-dependent coats. When the mesoderm becomes condensed in vivo, the cells lack coats in vitro. On differentiating, chondrocytes again elaborate large hyaluronan-dependent coats in vitro and an extensive matrix in vivo. Myoblasts exhibit coats prior to fusion but lose them on fusion. Ch, chondrocytes; Fb, fibroblasts; Mb, myoblasts; Mt, myotubes. (From Knudson & Toole, 1985, with permission).

in culture produce large, hyaluronan-dependent, pericellular coats (Knudson and Toole, 1985). When the chondrogenic and myogenic areas of the limb bud become condensed in vivo, i.e. these cells are separated by a smaller volume of matrix, the isolated mesodermal cells do not exhibit visible coats in culture. This change in the ability to express coats is accompanied by a large decrease in the ratio of hyaluronan to chondroitin sulfate-proteoglycan produced by the condensation-stage mesoderm, as compared to the pre-condensation mesoderm (Knudson and Toole, 1985).

During differentiation of chondrocytes, which are again separated by extensive spaces in vivo, large pericellular coats are re-expressed in culture. Chondrocyte coat structure is still dependent on hyaluronan even though proteoglycan is now a quantitatively more prominent component (Goldberg and Toole, 1984; Knudson and Toole, 1985). In contrast to chondrogenesis, fusion of myoblasts is accompanied by loss of pericellular coats (Orkin et al, 1985), apparently a necessary step towards differentiation since myoblasts cultured on a hyaluronan substratum fail to fuse (Kujawa et al., 1986). Thus there is a close correlation between the presence of large intercellular spaces in vivo and the expression of large hyaluronan-dependent coats in vitro (see Fig. 1).

CHANGES IN HYALURONAN-CELL INTERACTIONS DURING CONDENSATION

An important stage in differentiation of limb mesoderm is the above-mentioned condensation of cells that precedes final cytodifferentiation of muscle and cartilage. In parallel to the loss of ability to express

hyaluronan-dependent coats and the changes in glycosaminoglycan synthesis
that occur at this stage, membrane-associated binding sites for hyaluronan
are expressed (Knudson and Toole, 1987). Hyaluronan-binding sites are
known, in other systems, to be involved in endocytosis en route to
degradation of hyaluronan (Laurent et al., 1986; McGuire et al., 1987;
McGary et al., 1989), and so it is reasonable to suppose that the appearance
of hyaluronan-binding sites at the time of condensation represents the onset
or increase in receptor-mediated endocytosis of hyaluronan. Decreased
hyaluronan synthesis and coat production together with increased endocytosis
and degradation of hyaluronan would lead to a dramatic reduction in volume
of matrix between the mesodermal cells, thus allowing them to "condense".

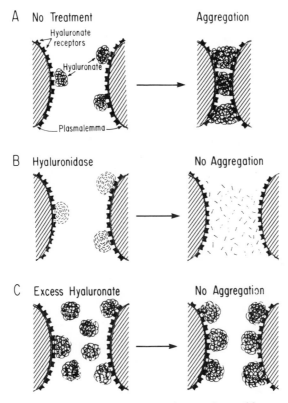

Figure 2. **Hyaluronan-mediated aggregation of cells**. Hyaluronan can
crossbridge cells via a calcium-independent, multivalent
interaction with receptors on adjacent cells. This aggregation
is inhibited by treatment of cells with hyaluronidase (B), or
excess hyaluronan (C); the latter causes saturation of the
receptors, inhibiting crossbridging. If hyaluronan receptors are
present but there is no hyaluronan on the surface of the cells,
as occurs with some lymphocytes (e.g. see Green et al., 1988;
Lesley et al., 1990), then addition of hyaluronan will induce
aggregation. If the receptors are partially occupied, as occurs
with some virally transformed cells (e.g. see Underhill and Toole,
1981; Green et al. 1988) and condensation-stage mesodermal cells
(see Knudson, 1990a), then the cells will spontaneously aggregate.
If the cells have large hyaluronan-dependent coats such as pre-
condensation-stage mesodermal cells (Knudson, 1990a), aggregation
will not occur. (From Toole, 1981, with permission).

217

In addition to the events leading to reduction in matrix volume, condensation may also involve direct cell interactions (e.g. see Knudsen et al., 1990; Bee and von der Mark, 1990). One such interaction is likely to be cross-bridging of cells via multivalent binding of hyaluronan to the cell surface binding sites that are expressed at this stage (Knudson and Toole, 1987). We have shown previously that Ca^{++}-independent self-aggregation of transformed cell lines is due to crossbridging by hyaluronan of binding sites on adjacent cells; removal of cell surface hyaluronan or saturation of the binding sites inhibits this crossbridging (Underhill and Toole, 1981; Underhill, 1982) (see Fig. 2). Knudson (1990a) has recently demonstrated that mesodermal cells from condensation stage limbs, but not cells from pre-condensation limbs, aggregate in vitro via such hyaluronan-mediated crossbridging.

Thus we propose that condensation may be explained in large part by the following events (see Fig. 3A vs B):
1. Receptor-mediated endocytosis of pre-existing pericellular hyaluronan, to permit condensation to begin;
2. Decreased hyaluronan synthesis and cessation of coat assembly, to permit condensation to continue;
3. Hyaluronan-mediated crossbridging of cells, to stabilize the condensate.

HYALURONAN-CELL INTERACTIONS IN CARTILAGE AND MUSCLE DIFFERENTIATION

Recent studies in our laboratory (Turner et al., 1990) have shown that interference with binding of hyaluronan, presumably at the surface of mesodermal cells, blocks chondrogenesis as assessed in the micromass culture

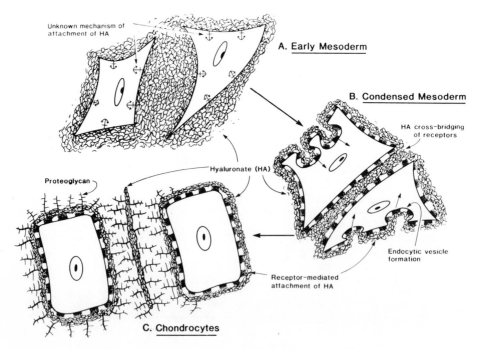

Figure 3. **The hypothesized role of hyaluronan-cell interactions in mesodermal condensation and chondrocyte differentiation.** (From Toole et al., 1989, with permission).

system of Solursh and colleagues (1978). In these studies, chondrogenesis was blocked by treatment of mesodermal cells either with the monoclonal antibody, MAb 4D4, to chick embryo hyaluronan-binding protein or with hyaluronan hexasaccharide that also disrupts binding of polymeric hyaluronan to binding protein. However, further experiments are necessary to distinguish whether hyaluronan and hyaluronan-binding proteins are directly involved in cell aggregation or differentiation in addition to their role in matrix assembly, discussed below.

Many studies indicate that hyaluronan plays an important role in matrix assembly by mature chondrocytes; for example, the well-documented interaction of hyaluronan with link protein and proteoglycan is central to the structure of cartilage matrix (Hascall and Hascall, 1981). Recent evidence indicates that retention of hyaluronan-proteoglycan aggregates in the pericellular matrix by binding to other hyaluronan-binding proteins is also important. As mentioned above, chondrocytes in culture exhibit pericellular coats that have provided a useful in vitro model for studying the pericellular matrix of chondrocytes. Using this system it has been shown, first, that chondrocyte coats are destroyed by treatment with hyaluronan-specific hyaluronidase, and this loss of coats is accompanied by loss of much of the proteoglycan associated with the chondrocyte cell layer, thus indicating that it was retained in the cell layer by interaction with hyaluronan (Goldberg and Toole, 1984; McCarthy and Toole, 1989). Second, production of these pericellular coats by chondrocytes is inhibited by hyaluronan hexasaccharide (Knudson, 1990b) or the antibody to hyaluronan-binding protein, MAb 4D4. The 4D4 antibody reacts strongly with embryonic cartilage but only after treatment of the tissue with hyaluronan-specific hyaluronidase, implying that the binding protein is also occupied by hyaluronan in vivo. The relationship of the proteins recognized by MAb 4D4 with other hyaluronan-binding proteins in cartilage (McCarthy and Toole, 1989; Crossman and Mason, 1990) is not yet known. Third, pericellular coats can be rebuilt around chondrocytes by addition of hyaluronan and proteoglycan to chondrocytes stripped of their coats by treatment with hyaluronidase, but this process is inhibited in the presence of hyaluronan hexasaccharide (Knudson, 1990b). Since hyaluronan hexasaccharides do not inhibit hyaluronan-link protein or hyaluronan-proteoglycan binding (Tengblad, 1981) but do inhibit hyaluronan interaction with the proteins recognized by MAb 4D4 (Banerjee and Toole, 1990), we conclude that hyaluronan-proteoglycan aggregates are retained in the pericellular matrix or at the chondrocyte surface by interaction with the hyaluronan-binding proteins recognized by MAb 4D4 (see Fig. 3C).

Whereas hyaluronan is clearly an important structural component of differentiated cartilage, there is no evidence that hyaluronan plays a central role in the structure of differentiated muscle. However, hyaluronan may be important in early events prior to cytodifferentiation. Exposure of myoblasts to a substratum to which hyaluronan is conjugated inhibits their differentiation and maintains the cells in a proliferative state (Kujawa et al., 1986). It is known that fibroblast-myoblast interactions are important in regulating basement membrane formation in muscle (Sanderson et al., 1986). We have shown in addition that chick embryo muscle fibroblasts in culture exhibit a hyaluronan-rich apical matrix that inhibits myoblast differentiation and a basal matrix that promotes differentiation (Welles and Toole, 1987). These findings suggest that matrix constituents, including hyaluronan, produced by neighboring mesodermal cells or fibroblasts are important in regulating myoblast differentiation. Myoblasts themselves, but not myotubes, exhibit hyaluronan-dependent coats (Orkin et al., 1985) and membrane-associated hyaluronan-binding sites (Knudson and Toole, 1987), characteristics that may be important in their response to cellular interactions prior to differentiation.

REGULATION OF HYALURONAN SYNTHESIS AND COAT FORMATION

A positive relationship may exist between hyaluronan synthesis and cell proliferation (Brecht et al., 1986), two prominent activities during the earliest stages of limb development. Basic FGF is a major growth factor in the early developing limb. The amount of this factor is highest in pre-condensation stage limb buds, when hyaluronan synthesis and cell proliferation are maximal, and decreases during subsequent condensation and differentiation of the mesoderm (Munaim et al., 1988; Seed et al., 1988). Basic FGF stimulates hyaluronan synthesis and hyaluronan-dependent coat formation in cultures of limb mesoderm (Munaim et al., 1990). Thus FGF may be important in the coordinated regulation of cell proliferation, hyaluronan synthesis and pericellular matrix assembly in the pre-condensation limb bud mesoderm. Diffusible growth factors, including FGF, have been implicated in the growth regulatory effects of the apical ectodermal ridge and zone of polarizing activity of the early limb bud (Bell and McLachlan, 1985; Aono and Ide, 1988). Therefore it will be of interest to determine whether these regions of the limb also influence hyaluronan-rich matrix assembly, and whether this potential effect contributes to their morphogenetic roles.

Interaction of limb mesoderm and ectoderm is important in the regulation of morphogenesis and differentiation in the limb bud. The subectodermal mesoderm remains non-condensed during premyogenic and prechondrogenic condensation, retaining mesenchymal morphology and large hyaluronan-rich spaces between cells (Singley and Solursh, 1981). This regionalization of condensed and non-condensed mesoderm appears to be under the control of the ectoderm since factors produced by the ectoderm prevent differentiation of nearby mesoderm to cartilage and cause the retention of mesenchymal characteristics, including hyaluronan-rich intercellular spaces (Solursh et al, 1981; Hurle et al., 1989; Solursh, 1990). We have found that ectodermal cells in culture produce a factor or combination of factors that stimulate hyaluronan synthesis and formation of hyaluronan-dependent coats in condensation-stage mesodermal cells (Knudson and Toole, 1988). Antibody raised against TGF-beta, but not antibodies against several other growth factors, inhibits these effects (Toole et al., 1989). TGF-beta itself stimulates hyaluronan-dependent coat formation and hyaluronan synthesis in these mesodermal cells but, unlike the ectodermal factor, it also stimulates chondroitin sulfate synthesis (Munaim et al., 1990). Carcinoma cells, but not normal adult epithelial cells, also produce a factor that stimulates hyaluronan synthesis in limb mesodermal cells (Knudson and Knudson, 1990). It seems likely that this factor would be similar to that produced by the limb ectoderm but its effect is not blocked by neutralizing antibody to TGF-beta (Knudson et al., 1989). We conclude from these studies that the ectoderm produces a factor that causes subjacent mesoderm to maintain a high rate of hyaluronan synthesis relative to the central condensed mesoderm; however its relationship to TGF-beta requires further investigation.

Thus a combination of regulatory factors, including factors related to FGF and TGF-beta, may regulate regional matrix production during condensation and other morphogenetic events in the limb bud. Cooperative effects between factors related to FGF and TGF-beta have been demonstrated in other morphogenetic systems (see Smith, 1989; Whitman and Melton, 1989), and these agents interact with and influence the composition of extracellular matrices. It is anticipated, then, that future research will continue to focus on their important role in regulation of early morphogenetic events in the limb.

REFERENCES

Aono, H., and Ide, H., 1988, A gradient of responsiveness to the growth-promoting activity of ZPA (zone of polarizing activity) in the chick limb bud, **Develop. Biol.** 128:136-141.

Aruffo, A., Stamenkovic, I., Melnick, M., Underhill, C.B., and Seed, B., 1990, CD44 is the principal cell surface receptor for hyaluronate, **Cell** 61:1303-1313.

Banerjee, S.D., and Toole, B.P., 1990, Monoclonal antibodies to chick embryo hyaluronan-binding proteins, submitted for publication.

Barnhart, B.J., Cox, S.H., and Kraemer, P.M., 1979, Detachment variants of Chinese hamster cells. Hyaluronic acid as a modulator of cell detachment, **Exp. Cell Res.** 119:327-332.

Bee, J.A., and von der Mark, K., 1990, An analysis of chick limb bud intercellular adhesion underlying the establishment of cartilage aggregates in suspension culture, **J. Cell Sci.** 96:527-536.

Bell, K.M., and McLachlan, J.C., 1985, Stimulation of division in mouse 3T3 cells by coculture with embryonic chick limb tissue, **J. Embryol. Exp. Morphol.** 86:219-226.

Brecht, M., Mayer, U., Schlosser, E., and Prehm, P., 1986, Increased hyaluronate synthesis is required for fibroblast detachment and mitosis, **Biochem. J.** 239:445-450.

Comper, W.D., and Laurent, T.C., 1978, Physiological function of connective tissue polysaccharides, **Physiol. Rev.** 58:255-315.

Crossman, M.V., and Mason, R.M., 1990, Purification and characterization of a hyaluronan-binding protein from rat chondrosarcoma, **Biochem. J.** 266:399-406.

Goldberg, R.L., and Toole, B.P., 1984, Pericellular coat of chick embryo chondrocytes: structural role of hyaluronate, **J. Cell Biol.** 99:2114-2122.

Green, S.J., Tarone, G., and Underhill, C.B., 1988, Aggregation of macrophages and fibroblasts is inhibited by a monoclonal antibody to the hyaluronate receptor, **Exp. Cell Res.** 178:224-232.

Hascall, V.C., and Hascall, G.K., 1981, Proteoglycans, in: **Cell Biology of Extracellular Matrix**, E.D. Hay, ed., Plenum, New York, pp 39-63..

Hurle, J.M., Ganan, Y., and Macias, D., 1989, Experimental analysis of the in vivo chondrogenic potential of the interdigital mesenchyme of the chick leg bud subjected to local ectodermal removal, **Develop. Biol.** 132:368-374.

Knudsen, K.A., Myers, L., and McElwee, S.A., 1990, A role for the Ca^{2+}-dependent adhesion molecule, N-cadherin, in myoblast interaction during myogenesis, **Exp. Cell Res.** 188:175-184.

Knudson, C.B., 1990a, Cell-cell adhesion of limb bud mesoderm mediated by hyaluronan, **Development**, in press.

Knudson, C.B., 1990b, Role of hyaluronan in the assembly of chondrocyte pericellular matrix, submitted for publication.

Knudson, C.B., and Knudson, W., 1990, Similar epithelial-stromal interactions in the regulation of hyaluronate production during limb morphogenesis and tumor invasion, **Cancer Lett.** 52:113-122.

Knudson, C.B., and Toole, B.P., 1985, Changes in the pericellular matrix during differentiation of limb bud mesoderm, **Develop. Biol.** 112:308-318.

Knudson, C.B., and Toole, B.P., 1987, Hyaluronate-cell interactions during differentiation of chick embryo limb mesoderm, **Develop. Biol.** 124:82-90.

Knudson, C.B., and Toole, B.P., 1988, Epithelial-mesenchymal interaction in the regulation of hyaluronate production during limb development, **Biochem. Intl.** 17:735-745.

Knudson, W., Biswas, C., Li, X-Q., Nemec, R.E., and Toole, B.P., 1989, The role and regulation of tumour-associated hyaluronan, **Ciba Found. Symp.** 143:150-169.

Kujawa, M.J., Pechak, D.G., Fiszman, M.Y., and Caplan, A.I., 1986, Hyaluronic acid bonded to cell culture surfaces inhibits the program of myogenesis, **Develop. Biol.** 113:10-16.

Laurent, T.C., Fraser, J.R.E., Pertoft, H., and Smedsrod, B., 1986, Binding of hyaluronate and chondroitin sulphate to liver endothelial cells, **Biochem. J.** 234:653-658.

Lesley, J., Schulte, R., and Hyman, R., 1990, Binding of hyaluronic acid to lymphoid cell lines is inhibited by monoclonal antibodies against Pgpl, **Exp. Cell Res.** 187:224-233.

McCarthy, M.T., and Toole, B.P., 1989, Membrane-associated hyaluronate-binding activity of chondrosarcoma chondrocytes, **J. Cell. Physiol.** 141:191-202.

McGary, C.T., Raja, R.H., and Weigel, P.H., 1989, Endocytosis of hyaluronic acid by rat liver endothelial cells. Evidence for receptor recycling, **Biochem. J.** 257:875-884.

McGuire, P.G., Castellot, J.J., and Orkin, R.W., 1987, Size-dependent hyaluronate degradation by cultured cells, **J. Cell. Physiol.** 133:267-276.

Meyer, F.A., 1983, Macromolecular basis of globular protein exclusion and of swelling pressure in loose connective tissue (umbilical cord), **Biochim. Biophys. Acta** 755:388-399.

Miyake, K., Underhill, C.B., Lesley, J., and Kincade, P.W., 1990, Hyaluronate can function as a cell adhesion molecule and CD44 participates in hyaluronate recognition, **J. Exp. Med.** 172:69-75.

Morris-Wiman, J., and Brinkley, L.L., 1990, Changes in mesenchymal cell and hyaluronate distribution correlate with in vivo elevation of the mouse mesencephalic neural folds, **Anat. Rec.** 226:383-395.

Munaim, S.I., Klagsbrun, M., and Toole, B.P., 1988, Developmental changes in fibroblast growth factor in the chicken embryo limb bud, **Proc. Natl. Acad. Sci. USA** 85:8091-8093.

Munaim, S.I., Klagsbrun, M., and Toole, B.P., 1990, Hyaluronan-dependent pericellular coats of chick embryo limb mesodermal cells: induction by basic fibroblast growth factor, **Develop. Biol.**, in press.

Orkin, R.W., Knudson, W., and Toole, B.P., 1985, Loss of hyaluronate-dependent coat during myoblast fusion, **Develop. Biol.** 107:527-530.

Sanderson, R.D., Fitch, J.M., Linsenmayer, T.F., and Mayne, R., 1986, Fibroblasts promote the formation of a continuous basal lamina during myogenesis in vitro, **J. Cell Biol.** 102:740-747.

Schor, S.L., Schor, A.M., Grey, A.M., Chen, J., Rushton, G., Grant, M.E., and Ellis, I., 1989, Mechanism of action of the migration stimulating factor produced by fetal and cancer patient fibroblasts: Effect on hyaluronic acid synthesis, **In Vitro Cell Develop. Biol.** 25:737-746.

Seed, J., Olwin, B.B., and Hauschka, S.D., 1988, Fibroblast growth factor levels in the whole embryo and limb bud during chick development, **Develop. Biol.** 128:50-57.

Singley, C.T., and Solursh, M., 1981, The spatial distribution of hyaluronic acid and mesenchymal condensation in the embryonic chick wing, **Develop. Biol.** 84:102-120.

Smith, J.C., 1989, Mesoderm induction and mesoderm-inducing factors in early amphibian development, **Development** 105:665-677.

Solursh, M., 1990, The role of extracellular matrix molecules in early limb development, **Seminars Develop. Biol.** 1:45-53.

Solursh, M., Ahrens, P.B., and Reiter, R.S., 1978, A tissue culture analysis of the steps in limb chondrogenesis, **In Vitro** 14:51-61.

Solursh, M., Singley, C.T., and Reiter, R.S., 1981, The influence of epithelia on cartilage and loose connective tissue formation by limb mesenchyme cultures, **Develop. Biol.** 86:471-482.

Tengblad, A., 1981, A comparative study of the binding of cartilage link protein and the hyaluronate-binding region of the cartilage proteoglycan to hyaluronate-substituted Sepharose gel, **Biochem. J.** 199:297-305.

Toole, B.P., 1981, Glycosaminoglycans in morphogenesis, in **Cell Biology of Extracellular Matrix**, E.D. Hay, ed., Plenum, New York, pp. 259-294.

Toole, B.P., 1990, Hyaluronan and its binding proteins, the hyaladherins, **Current Opinion Cell. Biol.** 2: in press.

Toole, B.P., Munaim, S.I., Welles, S., and Knudson, C.B., 1989, Hyaluronate-cell interactions and growth factor regulation of hyaluronate synthesis during limb development, **Ciba Found. Symp.** 143:138-149.

Turley, E.A., Bowman, P., and Kytryk, M.A., 1985, Effects of hyaluronate and hyaluronate binding proteins on cell motile and contact behavior, **J. Cell Sci.** 78:133-145.

Turner, R., Banerjee, S.D., and Toole, B.P., 1990, Role of hyaluronan-binding protein in chondrogenesis, submitted for publication.

Underhill, C.B., 1982, Interaction of hyaluronate with the surface of simian virus 40-transformed 3T3 cells: Aggregation and binding studies, **J. Cell Sci.** 56:177-189.

Underhill, C.B., and Toole, B.P., 1981, Receptors for hyaluronate on the surfaces of parent and virus-transformed cell lines. Binding and aggregation studies, **Exp. Cell Res.** 131:419-423.

Underhill, C.B., Green, S.J., Comoglio, P.M., and Tarone, G., 1987, The hyaluronate receptor is identical to a glycoprotein of M_r 85,000 (gp85) as shown by a monoclonal antibody that interferes with binding activity, **J. Biol. Chem.** 262:13142-13146.

Welles, S., and Toole, B.P., 1987, Muscle fibroblast extracellular matrix modulates myoblast differentiation in vitro, **J. Cell Biol.** 105:139a (abstract).

Whitman, M., and Melton, D.A., 1989, Growth factors in early embryogenesis, **Annu. Rev Cell. Biol.** 5:93-117.

Yu, Q., Banerjee, S.D., and Toole, B.P., 1990, Inhibition of pericellular coat production by antibody to hyaluronan-binding protein, submitted for publication.

ROLE OF THE TRANSFORMING GROWTH FACTOR-β (TGF-β) FAMILY, EXTRACELLULAR MATRIX, AND GAP JUNCTIONAL COMMUNICATION IN LIMB CARTILAGE DIFFERENTIATION

Robert A. Kosher, Eileen F. Roark, Stephen E. Gould, and Caroline N. D. Coelho

Department of Anatomy
University of Connecticut Health Center
Farmington, CT 06030

INTRODUCTION

The differentiation of limb mesenchymal cells into chondrocytes involves a sequential series of regulatory events mediated in part by extracellular matrix macromolecules, peptide growth factors, cell-cell and cell-matrix interactions, cytoskeletal reorganization, and intracellular second messengers such as cAMP. In the present manuscript we describe some of our current studies on the role of and the possible relationship between the TGF-β family of growth factors and extracellular matrix and cell surface molecules including the membrane-intercalated proteoglycan, syndecan in the regulation of chick limb cartilage differentiation. We also describe studies indicating that intercellular communication via gap junctions may be involved in the regulation of chondrogenesis.

STUDIES ON THE ROLE OF THE TGF-β FAMILY IN LIMB CHONDROGENESIS

TGF-β's constitute a large family of multifunctional peptide growth factors that regulate a variety of biological processes (Roberts et al., 1990). TGF-β1 and TGF-β2 are potent promoters of the chondrogenic differentiation of embryonic chick limb mesenchymal cells in vitro (Kulyk et al., 1989a). In fact, both of these forms of TGF-β promote chondrogenesis and cartilage-specific gene expression in low density subconfluent spot cultures of limb mesenchymal cells, which are situations in which little, or no cartilage differentiation normally occurs (Kulyk et al., 1989a). Only brief exposure to the growth factors at the initiation of culture is sufficient to stimulate chondrogenesis, suggesting they are acting at an early step in the process. We have therefore suggested that some form of TGF-β may be involved in regulating the onset of limb cartilage differentiation (Kulyk et al., 1989a).

Developmental Patterning of the Vertebrate Limb
Edited by J.R. Hinchliffe *et al.*, Plenum Press, New York, 1991

To determine if TGF-β's are expressed in a temporal and spatial fashion consistent with their role in regulating chondrogenesis, we have used cloned chick cDNA probes to examine the expression of TGF-β1 and TGF-β3 mRNAs during limb development and during the uniform, progressive chondrogenic differentiation distal subridge mesenchymal cells undergo in micromass culture (Roark et al., 1990). Northern blot analyses have shown that TGF-β3 mRNA is expressed at several early stages (stages 21-28) of chick limb development in which chondrogenesis is being initiated (Roark et al., 1990). Furthermore, the TGF-β3 gene is expressed at relatively high levels <u>in vitro</u> just before the onset of the critical cellular condensation process that characterizes the initiation of overt chondrogenesis (Roark et al., 1990). After the onset of chondrogenesis, TGF-β3 expression declines dramatically (Roark et al., 1990). In contrast to TGF-β3, TGF-β1 mRNA is not detectable either during early limb development or during chondrogenesis <u>in vitro</u> (Roark et al., 1990).

Since TGF-β3 is expressed in a temporal manner consistent with its involvement in regulating the onset of chondrogenesis, we have examined the effect of exogenous TGF-β3 on limb chondrogenic differentiation <u>in vitro</u>. TGF-β3 does indeed promote chondrogenesis, stimulating the accumulation of cartilage matrix, sulfated GAGs, and cartilage-specific mRNAs by limb mesenchymal cells in micromass culture (Roark et al., 1990). In order to determine if the expression of endogenous TGF-β3 is required for chondrogenic differentiation, in a preliminary experiment we have prepared an antisense oligonucleotide complementary to a 15-mer sequence of chick TGF-β3 mRNA located just 3' to the methionine start codon, and have examined the effect of this antisense oligonucleotide and the corresponding sense oligonucleotide on <u>in vitro</u> limb chondrogenesis (Roark and Kosher, in progress). Cultures treated with the anti-sense oligonucleotide against TGF-β3 exhibited a dramatic decrease in type II collagen mRNA accumulation compared to cultures treated with the sense oligonucleotide, and no cartilage proteoglycan core protein mRNA was detectable in the anti-sense oligonucleotide-treated cultures. Although we have not yet determined if this anti-sense oligonucleotide specifically inhibits endogenous TGF-β3 expression, this preliminary experiment is consistent with the possibility that TGF-β3 plays an important role in the onset of limb chondrogenesis.

Another member of the TGF-β family of growth factors that may be involved in limb chondrogenesis is bone morphogenetic protein-2A (BMP-2A). BMP-2A is a component of a protein extract derived from demineralized bone that promotes cartilage formation (and subsequently bone formation) when implanted into extraskeletal ectopic sites <u>in vivo</u> (Wozney et al., 1988). Lyons et al. (1989) have shown by <u>in situ</u> hybridization that BMP-2A transcripts are present in the condensing precartilaginous mesenchymal tissue of the mouse embryo that will give rise to the cartilaginous ribs and vertebrae. Therefore, we have examined the effect of exogenous BMP-2A on <u>in vitro</u> chick limb chondrogenesis, and have found it to be a potent stimulator of the process (Roark and Kosher, in progress). It is of interest that our studies (Roark et al., 1990) indicate that TGF-β3 is expressed at high levels just prior to the onset of condensation, and the studies of Lyons et al. (1989) indicate that BMP-2A is expressed at high amounts during the condensation phase of chondrogenesis. This suggests the possibility that the sequential expression of TGF-β3 and BMP-2A may be involved in regulating chondrogenesis, and raises the possibility that the expression of BMP-2A might be regulated by TGF-β3. It should be noted that in several systems various forms of TGF-β regulate the expression of genes for other forms of the molecule (Bascom et al., 1989).

The mechanism by which TGF-β's might regulate chondrogenesis is not known. The onset of the critical condensation phase of chondrogenesis appears to be mediated by interactions between several extracellular matrix molecules including fibronectin (see below), heparan sulfate proteoglycans (see below), tenascin (Mackie et al., 1987), and the mesenchymal chondroitin sulfate proteoglycan, PG-M (Kimata et al., 1986), and in various systems TGF-β's stimulate the expression of most of the matrix macromolecules that have been implicated in the condensation phase of chondrogenesis (Ignotz and Massague, 1986; Pearson et al., 1988; Border et al., 1990). Since TGF-β3 is expressed at high levels just before condensation, we suggest that it might regulate chondrogenesis by controlling the production of, or the levels of receptors for, the extracellular matrix molecules that are involved in mediating the crucial cell-cell or cell-matrix interactions that trigger chondrogenesis during condensation.

ROLE OF MATRIX MOLECULES INCLUDING SYNDECAN IN CHONDROGENESIS

As mentioned above, the onset of limb chondrogenesis is characterized by a transient cellular condensation process in which prechondrogenic mesenchymal cells become closely apposed to one another prior to depositing cartilage matrix, and during this condensation process intimate cell-cell and/or cell matrix interactions occur that are necessary to trigger chondrogenic differentiation. The onset of the critical condensation process may be initiated, at least in part, by a progressive decline in the accumulation of extracellular hyaluronate (Kosher et a., 1981; Knudson and Toole, 1985). A matrix molecule that appears to play a particularly important role in regulating condensation is the adhesive glycoprotein, fibronectin (FN). A striking transient increase in FN gene expression occurs during condensation and the onset of chondrogenesis in vivo and in vitro (Kulyk et al., 1989b); large amounts of FN are present along the surfaces of the closely apposed cells during the process (Kosher et al., 1982; Tomasek et al., 1982); and, monoclonal antibodies against the amino-terminal heparin-binding domain of FN impair the formation of prechondrogenic aggregates in vitro (Frenz et al., 1989). The latter observation suggests that an adhesive interaction between FN and heparin-like cell surface molecules may be involved in condensation formation, particularly since heparinase treatment also impairs prechondrogenic aggregate formation in vitro (Frenz et al., 1989).

We have recently performed a series of experiments to determine if limb mesenchymal cells undergoing chondrogenesis in vitro synthesize heparin-like cell surface molecules (Gould and Kosher, 1990). We have found that during condensation and the onset of chondrogenesis in vitro limb mesenchymal cells synthesize a hydrophobic PG with high affinity for Octyl-Sepharose that elutes as a single peak with high concentrations of Triton X-100 (Gould and Kosher, 1990). The GAG composition of this hydrophobic PG is 75% heparan sulfate (HS) and 25% chondroitin sulfate (CS) (Gould and Kosher, 1990). This hydrophobic HS/CS PG thus resembles syndecan, an integral membrane PG that links the cytoskeleton to extracellular matrix components including FN in several cell types (Bernfield and Sanderson, 1990).

To more critically determine if syndecan is expressed during chick limb development, we have screened a cDNA library prepared from poly(A)+ RNA isolated from stage 22, 26, and 28 chick limb buds with a mouse syndecan cDNA (Saunders et al., 1989) obtained from Dr. Merton Bernfield. We have isolated and partially sequenced a chicken cDNA clone which encodes a transmembrane and cytoplasmic domain that possesses a high degree of sequence similarity at the nucleotide and amino acid levels to mouse syndecan cDNA, and like mouse syndecan, encodes a protease-sensitive dibasic site adjacent to the extracellular face of the transmembrane domain and GAG attachment sites in its ecto-domain (Gould and Kosher, 1990). The chicken cDNA clone we have isolated thus appears to be a chicken cognate of mouse syndecan cDNA. Using this chicken cDNA as a probe, we have found that a striking progressive increase in the accumulation of "syndecan" mRNA occurs during condensation and the initiation of overt chondrogenesis in vitro (Gould and Kosher, 1990). This observation suggests the possible involvement of this molecule in the onset of chondrogenesis. Syndecan has the potential to play a role in this process by promoting cell aggregation perhaps in conjunction with FN (Frenz et al., 1989), and, by virtue of its intracellular link to the cytoskeleton, may also be involved in the cytoskeletal reorganization that may be involved in regulating chondrogenesis (Zanetti and Solursh, 1984). Syndecan also has the potential to bind, sequester, and modulate the activity of peptide growth factors (see Bernfield and Sanderson, 1990) that may be involved in regulating chondrogenesis.

It should be noted that a recent immunohistochemical study of Solursh et al. (1990) indicates that syndecan is expressed in the prechondrogenic distal mesenchyme of mouse limb buds. However, the expression of immunohistochemically-detectable syndecan essentially ceases in the central core of the limb and in high density micromass cultures prior to chondrogenic differentiation (Solursh et al., 1990). This observation contrasts to our studies which indicate that chick syndecan mRNA is expressed in high amounts during the initiation of chondrogenesis. Resolution of this apparent discrepancy will require further experimentation.

GAP JUNCTIONAL COMMUNICATION DURING LIMB CHONDROGENESIS

Regulatory events occurring during the critical condensation phase of chondrogenesis result in the expression of genes for cartilage-specific matrix molecules including type II collagen, cartilage proteoglycan core protein, and type IX collagen (Kosher et al., 1986a,b; Mallein-Gerin et al., 1988; Nah et al., 1988; Kulyk et al., 1990). A transient increase in cellular cAMP content occurs during condensation, and it has been suggested cAMP may be involved in triggering cartilage-specific gene expression (Kosher and Savage, 1980; Solursh et al., 1981; Kosher et al., 1986a; Leonard and Newman, 1987). It is of interest that cAMP stimulates the chondrogenesis of limb mesenchymal cells cultured at high (supraconfluent) densities, but fails to promote the chondrogenesis of cells cultured at low (subconfluent) densities (Rodgers et al., 1989). Therefore, we have suggested that cAMP might regulate chondrogenesis at least in part by facilitating cell-cell communication, perhaps gap junctional communication, during condensation and the onset of chondrogenesis (Rodgers et al., 1989). These observations combined with the fact that gap junctions are present between prechondrogenic mesenchymal cells in the condensing central core of the chick limb bud (Kelley and Fallon, 1978, 1983) prompted us to determine if gap junctional communication actually occurs during _in vitro_ chondrogenesis (Coelho and Kosher, 1990).

We have found that extensive cell-cell communication via gap junctions as assayed by the intercellular transfer of lucifer yellow dye occurs during condensation and the onset of chondrogenesis in high density micromass cultures prepared from the uniform population of chondrogenic precursor cells comprising the distal subridge region of stage wing buds (Coelho and Kosher, 1990). Furthermore, in heterogeneous micromass cultures prepared from the mesodermal cells of whole stage 23/24 limb buds, extensive gap junctional communication is limited to differentiating cartilage cells, while the nonchondrogenic cells of the cultures that are differentiating into the connective tissue lineage exhibit little, or no intercellular communication via gap junctions (Coelho and Kosher, 1990). These observations suggest the possibility that cell-cell communication via gap junctions may be involved in regulating limb cartilage differentiation.

We are currently investigating the relationship between cAMP and the gap junctional communication that occurs at the onset of limb chondrogenesis. In particular, we are testing the possibilities that cAMP might regulate chondrogenesis by promoting gap junction formation and/or gap junctional communication during condensation, or that gap junctional communication at the onset of chondrogenesis might facilitate direct intercellular of cAMP or some other low molecular weight molecule that regulates cartilage-specific gene expression.

SUMMARY

The chondrogenic differentiation of limb mesenchymal cells involves a sequential and complex series of regulatory events and transitions mediated by interactions between peptide growth factors, extracellular matrix macromolecules, cell surface receptors, the cytoskeleton, gap junctions, and intracellular second messengers. A hypothetical scheme outlining the possible regulatory transitions limb mesenchymal cells undergo during the course of their chondrogenic differentiation is presented in Fig. 1.

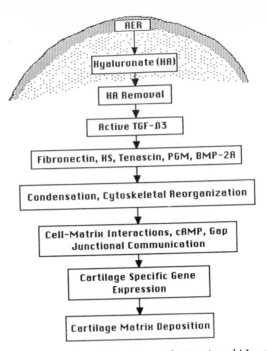

Fig. 1. A hypothetical scheme consistent with experimental
data outlining a possible sequence of regulatory events and
transitions involved in the differentiation of limb
mesenchymal cells into chondrocytes. The AER via molecules
(growth factors?) it synthesizes and secretes maintains the
mesenchymal cells of the limb subjacent to it in a labile,
undifferentiated condition (Kosher et al., 1979), perhaps in
part by causing the cells to synthesize and secrete
extracellular hyaluronate which maintains the cells spatially
separated thus preventing the cell-cell and cell-matrix
interactions necessary for subsequent chondrogenesis (Kosher
et al., 1981; Knudson and Toole, 1985). As cells leave the
influence of the AER, extracellular hyaluronate is removed
(Knudson and Toole, 1985), and an active form of the peptide
growth factor, TGF-β3 is expressed (Roark et al., 1990). TGF-
β3 promotes the production of extracellular matrix and cell
surface molecules including fibronectin (Kulyk et al., 1989b),
heparin-like cell surface molecules (Frenz et al., 1989) such
as syndecan (Gould and Kosher, 1990), tenascin (Mackie et al.,
1987), and PG-M (Kimata et al., 1986); and, TGF-β3 might also
promote expression of other growth factors such as BMP-2A
(Lyons et al., 1989; Roark and Kosher, in progress). An
interaction between these matrix, cell surface, and growth
factor molecules promotes condensation formation, cytoskeletal
reorganization (Zanetti and Solursh, 1984), and the critical
cell-cell and cell-matrix interactions that trigger
chondrogenic differentiation. During condensation cellular
cAMP levels increase and extensive cell-cell communication via
gap junctions occurs (Coelho and Kosher, 1990). These and
perhaps other regulatory events trigger the expression of
genes for cartilage-specific matrix proteins including type II
collagen (Kosher et al., 1986a), cartilage proteoglycan core
protein (Kosher et al., 1986b), and type IX collagen (Kulyk et
al., 1990). A somewhat similar sequence of events has been
proposed by Newman (1988).

This research was supported by NIH grants HD22896 and HD22610.

REFERENCES

Bascom, C. C., Wolfshohl, J. R., Coffey, Jr., R. J., Madisen, L., Webb, N. R., Purchio, A. R., Derynck, R., and Moses, H. L., 1989, Complex regulation of transforming growth factor β1, β2, and β3 mRNA expression in mouse fibroblasts and keratinocytes by transforming growth factor β1 and β2, Mol. Cell Biol., 9: 5508.

Bernfield, M. and Sanderson, R. D., 1990, Syndecan, a developmentally regulated cell surface proteoglycan that binds extracellular matrix and growth factors, Phil. Trans. R. Soc. Lond. B, 327: 171.

Border, W. A., Okuda, S., Languino, L. R., and Ruoslahti, E., 1990, Transforming growth factor-β regulates production of proteoglycans by mesangial cells, Kidney Int., 37: 689.

Coelho, C. N. D. and Kosher, R. A., 1990, Gap junctional communication during limb cartilage differentiation, Dev. Biol., submitted for publication.

Frenz, D. A., Jaikaria, N. S., and Newman, S. A., 1989, The mechanism of precartilage mesenchymal condensation: a major role for interaction of the cell surface with the amino-terminal heparin-binding domain of fibronectin, Dev. Biol., 136: 97.

Gould, S. E., Upholt, W. B., and Kosher, R. A., 1990, Molecular cloning of chicken syndecan and expression of the syndecan gene during chick limb cartilage differentiation, in preparation.

Ignotz, R. A. and Massague, J., 1986, Transforming growth factor-β stimulates the expression of fibronectin and collagen and their incorporation into the extracellular matrix, J. Biol. Chem., 261: 4337.

Kelley, R. O. and Fallon, J. F., 1978, Identification and distribution of gap junctions in the mesoderm of the developing chick limb bud, J. Embryol. Exp. Morph., 46: 99.

Kelley, R. O. and Fallon, J. F., 1983, A freeze-fracture and morphometric analysis of gap junctions of limb bud cells: initial studies on a possible mechanism for morphogenetic signalling during development, in: "Limb Development and Regeneration, Part A," J. F. Fallon and A. I. Caplan, eds., Allen R. Liss, Inc., New York.

Kimata, K., Oike, Y., Tani, K., Shinomura, T., Yamagata, M., Uritani, M., and Suzuki, S., 1986, A large chondroitin sulfate proteoglycan (PG-M) synthesized before chondrogenesis in the limb bud of chick embryo, J. Biol. Chem., 261: 13517.

Kosher, R. A., Gay, S. W., Kamanitz, J. R., Kulyk, W. M., Rodgers, B. J., Sai, S., Tanaka, T., and Tanzer, M. L., 1986a, Cartilage proteoglycan core protein gene expression during limb cartilage differentiation, Dev. Biol., 118: 112.

Kosher, R. A., Kulyk, W. M., and Gay, S. W., 1986b, Collagen gene expression during limb cartilage differentiation, J. Cell Biol., 102: 1151.

Kosher, R. A. and Savage, M. P., 1980, Studies on the possible role of cyclic AMP in limb morphogenesis and differentiation, J. Embryol. Exp. Morph., 56: 91.

Kosher, R. A., Savage, M. P., and Chan, S.-C., 1979, In vitro studies on the morphogenesis and differentiation of the mesoderm subjacent to the apical ectodermal ridge of the embryonic chick limb bud, J. Embryol. Exp. Morph., 50: 75.

Kosher, R. A., Savage, M. P., and Walker, K. H., 1981, A gradation of hyaluronate accumulation along the proximodistal axis of the embryonic chick limb bud, J. Embryol. Exp. Morph., 63: 85.

Kosher, R. A., Walker, K. H., and Ledger, P. W., Temporal and spatial distribution of fibronectin during development of the embryonic chick limb bud, Cell Differ., 11: 217.

Knudson, C. B. and Toole, B. P., 1985, Changes in the pericellular matrix during differentiation of limb bud mesoderm, Dev. Biol., 112: 308.

Kulyk, W. M., Coelho, C. N. D., and Kosher, R. A., 1990, Type IX collagen gene expression during limb cartilage differentiation, in preparation.

Kulyk, W. M., Rodgers, B. J., Greer, K., and Kosher, R. A., 1989a, Promotion of embryonic chick limb cartilage differentiation by transforming growth factor-β, Dev. Biol., 135: 424.

Kulyk, W. M., Upholt, W. B., and Kosher, R. A., 1989b, Fibronectin gene expression during limb cartilage differentiation, Development, 106: 449.

Leonard, C. M. and Newman, S. A., 1987, Nuclear events during early chondrogenesis: phosphorylation of the precartilage 35.5-kDa domain-specific chromatin protein and its regulation by cyclic AMP, Dev. Biol., 120: 92.

Lyons, K. M., Pelton, R. W., and Hogan, B. L. M., Patterns of expression of murine Vgr-1 and BMP-2a RNA suggest that transforming growth factor-β-like genes coordinately regulate aspects of embryonic development, Genes and Dev., 3: 1657.

Mackie, E. J., Thesleff, I., and Chiquet-Ehrismann, R., 1987, Tenascin is associated with chondrogenic and osteogenic differentiation in vivo and promotes chondrogenesis in vitro, J. Cell Biol., 105: 2569.

Mallein-Gerin, F., Kosher, R. A., Upholt, W. B., and Tanzer, M. L., 1988, Temporal and spatial analysis of cartilage proteoglycan core protein gene expression during limb development by in situ hybridization, Dev. Biol., 126: 337.

Nah, H.-D., Rodgers, B. J., Kulyk, W. M., Kream, B. E., Kosher, R. A., and Upholt, W. B., 1988, In situ hybridization analysis of the expression of the type II collagen gene in the developing chicken limb bud, Collagen Rel. Res., 8: 277.

Newman, S. A., 1988, Lineage and pattern in the developing vertebrate limb, Trends in Genetics, 4: 329.

Pearson, C. A., Pearson, D., Shibahara, S., Hofsteenge, J., and Chiquet-Ehrismann, R., Tenascin: cDNA cloning and induction by TGF-β, EMBO J., 7: 2677.

Roark, E. F., Greer, K., and Kosher, R. A., 1990, Studies on the possible role of TGF-β3 in the regulation of limb cartilage differentiation, in preparation.

Roberts, A. B., Flanders, K. C., Heine, U. I., Jakowlew, S., Kondaiah, P., Kim, S.-J., and Sporn, M. B., 1990, Transforming growth factor-β: multifunctional regulator of differentiation and development, Phil. Trans. R. Soc. Lond. B, 327: 145.

Rodgers, B. J., Kulyk, W. M., and Kosher, R. A., 1989, Stimulation of limb cartilage differentiation by cyclic AMP is dependent on cell density, Cell Differ. Dev., 28: 179.

Solursh, M., Reiter, R. S., Ahrens, P. B., and Vertel, B. M.,

1981, Stage- and position-related changes in chondrogenic response of chick embryonic wing mesenchyme to treatment with dibutyryl cyclic AMP, <u>Dev. Biol.</u>, 83: 9.

Solursh, M., Reiter, R. S., Jensen, K. L., Kato, M., and Bernfield, M., Transient expression of a cell surface heparan sulfate proteoglycan (syndecan) during limb development, <u>Dev. Biol.</u>, 140: 83.

Tomasek, J. J., Mazurkiewicz, J. E., and Newman, S. A., 1982, Nonuniform distribution of fibronectin during avian limb development, <u>Dev. Biol.</u>, 90: 118.

Wozney, J. M., Rosen, V., Celeste, A. J., Mitsock, L. M., Whitters, M . J., Kriz, R. W., Hewick, R. M., and Wang, E. A., 1988, Novel regulators of bone formation: molecular clones and activities, <u>Science</u>, 242: 1528.

Zanetti, N. C. and Solursh, M., 1984, Induction of chondrogenesis in limb mesenchymal cultures by disruption of the actin cytoskeleton, <u>J. Cell Biol.</u>, 99: 115.

VASCULAR INVOLVEMENT IN CARTILAGE AND BONE DEVELOPMENT

David J. Wilson

Department of Dental Surgery, School of Clinical Dentistry
The Queen's University of Belfast, Northern Ireland

Introduction

The very early embryonic vascular system develops from mesodermal
blood islands and is laid down prior to the onset of circulation.
According to Risau (1990), all subsequent angiogenic events seem to be
regulated by the capillary requirements of surrounding tissues. Whether
such a view holds true for the development of cartilage (and subsequently
bone) during early limb development is controversial, with two opposing
views being expressed. The first suggests that the developing
vasculature dictates the spatial pattern of cell differentiation in the
limb, whilst the other suggests that the vasculature is passive, simply
responding to the requirements of tissues as they differentiate. In this
paper these two views of the limb vasculature will be assessed with
reference to factors which may govern blood vessel growth in the
developing limb.

Vascular pattern and differentiation

In the chick, the early limb bud appears as a bulge in the flank at
stage 16-17 (Hamburger and Hamilton, 1951) and is vascularised by a
capillary bed originating from the dorsal aorta, and the primary body
wall plexus. As limb outgrowth proceeds there is an increase in the size
of the bed with new capillaries sprouting form existing ones (Wilson,
1983). By stage 21-22 a basic vascular pattern has become established, a
single central vessel (the primary subclavian artery) carries blood into
the limb, it radiates into a network of smaller bore capillaries which in
turn drain into two marginal vessels at the anterior and posterior
borders of the limb (Caplan and Koutroupas, 1973; Drushel, Pechak and

Caplan, 1985; Feinberg, Latker and Beebe, 1986; Wilson and Orr-Urtereger, 1986). By stage 22-3, the proximal part of the subclavian artery has come to lie more posteriorly and ventrally, and this coincides with the onset of formation of the humerus (Wilson, 1986). The vascular pattern is essentially fountain-like, with the terminal capillary spray supplying the dorsal, ventral and distal parts of the limb, and draining into the now larger and discrete anterior and posterior marginal veins. By stage 25, the precursor of the adult wing pattern has been established (Drushel, Pechak and Caplan, 1985) and it is interesting to note that the vascular pattern at this stage is not significantly different from the pattern seen in the hatchling chick (Levinsohn, Packard, West and Hootnick, 1984).

In terms of differentiation, there is general agreement that the limb vessels remain little more than capillaries during the early stages of cartilage and muscle cytodifferentiation, and during the establishment of the vascular pattern (Drushel et al, 1985; Wilson and Orr-Urtereger, 1986). Vascular differentiation starts in the proximal portion of the subclavian at stage 26-7, when mesenchymal apposition to the endothelial cells increases, the basal lamina becomes more conspicuous and the nuclear portion of the endothelial cells protrude into the lumen. The changes in the artery become more pronounced at later stages, but are made more conspicuous by the lack of change in the other vessels of the limb, which even at stage 35 still retain an 'early' morphology. Pericyte-type cell apposition to the future arterial vessels coincided with a decrease in tritiated thymidine labeling (Wilson and Orr-Urtereger, 1986). This was interpreted at the time as possible 'contact inhibition' of endothelial cell mitosis. More recent evidence suggests that pericytes can activate latent TGF-beta upon cell-cell contact with endothelial cells, and thus may act as a short range mechanism for inhibition of endothelial cell proliferation (Antonelli-Orlidge, Saunders, Smith and D'Amore, 1989).

Vascular patterning and onset of skeletal development

It has been argued that during early limb development there is a vascular pre-pattern which has a causal relationship with the spatial differentiation of cartilage and muscle in the limb (Caplan and Koutroupas, 1973; Hunter and Caplan, 1983; Jargiello and Caplan, 1983; Drushel et al. 1985). Central to this argument is the idea that a vascular pre-pattern predates overt differentiation of cartilage and muscle. Caplan and Koutroupas (1973) suggested that differential vascularisation of the stage 23-4 wing bud establishes gradients of low molecular weight metabolic substances which are responsible for

236

patterning an inner core of cartilage and outer blocks of muscle. These authors based much of their argument on the effects in vitro of nicotinamide on cartilage and muscle differentiation, and the effects of nicotinamide analogues (especially 3-acetyl-pyridine) in ovo, which caused muscle hypoplasia in the developing limbs. However, the interpretation of their findings appears to be inconsistant with the results of McLachlan, Bateman and Wolpert (1976), who found that 3-AP induced muscle hypoplasia was preceded by complete peripheral nerve destruction - a more likely cause of the muscle degeneration. Similarly, Hwang, Byrne and Kitos (1988) have shown that NAD content of limb bud cells in vitro did not vary greatly as a function of oxygen availability, and that high oxygen tension is compatible with some aspects of chondrogenic differentiation of limb bud cells.

Despite these apparent inconsistencies, the idea that a vascular pre-pattern governs the establishment of the skeletal pattern has still been advanced (Feinberg, Latker and Beebe, 1986; Hallman, Feinberg, Latker, Sasse and Risau, 1987), with the central issue being whether regression of blood vessels precedes cartilage differentiation. These studies have employed labelling techniques such as radiosulphate incorporation or novel monoclonal antibodies to identify the onset of cartilage matrix synthesis with correlative vascular labelling. The results of the first of these studies are, as the authors admit, only suggestive of a cause and effect relationship between vascular regression and chondrogenesis. In the other study, the antibody used 'against a cartilage-specific protein which is expressed by chondrocytes earlier than collagen type-II and keratan sulphate' is questionable. For example, the temporal correlation between avascular zones and antibody staining pattern at stage 27 in the radius and ulna is not surprising: simple Alcian green staining of whole mount limbs at this stage clearly shows that chondrogenesis in these elements is well advanced at this stage (Summerbell, 1976).

The alternative view, that the development of avascularity seen during cartilage differentiation can been attributed to the simple exclusion of the vessels by the process of condensation formation (Wilson, 1986), cannot be ruled out as a possible mechanism. Currently ultrastructural examinination of the relationship between limb blood vessels and the process of mesenchymal condensation is underway using Ruthenium Hexammine Trichloride-fixed limb tissue (Wilson and Archer, in preparation). This fixative improves preservation of extracellular matrix components, and preliminary results show evidence of compression/ occlusion of vessels in parts of the mesenchyme where condensation is occuring (Fig.1a,b). However, the absence of blood vessels from the precartilage condensations would not provide a clear answer to the issue of vascular cause or effect in patterning cartilage differentiation.

Growth factors such as EGF that have been shown to promote embryonic angiogenesis (Stewart, Nelson and Wilson, 1989), together with antiangiogenic agents such as retinoids (Oikawa, Hirotani, Nakamura, Shudo, Hiragun and Iwaguchi, 1989) may prove decisive in resolving this controversy. Application of such molecules so as to locally increase or decrease the limb vasculature may provide a definitive answer to the question of whether there is a causal link between the vasculature and the pattern of cartilage/muscle differentiation.

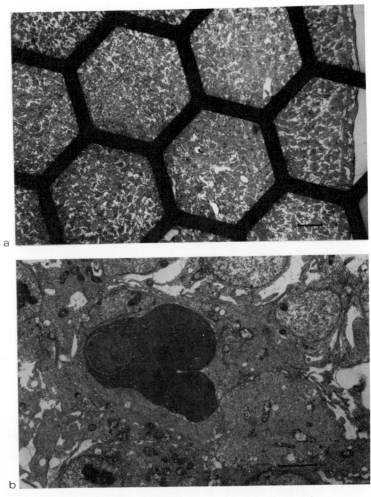

Figure 1. (a) Low power TEM of stage 25 wing bud showing mesenchymal condensation. Bar represents 20μm. (b) Vascular occlusion apparent at the edge of the condensation. Bar represents 2μm.

Vascular involvement in bone development

Recently several elegant accounts of the relationship between the limb vasculature and endochondral ossification have appeared (Pechak, Kujawa and Caplan, 1986a,b; Caplan and Pechak, 1987). Vascular invasion of the cartilage rudiment first occurs in the mid-diaphyseal region. It is coincidental with the differentiation of osteoblasts on the inner aspect of the perichondrium and the appearance of osteoid adjacent to hypertrophying cartilage.

What remains unclear is the signalling mechanism responsible for the blood vessel invasion. It has been proposed that the periosteum, when it forms, acts as a diffusion barrier that restricts nutrient flow from the vasculature to the cartilage core, and that such a nutrient deprivation could result in chondrocyte hypertrophy, with chondrocytes being 'starved' to death (Pechak et al., 1986a). As a result, cartilage secretion of anti-angiogenesis factors would stop thus allowing vascular invasion of the cartilage to occur. However work on angiogenesis associated with the devleopment of intramembranous bone (which forms directly from a mesenchymal condensation, without a cartilage model phase) suggests that differentiating osteoblasts may actively recruit blood vessels by producing angiogenic factors (Thompson, Owens and Wilson, 1989) in a similar manner to developing brain (Risau, 1986) and kidney (Risau and Ekblom, 1986). If one accepts this suggestion, it is possible to argue that the same stimulus for vascular invasion is present during both endochondral and intramembranous bone development. In endochondral ossification, combined osteoblast production of an angiogenesis factor within the forming periosteum and cessation of antiangiogenic factor secretion during cartilage hypertrophy/death results in vascular invasion, whilst in intramembranous ossification direct differentiation of mesenchymal cells into osteoblasts leads to the secretion of angiogenic factors which promote vascular invasion of the condensation. This hypothesis is currently being examined using both in vivo angiogenesis assays (Stewart, Nelson and Wilson, 1990) and in vitro endothelial migration assays.

Conclusion

Angiogenesis is directed. In other words, new blood vessel growth occurs in response to a stimulus. It is a complex multistep process which occurs in both physiological and pathological situations, for example angiogenesis associated with wound healing (Knighton et al, 1983) and neoplasia (Folkman, 1974). There are a wide range of factors which either directly or indirectly induce or inhibit angiogenesis, some of which have demonstrable effects on embryonic angiogenesis. Thus it seems to be reasonable to accept that angiogenic events in the limb are regulated by the demands of the surrounding tissue, and that the vessels are not in themselves directing that demand.

References

Antonelli-Orlidge, A., Saunders, K. B., Smith, S. R., and D'Amore, P. A., 1989, An activated form of transforming growth factor-beta is produced by cultures of endothelial cells and pericytes, Proc Nat Acad Sci., 86:4544

Caplan, A. I., and Koutroupas, S., 1973, The control of muscle and cartilage development in the chick embryo: the role of differential vascularisation, J Embryol exp Morphol., 29:571.

Caplan, A. I., and Pechak, D. G., 1987, The cellular and molecular biology of bone formation, in: "Bone and Mineral Research", W.A. Peck, ed., Elsevier Science Publishers, New York.

Drushel, R. F., Pechak, D. G., and Caplan, A. I., 1985, The anatomy, ultrastructure and fluid dynamics of the developing vasculature of the embryonic chick wing bud, Cell Diff., 16:13.

Feinberg, R. N., Latker, C. H., and Beebe, D. C , 1986, Localised vascular regression during limb morphogenesis in the chicken embryo. I. Spatial and termporal changes in the vascular pattern, Anat Rec., 214:405.

Folkman, J., 1974, Tumour angiogenesis, Adv Cancer Res, 19: 331-358.

Hallmann, R., Feinberg, R. N., Latker, C. H., Sasse, J., and Risau, W., 1987, Regression of blood vessels precedes cartilage differentiation during chick limb development, Differentiation, 34:98.

Hamburger, V., and Hamilton, H. L., 1951, A series of normal stages in the development of the chick embryo, J Morphol., 88:49.

Hunter, S and Caplan, A. I., 1983, Control of cartilage differentiation. in "Cartilage: Development, Differetiation and Growth", B.K Hall, ed., Academic Press, New York.

Hwang, P. M., Byrne, D. H., and Kitos, P. A., 1988. Effects of molecular oxygen on chick limb bud chondrogenesis, Differentiation, 37:14.

Jargiello, D. M.,and Caplan, A. I., 1983, The establishment of vascular-derived micro-environments in the developing chick wing, Dev Biol., 97:364.

Knighton, D. R., Hunt, T. K., Scheuenstuhl, H., Halliday, B. J., Werb, Z., and Carr, J., 1983, Oxygen tension regulates the expression of angiogenesis factors by macrophages, Science, 221:1283.

Levinsohn, E. M., Pakard, D. S., West, E. M., and Hootnick, D. R., 1984, The arterial anatomy of the hatchling chick, Am J Anat., 169:377.

McLaughlin, J., Bateman, M., and Wolpert, L., 1976, Effect of 3-Acetyl-pyridine on tissue differentiation of the embryonic chick limb. Nature, 264:267.

Oikawa, T., Hirotani, K., Nakamura, O., Shudo, K., Hiragun, A., and Iwaguchi, T., 1989, A highly potent antiangiogenic activity of retinoids. Cancer Lett., 48:157.

Pechak, D. G., Kujawa, M. J and Caplan, A. I 1986a, Morphological and histochemical events during first bone formation in embryonic chick limbs, Bone, 7:441.

Pechak, D. G., Kujawa, M. J., and Caplan, A. I., 1986b, Morphology of bone development and bone remodelling in embryonic chick limbs, Bone, 7:459.

Risau, W., 1986, Developing brain produces an angiogenesis factor, Proc Nat Acad Sci., 83:3855.

Risau, W and Ekblom, P., 1986, Production of a heparin-binding angiogenesis factor by the embryonic kidney, J Cell Biol., 103:1101.

Risau, W., 1990, Angiogenic growth factors, Prog Growth Factor Res., 2:71.

Stewart, R. A, Nelson, J and Wilson, D. J.,1989, Epidermal growth factor promotes chick embryonic angiogenesis Cell Biol Int Rep., 13:957.

Stewart, R. A, Nelson, J and Wilson, D. J.,1990 Growth of the chick area in ovo and in shell-less culture. J Anat., 172:81.

Summerbell, D., 1976, A descriptive study of the rate of elongation and differentiation of the developing chick wing, J Embryol exp Morph., 35:241.

Thompson T.J., Owens P. D. A., and Wilson D.J., 1989, Intramembranous osteogenesis and angiogenesis in the chick embryo. J Anat., 166:55.

Wilson, D. J., 1983, The origin of the endothelium in the developing marginal vein of the chick wing bud. Cell Diff., 13:63.

Wilson, D. J.,1986, Development of avascularity during cartilage differentiation in the embryonic limb - An exclusion model. Differentiation, 30:183.

Wilson, D. J., and Orr-Urtereger, A., 1986, Aspects of vascular differentiation in the developing chick wing. Acta Histochem, 32S:1

PATTERNING OF CONNECTIVE TISSUES IN THE REGENERATING
AMPHIBIAN LIMB

Nigel Holder

Anatomy and Human Biology Group, Division of Biomedical
Sciences, King's College, Strand, London WC2R 2LS

INTRODUCTION

A number of recent experiments have focussed attention on the role of connective tissues in the patterning process during vertebrate limb development and regeneration. During amphibian limb regeneration such studies have concentrated on the connective tissue dermis, the mesodermal component of the skin. Original experiments in which the limb skin is rotated (Carlson 1974, 1975), made symmetrical with respect to the principal limb axes (Tank 1979, Maden and Mustafa 1982, Slack 1980, 1983) or employed to create mismatched tissue implants (Tank 1981, Rollman-Dinsmore and Bryant 1984) all demonstrate that the cells of the dermis carry positional information that is necessary for the normal patterning process (Bryant 1978, Bryant et al 1981). Working from this baseline more direct evidence for this assertion has come from experiments in which cell markers have been used to examine directly the behaviour of dermal connective tissue cells during regeneration. Such experiments have demonstrated the migration patterns and relative contributions of these cells during blastema formation following limb amputation (Muneoka et al 1986), and the results are consistent with a controlling role for dermal connective tissue cells during limb patterning.

Experiments performed on the developing chick limb are also consistent with this view. In the absence of muscles, connective tissues can organize themselves into recognizable pattern elements (Kieny and Chevallier 1979). The connective tissues in muscleless limbs, which derive from the somatopleure (Chevallier et al 1977), are also capable of organizing grafted myogenic cells into recognizable muscles (Chevallier and Kieny 1982).

In this paper we discuss two experiments which analyze further the ability of dermal fibroblasts of the axolotl limb to organize into tissue patterns. The first of these experiments involves the creation of muscleless limb regenerates and the analysis of tendon and muscle-associated fibroblast condensations in the digit and carpal regions (Holder 1989). The second experiment involves a detailed analysis of the generation of cartilage element shape and form in a range of limb types including normal limbs, limbs symmetrical with respect to the dorso-ventral axis and muscleless limbs. The results of these two experiments strongly support the view that connective tissue cells have a remarkable self-organizing capability during limb regeneration and that they are a major controlling influence on pattern formation.

Developmental Patterning of the Vertebrate Limb
Edited by J.R. Hinchliffe *et al.*, Plenum Press, New York, 1991

Connective tissue patterns in muscleless limb regenerates

Muscleless limb regenerates were generated by amputating limb stumps in which all tissues except the skin had been X-irradiated (Dunis and Namenworth 1977, Lheureux 1983). Such regenerates produce completely muscleless regenerates in 40% of cases, the remaining regenerates having partly (26%) or completely (23%) formed muscles (Holder 1989). The analysis of connective tissues in such limbs concentrated on the cartilage, tendons and muscle-associated fibroblast condensations. With respect to the cartilage patterns were monitored in the formation of the radius and ulna, carpus and metacarpals and phalanges. With the exception of minor variations in a minority of cases, the recognizable arrangement of cartilage elements formed, producing the standard limb skeleton.

The tendon patterns were examined with particular reference to the digits and the carpus. In this region the tendons comprise a series of structures in a complex pattern. In the wrist region dorsal and ventral aponeurotic connective tissue sheets are attached to a series of distally projecting tendons which attach in various positions to the metacarpals and phalanges. In the muscleless limbs these sheets and tendons form in their normal locations and insert at their appropriate positions. At more proximal levels, those tendons associated with muscles and bones in the forearm did not form into a recognizable pattern, suggesting that the tendons of the digits and carpus are organized on a different principle. This may have something to do with their interactions with the aponeurotic sheets of the carpus and their function as an integrated complex during movements of the fingers.

With regard to the fibroblasts associated with the normal locations of muscles the experimental limbs revealed a striking pattern of identifiable cell groups in precise locations. During normal muscle morphogenesis the myotubes are surrounded by fibroblasts which create the endo-, epi- and perimysia. The distribution of condensations of connective tissue cells in the identifiable locations and patterns indicates strongly that the positions and shapes of muscles are normally controlled by the cells which form the mysial layers. This result is entirely consistent with the results of the experiments with muscles from chick limbs mentioned in the Introduction.

Taken together it is evident that connective tissues in the regenerating and developing limb show remarkable degrees of self-organization into defined tissue patterns. A diverse group of structures can form, the dermal fibroblasts are able to differentiate into tendons, cartilages and mysial condensations. It is likely, therefore, that the connective tissues are responsible for controlling the pattern of other tissues in the limb and direct evidence supports this notion with respect to muscle and nerves (Chevallier and Kieny 1982, Lewis et al 1981).

What controls the formation of detailed anatomical features of cartilage elements?

Cartilage elements, the precursors for limb bones, are part of the pattern provided by dermal fibroblasts when these are the only source of cells in the regenerate (Holder 1989, Dunis and Namenworth 1977). Each cartilage element in the limb is uniquely identifiable by its relative position, relative size and by its shape and form. An experienced anatomist can identify each element in isolation. The shape and form of each element varies; they are traditionally described as long bones or the smaller elements of the carpus and tarsus. An important element of the unique character of an individual element is, therefore, its growth profile. Little is known about the control of the processes which contribute to cartilage element growth, cell division, cell hypertrophy and the secretion of extracellular matrix. Even less is known about the control of cartilage element shape and form. For example, what is the relationship between pattern formation mechanisms and cartilage form; do the attachment of tendons and the actions of muscles alter cartilage shape; what is the role of the perichondrium, the constraining layer of connective tissue cells which surrounds each individual element?

In an attempt to begin to answer these basic questions we have derived a method for describing cartilage element shape and form and have used it to characterize transitions in cartilage element structure under certain experimental conditions which affect limb symmetry and which analyze the possible role of muscle action in cartilage morphogenesis.

A description of cartilage element shape and form

In order to derive a simple description of cartilage shape we decided to focus on the carpus of the axolotl forelimb and the distal regions of the radius and ulna. The carpus of the axolotl contains eight elements, each individually identifiable. There is a proximal row of three elements, from anterior to posterior these are the radiale, intermedium and ulnare; two central elements, the prepollicus and centrum; and three distal elements, basal carpals 1/2, 3 and 4. The distal ends of the radius and ulna articulate with the three elements of the proximal carpal row.

The numerical analysis of element shape was performed on representative transverse sections cut through the hand and distal forearm at 10 μm from limbs embedded in wax. Camera lucida drawings were made from appropriate sections containing regions of elements to be studied further. To ensure reproducibility of results the same proximo-distal level was analyzed for any particular element. A line was then drawn parallel to the dorso-ventral axis of the drawing of the transverse section for individual elements and the element divided into equal-sized parallel segments with lines drawn at right angles to the initial reference line (Figure 1). For each level the mid-point of the intersecting line was measured as it passed across the limits of the element. The distance (d in Figure 1) was then measured from this mid-point to the original reference line. The mean of this distance (d) for each segment was then calculated to give a line parallel to the reference line which divides the outline of the element into two regions of equal areas (Figure 1).

This analysis was performed on three examples for each of the following elements; bc 1/2, prepollicus (pp), centrum, and distal ulna. These four were selected because two of them, bc 1/2 and distal ulna, are clearly asymmetrical in shape with respect to the dorso-ventral axis whereas the pp and centrum are almost perfectly symmetrical with respect to this axis.

Analysis of cartilage element shape with respect to pattern formation in the dorso-ventral axis

To approach the question of the relationship between pattern formation and form the shape of the four cartilage elements was assessed in limbs with abnormal symmetry in the dorso-ventral axis. These were obtained from serial sectioned sets of double dorsal or double ventral regenerates obtained from previously published experiments. These were generated either by surgically creating double dorsal or double ventral stumps (Burton et al 1986), surgically creating mixed handed stumps (Holder and Weekes 1984) or producing supernumerary limbs following 180° ipsilateral blastemal rotations (Maden and Mustafa 1982b).

The basic question to be answered by this analysis is, does the dorso-ventral symmetry of patterning information in the forming blastemas from these experimental limb stumps generate symmetrical cartilage elements? The analysis also allows some comment on the true location of the dorso-ventral axial midline in these limbs, although it must be stressed the results say nothing about the basic patterning mechanisms per se. In order to determine whether symmetrical cartilage elements form in these experimental limbs, theoretical double dorsal and double ventral bc 1/2, pp, centrum and distal ulna were drawn, using the mean line passing through the normal elements (refer to Figure 1). In each case the elements seen in the symmetrical regenerates matched closely those of the theoretical outlines. For the asymmetrical elements, bc 1/2 and distal ulna this involved a dramatic alteration in shape, whereas the normally symmetrical examples (pp and centrum) remained symmetrical in outline, as would be predicted.

Having established that cartilage element shape may be influenced by pattern forming mechanisms, some assessment of the reality of the location of the dorso-ventral axial midline was required because the theoretical symmetrical outlines of the four elements analyzed were sensitive to the arbitrary placement of the initial reference line. For this reason reference lines were drawn at various angles relative to the initial outlines of the selected elements and the theoretical symmetrical outlines derived as before. In each case the shape of the theoretical symmetrical elements altered dramatically and for both the observed symmetrical bc 1/2 and the distal ulna the closest match was clearly that for the location of the original line selected to lie along the mid dorso-ventral axial line. This position is therefore the best estimate for its location.

The effect of muscle action on cartilage element shape

Despite the remarkable and predictable transition in cartilage element shape in symmetrical limbs, it remains a possibility that tendon attachment and muscle action can influence the generation of asymmetry, because such symmetrical limbs contain symmetrical tendon and muscle patterns. To analyze this possibility, the shape of the selected four cartilage elements was assessed in the muscleless limbs described in the first part of this paper. Exactly the same procedure was used to generate the necessary drawings and numerical data as was described for symmetrical elements. The results are of primary interest for the asymmetrical elements, bc 1/2 and distal ulna, because it is for these that muscle action may be responsible for distorting an initially symmetrical shape. For both of these elements they remained asymmetrical in muscleless limbs and have shapes closely comparable to those seen in normal limbs. This was also true for the pp and centrum. It can be concluded therefore that muscle action plays little part in generating the basic shapes of cartilage elements during morphogenesis.

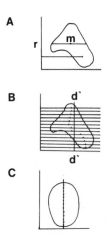

Figure 1. The quantitative analysis of symmetry illustrated in A and B relative to a transverse section of the distal ulna. In A the initial reference line is marked r and the mid-point of the element is shown by the point m. d is the distance from each point m for all parallel levels, as indicated in B. The mean value of the population of d values were calculated to produce a line d' which separates the outline into two regions of equal size, as shown in B. The variance ratio of d values for each population as compared with d' gives a measure of symmetry. If the element is initially symmetrical the values m will vary little from the mean value d' as shown in C, which is a transverse section of the centrum.

CONCLUSIONS

The experiments with muscleless limbs, taken together with previous work on developing chick limbs, strongly indicate that connective tissues are of primary importance to the patterning process. Not only do such cells ensheath and mould muscles into recognizable individual structures, but these cells, in the form of dermal fibroblasts, have the capacity to differentiate into cells making up various structures of the connective tissue family: tendons, dermis and cartilages, and mysial layers, which are organized into a recognizable limb pattern.

With respect to one component of this pattern, the cartilage elements, a detailed analysis of their shape and form indicates that these are derived by properties of cells inherent to the elements themselves, that are in some way an expression of the positional information necessary to derive the original limb pattern. It remains to be established how subtle alterations in the properties of the cartilage cells or the perichondrial cells which surround them lead to the creation of elements with a precise shape and form.

REFERENCES

Bryant, S.V., 1978, Pattern regulation and cell commitment in amphibian limbs, in "Clonal Basis of Development", Symp. Soc. Dev. Biol. 36, 63-82.

Bryant, S.V., French, V. and Bryant, P.J., 1981, Distal regeneration and symmetry. Science 212, 993-1002.

Burton, R., Holder, N. and Jesani, M., 1986, The regeneration of double dorsal and double ventral limbs in the axolotl. J. Embryol. exp. Morph. 94, 29-46.

Carlson, B.M., 1974, Morphogenetic interactions between rotated skin cuffs and underlying stump tissues in regenerating axolotl forelimbs. Dev. Biol. 39, 263-285.

Carlson, B.M., 1975, The effects of rotation and positional change of stump tissues upon morphogenesis of the regenerating axolotl limb. Dev. Biol. 47, 269-291.

Chevallier, A. and Kieny, M., 1982, On the role of connective tissue in the patterning of the chick limb musculature. Wilhelm Roux Arch. dev. Biol. 191, 277-280.

Chevallier, A., Kieny, M. and Mauger, A., 1977, Limb-somite relationship: origin of the limb musculature. J. Embryol. exp. Morph. 41, 245-253.

Dunis, O. and Namenworth, M., 1977, The role of grafted skin in the regeneration of X-irradiated axolotl limbs. Dev. Biol. 56, 97-109.

Holder, N., 1989, Organization of connective tissue patterns by dermal fibroblasts in the regenerating axolotl limb. Development 105, 585-593.

Holder, N. and Weekes, C., 1984, Regeneration of surgically created mixed-handed axolotl forelimbs: pattern formation in the dorsal-ventral axis. J. Embryol. exp. Morph. 82, 217-239.

Kieny, M. and Chevallier, A., 1979, Autonomy of tendon development in the embryonic chick wing. J. Embryol. exp. Morph. 41, 245-253.

Lewis, J.H., Chevallier, A., Kieny, M. and Wolpert, L., 1981, Muscle nerve branches do not develop in chick wings devoid of muscle. J. Embryol. exp. Morph. 64, 211-232.

Lheureux, E., 1983, The origin of tissues in the X-irradiated regenerating newt limb, in "Limb Development and Regeneration", Prog. clin. biol. Res. 110A, J. Fallon and A. Caplan, eds., A.R. Liss, New York.

Maden, M. and Mustafa, K., 1982a, Axial organization of the regenerating limb: asymmetrical behaviour following skin transplantation. J. Embryol. exp. Morph. 70, 197-213.

Maden, M. and Mustafa, K., 1982b, The structure of 180-degree supernumerary limbs and a hypothesis of their formation. Dev. Biol. 93, 257-265.

Muneoka, K., Fox, W. and Bryant, S.V., 1986, Cellular contribution from dermis and cartilage to the regenerating limb blastema in axolotls. Dev. Biol. 116, 256-260.

Rollman-Dinsmore, C. and Bryant, S.V., 1984, The distribution of marked dermal cells from localized implants in limb regenerates. Dev. Biol. 106, 275-281.

Slack, J.M.W., 1980, Morphogenetic properties of the skin in axolotl limb regeneration. J. Embryol. exp. Morph. 58, 265-288.

Slack, J.M.W., 1983, Positional information in the forelimb of the axolotl: properties of posterior skin. J. Embryol. exp. Morph. 73, 233-247.

Tank, P.W., 1979, Positional information in the forelimb of the axolotl: experiments with double-half tissues. Dev. Biol. 73, 11-24.

Tank, P.W., 1981, The ability of localized implants of whole or minced dermis to disrupt pattern formation in the regenerating forelimb of the axolotl. Am. J. Anat. 162, 315-326.

THE INTERDIGITAL SPACES OF THE CHICK LEG BUD AS A MODEL FOR
ANALYSING LIMB MORPHOGENESIS AND CELL DIFFERENTIATION

J. M. Hurle[1], D. Macias[2], Y. Gañan[2], M.A. Ros[1] and
M. A. Fernandez-Teran[1]

[2] Dpto. Ciencias Morfologicas. U. Extremadura Spain
[1] Dpto. Anatomia y Biologia Celular. U. Cantabria. Spain
Facultad de Medicina. 39011 Santander (Spain)

1.- Cell death as a developmental strategy for morphogenesis The interdigital necrotic areas

Normal embryonic development is not a linear process of cell and tissue growth and differentiation (Hurle, 1988). There are many examples in which an early embryonic structure undergoes a reshaping process involving a substantial phenomenon of cell death. In some cases the tissue loss affects to fully formed organs which play a transitory functional role in the developing individual.This is the case of the loss of the tail in the developing anura amphibia during metamorphosis. In other cases cell death appears closely linked to morphogenesis an appears to play an sculpturing role for the achievement of the final shape of a developing structure. This appears to be the case in "interdigital cell death" occurring during morphogenesis of the digits of the amniote embryos.

The early primordium of the autopodium in the amniotes is a paddle shaped structure consisting of a mesodermal core covered by ectoderm. As distinct to the amphibians in which the digits grow at the tip of the developing limb (Cameron & Fallon, 1977), in the amniote the formation of the digits involves the formation of chondrogenic digital rays followed by the loss by cell death of the intervening interdigital tissue. In the chick leg bud interdigital cell death begins at stage 30-31 (day 7-7.5) and by day 9 the digits are fully formed and free.

In the last few years we have studied the fate of the interdigital mesoderm when it is diverted from the necrotic program(Hurle & Gañan, 1986;1987 ; Hurle et al., 1989). Since interdigital cell death takes place much later than the stages in which normal patterning of the limb is established, this study was aimed to analyse the histogenetic and morphogenetic potentiality of the limb mesodermal cells prior to their commitment to death.

In all the experiments we have employed the third interdigital space of the leg bud of white Leghorn chick embryos.

Developmental Patterning of the Vertebrate Limb
Edited by J.R. Hinchliffe *et al.*, Plenum Press, New York, 1991

2.- In vitro analysis of the cartilage forming capacity of the interdigital mesoderm prior to death

By day 5 of development (stage 27) the chick embryonic leg bud has a prominent autopodial plate in which the digital rays and the interdigital spaces are clearly identifiable. This is 48 h before the onset of cell death. In a series of experiments we have analysed the prospective fate of the interdigital tissue when cultured in vitro either as dissociated cells in high-density micromass cultures or as tissue explants.

Micromass cultures.- High-density micromass cultures were prepared according to the usual procedure (Ahrens et al., 1977). Briefly, the third interdigit was removed from leg buds ranging from stage 28 to 31 (Hamburger and Hamilton, 1951) and incubated for 16 or 20 min in 2% trypsin in calcium and magnesium free saline.The interdigital mesenchyme was then disaggregated and the cells were resuspended at a concentration of $2x10^4$ per μl in Ham F12 (GIBCO) supplemented with 10% fetal calf serum, penicillin (50 u/ml) and streptomycin (25μg/ ml). 10μl drops of cell suspension were plated out in 35-mm tissue culture dishes and the cells allowed to attach for 1 h before flooding with 1ml of medium. Cultures were incubated in an atmosphere of 5% CO_2, 95% air in a humidified incubator, for 3 or 4 days.

Stage 28 interdigital mesoderm exhibits an intense chondrogenic response under micromass culture conditions. By 24 h of culture discrete areas of increased cell density are recognizable by phase-contrast microscopy. The number and size of these areas of cell aggregation increase in the following days giving rise to chondrogenic alcian blue-positive nodules at day 3. By day 4 of culture, most of the micromass appears chondrogenic due to the coalescence of the cartilage nodules (Fig. 1a). At stage 29 the cartilage-forming capacity of the interdigital mesoderm remains intense although the number of nodules detected in each micromass after 3 or 4 days of culture appears reduced with respect to micromasses of stage 28 (Fig. 1b). By stages 30 and 31 the chondrogenic response of the interdigital cells drops drastically. As can be seen in figure 1c, only 2 or 3 nodules are detected in each micromass after 4 days of culture. A considerable increase of chondrogenesis is obtained when the cell concentration of the micromass is increased two or four-fold, but even in these conditions the number of nodules per micromass after 4 days of culture is only of 12 to 16 (Fig. 1d). At stage 32 no chondrogenic activity was detected even by increasing the cell concentration. Histological examination of these cultures revealed a significant increase of cell death and phagocytes in the older stages while the surviving cells exhibit a fibroblastic appearance.

Organ cultures.- For this purpose the interdigital spaces were dissected free from leg buds of stage 29 to 31 and cultured at 37ºC for 3 days immersed in culture medium. The explants were placed in the culture dishes in a small drop of medium and allowed to attach for 1h in humidified atmosphere before flooding with 1 ml of medium. The culture conditions were the same as those of the micromass cultures.

After 3 days of culture 83% of the interdigital explants of stage 29 showed a prominent alcian blue-positive cartilage.In

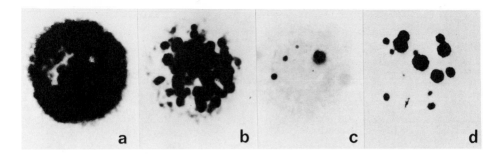

Fig. 1.- Alcian blue stained whole mounts of interdigital
mesenchyme after 4 days of micromass culture. a)Stage 28
H.H. The whole culture is chondrogenic. b)Stage 29 H.H.
Chondrogenic nodules are confluent. c)Stage 30 H.H. Only
isolated nodules appear. d)Stage 31 H.H. Discrete nodules
appear only after increasing the micromass concentration
twice. 8X.

most cases the cartilage appeared as a rod with rounded tips
ranging from 0.4mm to 0.9 mm long (Fig. 2a). In some cases one
of the tips of the cartilage exhibited a bifurcated appearance.
These cartilages were usually thicker than the rod shaped ones.
In no case we detected the formation of a joint. Neither did
the morphology of the cartilage resemble a phalange.
Examination of the explants with scanning electron microscopy
revealed that the mesodermal core of the explants
expanded more than the ectodermal jacket thus leaving a
significant area of the interdigit devoid of ectodermal cover
(Fig. 2b). Removal of a small portion of the interdigital
ectoderm before explanting increased the incidence of
chondrogenesis to 94% of the explants.

In the explants of stages 30 and 31 formation of cartilages
was also obtained but they were smaller and rounded rather than
elongated. Stage 32 explants always underwent
disintegration.This lack of chondrogenesis was not affected by
ectoderm removal at the moment of explanting.

A significant feature worth mentioning was that the
incidence of interdigital chondrogenesis was reduced
drastically when the explants contained an adjacent digit. At
stage 29 isolated interdigits developed a cartilage in 83% of
the explants, while when the explants contained an adjacent
digit chondrogenesis was only detected in 45% of the explants.
The size of the cartilage also showed a reduction when the
explants contained an adjacent digit.

These results show that the interdigital mesoderm consists
of a chondrogenic cell population and that their chondrogenic
potentiality is lost concomitantly with the establishment of
the cell death program. The results also suggest that the
interdigital tissue in vivo may be subjected to an
antichondrogenic influence. Our results along with the studies
of Solursh and coworkers (Solursh, 1984; Solursh et al.,
1981; Solursh & Reitter, 1988; Zanetti & Solursh, 1986)
suggest that the ectoderm is the most likely structure
responsible for the antichondrogenic effect but it is also
possible that the developing digits exerts some kind of lateral

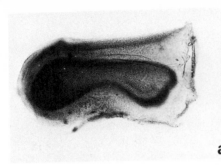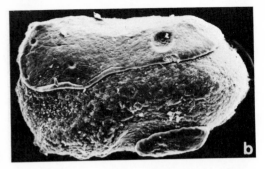

Fig. 2a.- Alcian-blue staining showing the presence of a prominent cartilage in the organ culture when the whole interdigit was cultured for 3 days after removing the marginal. ectoderm. 125X.

Fig. 2b.- SEM micrograph of the whole interdigit culture reveals a significant area of the interdigit devoid of ectodermal cover. 170X.

inhibitory effect on their adjacent interdigits. This lastpossibility was supported by the culture experiments explanting interdigits with and without an adjacent digit.

3.- In vivo induction of interdigital chondrogenesis

Experimental approaches devised to reproduce in vivo the chondrogenic potentiality of the interdigital mesenchyme detected in vitro were based on interdigital ectoderm removal.

Since the ectoderm appears to exert a major antichondrogenic effect on the limb mesenchyme (Solursh, 1984) in a series of experiments we tried to reduce or abolish such antichondrogenic influence by removing locally the interdigital ectoderm. Results were positive both when the ectoderm was removed surgically (Hurle & Gañan, 1986) or enzymatically. Surgical removal of a small piece of the marginal interdigital ectoderm at stage 29 resulted in the formation of interdigital cartilages in 47 % of the cases detectable 48 h after the operation by whole-mount alcian blue staining (Fig. 3).

Enzymatic elimination of the ectoderm was achieved by local microinjection of 2% trypsin (GIBCO))into the interdigit of stage 29 embryos. The trypsin solution was backfilled into glass micropipettes that were connected to an oil pressure source. After windowing the egg and exposing the leg bud, the micropipette was inserted into the third interdigit using a micromanipulator and 1-2.5 μl of solution was injected into the mesenchymal core of the interdigit. Control embryos were microinjected with the same volume of PBS alone. After microinjection the eggs were returned to the incubator and the embryos were sacrificed at the desired stage and examined by histological procedures. The efficiency of local trypsin microinjection for ectoderm removal was assessed 24h after microinjection by examining the limbs by scanning electron microscopy. As can be seen in figure 4, a wide area denuded of ectoderm was detectable in the experimental but not in the control embryos 24 h after microinjection. Interdigital chondrogenesis by this procedure was observed in 31% of the surviving embryos. Cartilages were detectable by alcian blue

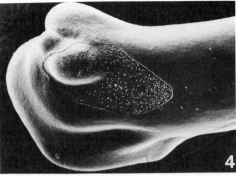

Fig. 3.- Alcian-blue staining showing the presence of an ectopic cartilage in the interdigit of leg bud 72 h. after ridge removal at stage 29 H.H. 55X.

Fig. 4.- SEM micrograph showing the area denuded of ectoderm in an experimental leg 24 h. after microinjection of 2 % trypsin. 85X.

staining 48 h after microinjection. As distinct to surgically induced chondrogenesis which consisted always of a single cartilage nodule located in the distal interdigit,enzymatically induced interdigital cartilages were often multiple (up to 3) lying along the proximodistal axis of the interdigit.

The most precocious histochemical change detected in the interdigits subjected to those experimental procedures was the formation of a cell aggregate under the wounded interdigital surface reacting positively with Peanut lectin. This feature was detected 15 h after ectoderm removal and is in agreement with previous histochemical analysis of chondrogenesis in vitro (Althouse & Solursh, 1987). The specific detection of mRNA for collagen type II in these aggregates by in situ hybridization was not as precocious as lectin labelling. For this purpose we used a 35S-labeled cDNA probe (a generous gift of Dr. William Upholt; University of Connecticut Health Center). Analysis of the autoradiographs showed an intense labelling of the aggregates only 20 h after ectoderm removal (Fig. 5).

4.- Morphogenetic fate of the ectopic interdigital cartilages

A major task of this study was the analysis of the morphogenetic fate of the experimentally induced interdigital cartilages. It must be taken into account that in these experiments interdigital cartilages are induced long after theperiod in which the skeletal elements of the limb have been specified (Honig & Summerbell, 1985). This feature makes the model particularly interesting in detecting local tissue interactions as factors accounting for tissue patterning as proposed by the mechano-chemical theory (Oster et al., 1983, 1985).

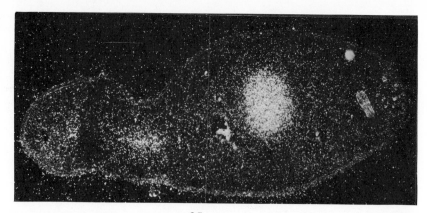

Fig.5.- Hybridization with (^{35}S)-labelled type II collagen cDNA
probe to a leg bud 20 h. after ridge removal at stage 29
H.H. In addition to the intense accumulation of silver
grains over the cartilage of the third digit, note another
detectable accumulation of silver grains over the ectopic
chondrogenesis. 42X.

Our results show that in a high percentage of the cases the
interdigital cartilages in the course of development form a
rudimentary digit.The extradigits were identifiable by their
shape 3 days after experimental manipulation (surgical ectoderm
removal or trypsin microinjection) and consisted of two distal
phalanges separated by a developing joint (Fig.6a).By day 6
after the operation in addition to a well defined digital
morphology, a set of flexor and extensor tendons were
identifiable in tissue sections (Hurle et al., 1990). Tendons
were always connected with those of the neighbouring digits by
fibrous expansions, and formed once the normal tendons of the
neighbouring digits were clearly established.

By three days after the operation, digit formation was
identified in 44% of the embryos with surgically induced
chondrogenesis. When chondrogenesis was induced by trypsin
microinjection, digit formation was detected in 20% of the
cases.However a distinctive feature between trypsin-induced
extradigits and those induced surgically, was the abundance of
abnormalities in the shape of the digits induced by trypsin
treatment. Abnormalities were present in 60% of the cases and
consisted of fusion of adjacent digits, abnormal oblique
alignment or bifurcation of the proximal phalange of the
extradigits (Fig. 6b). In both types of experiments the
interdigital cartilages which failed to form a digit appeared
as a rounded or ovoid cartilage without any distinct
morphological features.

Differences in size between cartilages with and without
digital morphology was analyzed by measuring the surface area
of the ectopic skeletal rudiment in whole-mount alcian blue
stained specimens using a semiautomatic image analyzer
(Videoplan, Kontron).The size of the unspecific cartilages
incomparison with those exhibiting digit morphology was
considerably reduced (p< 0.00001) (Fig. 7). Cartilages with a
surface area value below 0.06mm^2 corresponded only to isolated
cartilages. Values of dorsal surface area over 0.08 mm^2

a   b

Fig. 6a.- Whole-mount cartilage staining showing the two distal segments of an extra-digit in the experimental chick leg 3 days after ridge removal at stage 29 H.H. Dorsal view. 55X.

Fig. 6b.- Whole-mount cartilage staining showing the presence of an extra-digit with bifurcation of the proximal phalange after microinjection of 2 % trypsin in the leg bud. Ventral view. 100X.

corresponded only to extradigits. Values between 0.06 mm^2 and 0.08 mm^2 corresponded either to digits or ovoid unspecific cartilages. This feature suggest that the transition from an ectopic cartilage, lacking any particular shape, to the well defined extradigit requires the achievement of a critical size. If this is true it may be suggested that a cascade of morphogenetic events (joint formation, tendon formation) is triggered by the establishment of a cartilage nodule with a threshold size, thus emphasing the importance of local tissue interaction for morphogenesis .

A further feature analyzed in this model of extradigit formation was the influence on digit morphology of their relative position along the proximodistal axis of the interdigit. This feature was analyzed to assess how far the formation of extradigits could be explained by the existence of a precise memory in the interdigital mesoderm of an early positional specification (Wolpert, 1989; Wolpert & Hornbruch, 1990).To establish the positional parameter we compared the location of the interphalangeal joint of the extradigit with that of the joint between phalanges 2 and 3 of digit III. According to this reference "distal" and "proximal" extradigits were distinguishable (Fig. 8). The joint of the distal extradigits was always aligned with the joint between phalanges 4-5 of digit IV. In these cases the distal phalange of the extradigit was parallel to the distal phalange of digit IV. In proximal extradigits the interphalangeal joint was always aligned with the joint between phalanges 2-3 of digit III. It is interesting to emphasize that the morphology of the distal phalanges in both cases was similar and that truncated extradigitslacking a distal phalange were never found. These facts suggest that while the morphogenetic process follows a well-defined proximodistal gradient it has some autonomy in relation to its proximodistal position in the autopodium. This feature tends to disprove the existence of a precise memory of positional values in the mesodermal cells destined to die supporting again the importance of local properties of the interdigital tissue subjected to experimental manipulation. Further support to this hypothesis comes from

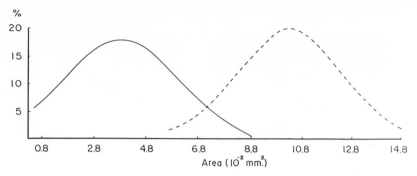

Fig. 7.- Size distribution curve of the chondrogenic areas of
the ectopic cartilages(———) and extra-digit (- - -).

experiments on chondrogenesis induced by trypsin
microinjection. In these cases the shape of the interdigit is
altered by the volume injected and it is correlated with a
reduced incidence of extradigits which in addition if formed
show an increased incidence of morphological abnormalities.

5.-Extradigit formation and the "chemical" versus the
"mechano-chemical" theory for limb patterning

Two major hypothesis have been proposed to explain limb
patterning and morphogenesis. The notion of "Positional
Information" set up by Wolpert (1971;1989) and other comparable
theories (see Meinhardt, 1982 and Wolpert & Hornbruch,
1990)proposes that embryonic patterns are specified by the
distribution of chemical (morphogens) concentrations. In this
view, embryonic cells behave (i.e. change of cell shape,
differentiation, migration, etc.) according to the information
received from variable concentrations of morphogen.
Morphogenesis, according to this view, is a secondary process
determined once the chemical pattern is established. The
discovery of a zone of polarizing activity (ZPA) (Gasseling &

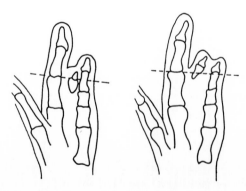

Fig. 8.- Schematic drawing showing the presence of extradigits
after ridge removal at stage 29 H.H. Note the joint of the
distal extradigit is aligned with the joint between
phalanges 4-5 of digit IV, and the joint of the proximal
extradigit is aligned with the joint between phalanges 2-3
of digit III.

Saunders, 1964) in the posterior margin of the early limb bud and subsequent studies on its biological role (Tickle, 1980) provided good experimental support for this hypothesis. Furthermore, there is now increasing evidence suggesting that retinoic acid might well be the hypothetical morphogen for the anteroposterior patterning of the limb bud (Smith et al., 1989). In a rather different approach, the "mechanochemical interaction" viewpoint (Oster et al., 1983;1985;1988) proposes that the embryonic cells and extracellular matrix interact continuously to produce definite patterns of tissue behaviour leading to morphogenesis. Each patterning process begins as a local interaction between the tissue components of the developing structure. In limb bud chondrogenesis the interaction is initiated when the mesodermal tissue achieves a threshold level of differentiation and progresses following modifications which can be predicted by mathematical laws on the basis of a definite geometry of the embryonic anlagen. Support to this hypothesis comes mostly from theoretical approaches while experimental evidence of its role in normal development is scarce (Alberch & Gale, 1983).

The formation of extradigits by the interdigital mesoderm reported in this study provides an interesting test of the two hypotheses. Specification signal mechanisms acting on limb development have ceased their function by the time we induce extradigit formation (Honig & Summerbell, 1985). Furthermore, previous studies made in our laboratory revealed that grafts of ZPA into the interdigit, or retinoic acid microinjection do not affect extradigit formation. All these facts point to a major role of local tissue interaction in the determination of the extradigits. Moreover,the possibility of a positional memory in the cells destined to die appears unlikely since there is a certain degree of variation in the position of the extradigits and the presumed information that we might expect is retained by these cells is that for the establishment of cell death. All these facts make more it likely that the distal part of the autopodium has structural and geometrical properties compatible with digit formation.

In conclusion our results tend to show that formation of the digits in the autopodium is due to the balance between a natural tendency of the local mesenchyme to differentiate into cartilage and an inhibitory antichondrogenic effect of the ectodermal covering as previously suggested by Solursh (1984) (see also Ide, 1990). According to this hypothesis early regulatory mechanisms responsible for morphogenesis, such as genes or their related morphogens, may operate by determining an autopodial plate, consisting of cells of the appropriate lineage, having a critical volume and geometry which accounts for the balance between chondrogenic promoters and antichondrogenic factors, thus resulting in the final pattern of the autopodium. In this view, the interdigital necrotic areas may reflect the elimination of chondrogenic cells which while required for the formation and growth of the early autopodial plate are later excluded from their natural pathways of differentiation due to their location in areas of high antichondrogenic influence.

ACKNOWLEDGEMENTS
    This work was supported by a grant from the DGICYT (PS87-0095)

References

Ahrens P. B.,Solursh, M.,Reiter, R. S.(1977).Stage-related capacity for limb chondrogenesis in cell cultures.Dev.Biol.,60: 69-82.

Alberch, P., & Gale, E. A. (1983). Size dependence during the development of the amphibian foot. Cochicine-induced digital loss and reduction. J. Embryol. exp. Morph., 76: 177-197.

Althouse A., L., & Solursh, M. (1987). The detection of a precartilage, blastema specific marker. Dev. Biol.,120: 377-384.

Cameron J. A., & Fallon, J. F. (1977). The absence of cell death during development of free digits in amphibians. Dev.Biol.,55: 331-338.

Gasseling, M.T., & Saunders, J. W. Jr. (1964). Effect of the "Posterior Necrotic Zone" of the early chick wing bud on the pattern and symmetry of limb outgrowth.Amer. Zool.,4:303-304.

Hamburger V. & Hamilton, H. L. (1951). A series of normal stages in the development of the chick embryo. J. Morphol., 88: 49-92.

Honig, L., S. & Summerbell, D. (1985). Maps of strength of positional signalling activity in the developing chick wing bud. J. Embryol. exp. Morph., 87: 163-174.

Hurle J. M. (1988). Cell death in developing systems. Meth.Achiev.exp. Pathol.vol. 13:Kinetics and Patterns of Necrosis (G. Jasmin ed.) pp 55-86. Karger.Basel.

Hurle J. M., & Gañan, Y. (1986). Interdigital tissue chondrogenesis induced by surgical removal of the ectoderm in the embryonic chick leg bud. J. Embryol. exp.Morph.,94:231-244.

Hurle J. M., & Gañan, Y. (1987). Formation of extra-digits induced by surgical removal of the apical ectodermal ridge of the chick embryo leg bud in the stages previous to the onset of interdigital cell death.Anat. Embryol.,176:393-399.

Hurle J. M., Gañan, Y, & Macias, D. (1989). Experimental analysis of the in vivo chondrogenic potential of the interdigital mesenchyme of the chick leg bud subjected to local ectodermal removal. Dev. Biol., 132: 368-374.

Hurle, J. M., Ros, M. A., Gañan, Y., Macias, D., Chritchlow, M., & Hinchliffe, J. R. (1990). Experimental analysis of the role of ECM in the patterning of the distal tendons of the developing limb bud. Cell Differ. Develop.,30: 97-108

Ide, H. (1990).Growth and differentiation of limb bud cells in vitro: Implications for limb pattern formation. Develop.Growth Differ., 32: 1-8.

Meinhardt, H. (1982). Models of biological pattern formation. Academic Press. London.

Oster, G.F., Murray, J. D., & Harris, A. K. (1983). Mechanical aspects of mesenchymal morphogenesis. J. Embryol. exp.Morph. ,78: 83-125.

Oster, G.F., Murray, J. D., & Maini, P. K. (1985). A model for chondrogenic condensations in the developing limb: the role of extracellular matrix and cell tractions. J.Embryol.exp.Morph.,89: 93-112.

Oster, G.F., Shubin, N., Murray, J. D. & Alberch, P. (1988). Evolution and morphogenetic rules: the shape of the vertebrate limb in ontogeny and phylogeny. Evolution.,42: 862-884.

Smith, S. M., Pang, K., Sundin, O., Wedden, S. E., Thaller, C. & Eichele, G. (1989). Molecular approaches to vertebrate limb morphogenesis. Development suppl., 121-131.

Solursh, M. (1984). Ectoderm as a determinant of early tissue pattern in the limb bud. <u>Cell Differ</u>.,15: 17-24.

Solursh, M., & Reiter, R. S. (1988). Inhibitory and stimulating effects of limb ectoderm on in vitro chondrogenesis. <u>J.exp.Zool</u>., 248: 147-154.

Solursh, M., Singley, C. T., & Reiter, R. S. (1981). The influence of epithelia on cartilage and loose connective tissue formation by limb mesenchyme cultures. <u>Dev. Biol</u>., 86: 471-482.

Tickle C. (1980). The polarising region and limb development. In: <u>"Development of Mammals vol. 4 "</u> (M. H. Johnson ed.) pp. 101-136. Elshevier/North Holland Biomed. Press. Amsterdam.

Wolpert L (1971). Positional information and pattern formation. <u>Curr. Top. Dev. Biol</u>. 6: 183-224.

Wolpert, L. (1989). Positional information revisited. <u>Development</u> supp. 3-12.

Wolpert L., & Hornbruch, A (1990). Double anterior chick limb buds and models for cartilage rudiment specification. <u>Development</u>.,109: 961-966.

Zanetti, N., & Solursh, M. (1986). Epithelial effects on limb chondrogenesis involve extracellular matrix and cell shape. <u>Dev.Biol</u>., 113: 110-118.

A "PACKAGING" MODEL OF LIMB DEVELOPMENT

Sergei L. Znoiko

Koltzov Institute of Developmental Biology
USSR Academy of Science
MOSCOW, USSR

This model is based on a hypothesis about the early separation of chondrogenic and non-chondrogenic subpopulations of cells in the limb bud (LB). It is proposed that there exists a morphogenetic mechanism leading to "packaging" of the chondrogenic subpopulation of cells into a properly arranged non-chondrogenic subpopulation of cells during the formation of every skeletal element. This model makes use of several facts and ideas discovered/proposed earlier.

For the convenience of presentation, mesenchyme cells participating in limb development can be operationally divided into two types:

Cells of Type 1 display chondrogenic potential (they carry on their surface corresponding tissue-specific molecules of adhesion).

Cells of Type 2 represent pluripotent cells of limb buds, carrying on their surface adhesion molecules characteristic of embryonic cells. Of course, intermediate cell types are possible which possess combinations of such adhesion molecules in different quantitative proportions.

According to our model all these cell types are arranged in the limb bud in correspondence with their adhesion properties according to the rules proposed by Steinberg (1963). The process of cell sorting with respect to their adhesion properties begins in the prospective limb area and continues during limb outgrowth. Cells of chondrogenic differentiation are concentrated in the inner portion of LB. Cells of the second type of tissue move or stay in the surface direction under the ectodermal epithelium. Other cells occupy an intermediate position between these two.

As demonstrated by Cairns (1975) and Kwasigroch et al (1975), cells of LBs obtained from different proximal-distal (P-D) and anterior-posterior (A-P) levels differ in terms of adhesion. According to their data, cells with maximal adhesion are found in the posterior dorsal region of the distal portion of the LB, whereas minimal adhesion corresponds to the anterior ventral region in the proximal part.

Probably the main difference between the chondrogenic and non-chondrogenic subpopulation of cells is associated with their different capacity to form capillaries, and this in turn creates a gradient of oxygen tension between them. As proposed by Caplan and Koutroupas (1973), these

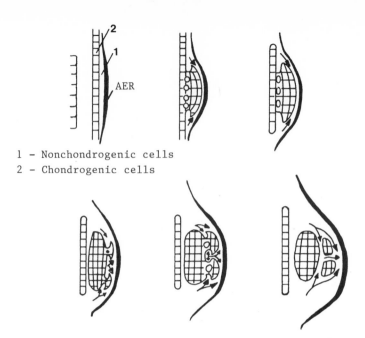

1 - Nonchondrogenic cells
2 - Chondrogenic cells

Diagram illustrating the "packaging model" of limb morphogenesis
(see text for details).

differences may be decisive for creating the direction of subsequent
differentiation of these tissues in the course of limb development. The LB
ectoderm could be involved in this process (Kosher et al 1979).

The phase of LB outgrowth is characterized by active cell proliferation
in the distal mesodermal part which is controlled by AER. On the other
hand, cells of the limb mesoderm maintain the AER in an active state. These
mutual influences of the two tissues are retained to the end of development.

We believe that the relative rate of proliferation and relative
movement (controlled by cell adhesion and proliferation) of chondrogenic and
non-chondrogenic subpopulations of cells within blastema are critical for
the size and shape of each skeletal element. For example, if the rate of
growth and migration of chondrogenic cells in the distal direction is
relatively higher than in non-chondrogenic cells, then cartilage boundaries
are laid down along the anterior-posterior (A-P) and dorsal-ventral (D-V)
axes of the limb. When the relative rates of growth and distal migration
are higher in non-chondrogenic subpopulations, then the distal boundary of
the prospective cartilage is formed and this results in the formation of the
joint. Separation of cartilage elements belonging to a row of cartilages
(radius) could take place in this way. In order to explain the appearance
of a new row of cartilage elements, Connelly and Bookstein (1983) propose a
model about differential growth of the LB along the A-P axis; they also
present experimental data in favour of this model. Their data provide
evidence about the increased incorporation of H^3-thymidine in the regions of
branching (appearance of a new row of cartilages). Asymmetry of
proliferation and perhaps adhesion of cells along the A-P axis can from time
to time lead to events that are described above for the longitudinal axis of
the regenerate.

Recently, new and intriguing data has been published about the gradient
of expression of homeobox containing genes in the developing limb bud. The

maximal number of these genes is expressed in posterior dorsal and distal parts of the limb bud which is in correspondence with the existing gradient of adhesion and. in the case of the axolotl limb. it also corresponds to the gradient of proliferation. Possible coincidence in time and space between the expression of homeobox-containing genes on the one hand, and the existence of zones of active morphogenesis on the other, leads to a hypothesis about participation of these genes in the control of spatio-temporal parameters of development in general and perhaps also of limb development in particular.

We believe that a similar mechanism may also operate in limb regeneration.

REFERENCES

CAIRNS. J. M., (1975). The function of the ectodermal apical ridge and distinctive characteristics of adjacent distal mesoderm in the avian wing bud. J. Embryol. exp. Morph. 34: 155-169.

CONNELLY, T. G., and BOOKSTEIN, F. L., (1983). Method for 3-dimensional analysis of patterns of thymidine labelling in regenerating and developing limbs. In "Limb Development and Regeneration. Part A". New York: Alan R. Liss Inc. pp 525-536.

KOSHER, R. A., SAVAGE. M. P., and CHAN S-C., (1979). In vitro studies on the morphogenesis and differentiation of the mesoderm subjacent to the apical ectodermal ridge of the embryonic chick limb bud. J. Embryol. exp. Morph. 50: 75-97.

KWASIGROCH. T. E., and KOCHAR, D. M., (1975). Locomotory behaviour of limb bud cells. Exptl. Cell Res. 95: 269-278.

STEINBERG, M. S., (1963). Reconstruction of tissues by dissociated cells. Science. 141: 401-411.

THE SOMITE-MUSCLE RELATIONSHIP IN THE AVIAN EMBRYO

Bodo Christ, Hans-Henning Epperlein, Helmut Flöel, Jörg Wilting

Anatomisches Institut
Albert-Ludwigs-Universität Freiburg
Freiburg, Federal Republic of Germany

INTRODUCTION

The trunk, limb, and tongue muscles originate from the somites (Christ et al. 1974, 1986; Chevallier et al. 1976). The muscle-forming cells (MF cells) are determined at very early stages of development (Krenn et al. 1988). The ways of MF cell separation from the dermomyotome are different with regard to the muscle groups examined. The intrinsic muscles of the back arise from the myotomes (Christ et al. 1986) which represent epithelially arranged derivatives of the somites. The abdominal muscles develop from the ventrally situated epithelial somite buds consisting of both dermatome and myotome cells (Christ et al. 1983). The limb and tongue muscles differentiate from MF cells that have migrated from the dermomyotomes (resp. dermatomes) to their destination as still replicating fibroblast-like cells. The modalities of MF cell separation seem to depend on the presence of different dermomyotome compartments which possibly arise by inter-actions between the somites and the environment (Jacob et al. 1974).
In this work earlier studies on MF cell migration and myotome formation (Christ et al. 1978; Jacob et al. 1978, 1979; Kaehn et al. 1988) are followed up. Using im-munohistochemistry, scanning electron microscopy, and microsurgery we investigate 1) the distribution of extracellular matrix (ECM) components around the somites during MF cell separation and 2) the ways of MF cell separation after neural tube extirpation or somite rotation.

RESULTS

At the limb level of chick embryos, MF cells begin to separate at the third day of development (Figs. 1-3). The formation of the myotome, which arises from the dermomyotome, starts in the cranio-medial corner of the somites (Fig. 3). Simultaneously, MF cells for the limb leave the lateral edge of the dermo-myotome. Migration commences in the anterior part of the somites and proceeds in a cranio-caudal fashion (Figs. 2, 3). There are two groups of MF cells. One of them migrates laterally and is still able to undergo mitosis, whereas the cells of the other group are postmitotic and situated within an epithelial layer (Christ et al. 1986). MF cells migrating laterally pass through the cell-free space above the Wolffian duct. The cells are oriented transversally and situated within an ECM network, the fibrils of which are also oriented laterally (Figs. 4,5). With MF cell migration which starts in the cranial part of each somite, ECM fibrils occur pre-dominantly in the area of migration. At later stages, ECM fibrils span the space between the whole lateral edge of the somites and the wing mesoderm of the pro-spective limb bud. The accumulation of ECM-material in the cranial region

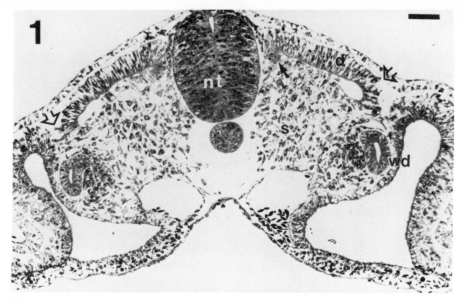

Fig. 1. Transverse section of a chick embryo (HH stage 15) at the wing level. Somites are differentiated into dermatome (d), myotome (arrow), and sclerotome (s). Note the emigrating muscle forming (MF) cells at the lateral edge of the dermatome (arrowheads). nt, neural tube; wd, Wolffian duct. Bar: 70 μm.

corresponds to earlier findings (Kaehn et al., 1988), in which the cranial edge of the dermomyotome was shown to be the source of the myotome cells. Thus, a possible morphogenetic signal for lateral MF cell migration and myotome formation might be localized within the cranial part of each somite.

The lateral migration of MF cells (Fig. 6) is preceded by an opening up of the fibronectin-rich basement membrane (BM). Results from our somite-rotation experiments (discussed later) suggest that an opening up of the BM is mainly stimulated by the environment of the limb anlage. After breaking up of the BM, lateral migration of MF cells proceeds into the mesoderm of the limb anlage on routes through a fibronectin-rich environment (Fig. 7). Conversely, anti-tenascin staining of the somite BM is largely lacking at the ventro-lateral part of the dermomyotome where MF cells start to migrate laterally (Fig. 8). These results suggest that fibronectin which is important for cell migration and adhesion (Greenberg et al., 1981; Hynes, 1981), seems to support lateral migration of MF cells whereas tenascin which does not support cellular adhesion and migration does not hinder MF cell migration (Chiquet-Ehrismann et al., 1986; Epperlein et al., 1988).

In order to understand how somite compartments are established and how the separation of MF cells is organized along the embryonic axis and with respect to limb development, microsurgical experiments were carried out.

In the first series of experiments fragments of the neural tube were removed in 2 d old chick embryos. Following extirpation of the neural tube, both paraxial mesodermal plates fuse and form unpair somites (Fig. 9). These somites are large but have the same sequence as those of unoperated embryos. Further segregation of the unpair somites into dermatome, myotome and sclerotome proceeds as normally (Fig. 10). However, the myoblasts of the myotome are irregularly arranged and may become transversally oriented (Jacob et al., 1974). Finally, the myotome cells desintegrate and disappear. MF cells for the limb leave the lateral dermomyotome edge in a normal way and continue their migration even into the limb region.

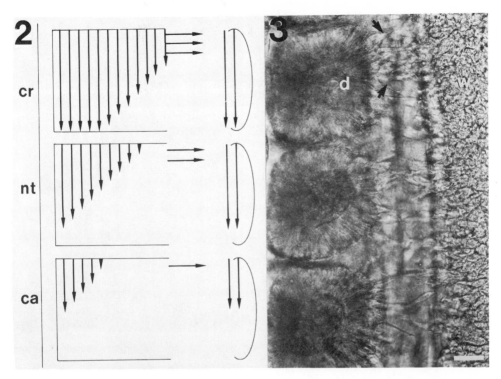

Figs. 2 and 3. Separation of MF cells. Fig. 2, schematic representation of myotome formation (vertical arrows) and MF cell migration (horizontal arrows) at a limb bud level. On the left, somites from the ventral aspect; on the right, somites in a sagittal section. Note cranial (cr) to caudal (ca) differentiation gradient. Fig. 3, MF cells (arrows) migrating from the lateral edges of the dermatome (d) to the wing anlage (w) passing through the cell-free space above the Wolffian duct. Migration of MF cells commences at the rostral part of the somite (compare also Figs. 2 and 5. nt, neural tube. Bar: 30 μm.

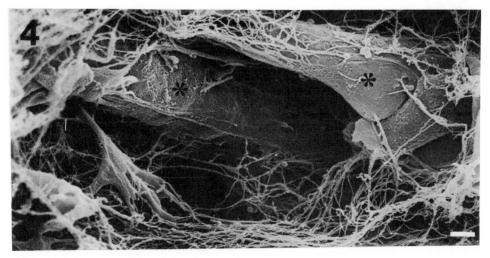

Fig. 4. Scanning electron micrograph (SEM) of laterally migrating MF cells (asterisks). Somites to the left: outside the figure; note transversally oriented ECM fibrils. Bar: 1 μm.

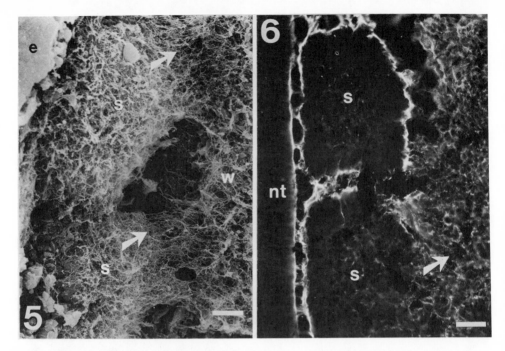

Fig. 5. SEM micrograph of two brachial somites (s) and adjacent wing anlage (w) after removal of the dorsal ectoderm (e). ECM fibrils are prominent in the rostral part of the somite (arrows), where MF cells commence migration. Bar: 20 µm.

Fig. 6. Fluorescence micrograph of a frontal cryostat section through a chick embryo (limb level) stained with anti-fibronectin (FN). Note migrating MC cells (arrow). nt, neural tube; s, somite. Bar: 25 µm.

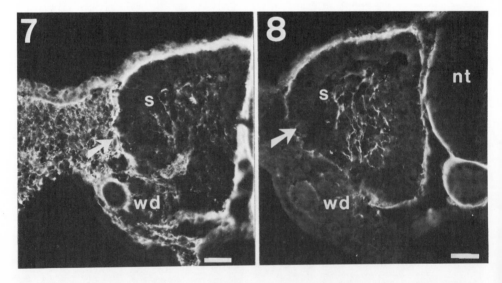

Fig. 7 and 8. Fluorescence micrographs of transverse cryostat sections through a chick embryo (limb level). Fig. 7, staining with anti-FN; Fig. 8, staining with anti-tenascin (TN). Note difference in anti-FN and anti-TN immunostaining with respect to the area of emigrating MF cells (arrow). nt, neural tube; s, somite; wd, Wolffian duct. Bars: 20 µm.

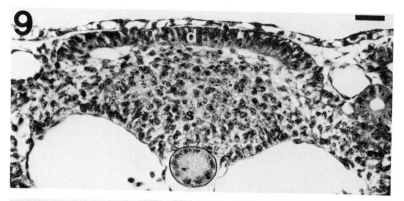

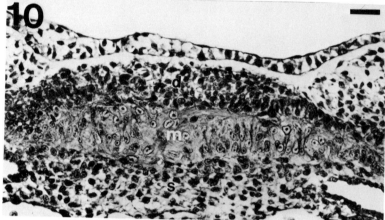

Figs. 9 and 10. Transverse sections of chick embryos from which the neural tube was removed at 2nd day of development (Fig. 9, reincubation period: 1d, Fig. 10, 2d). The plates of the paraxial mesoderm have fused and unpair somites have developed. The myotome cells (Fig. 10) are irregularly arranged. d, dermatome; m, myotome; s, sclerotome. Bars: 25 μm.

In the second experimental series, several somites or parts of unsegmented paraxial mesoderm of chick embryos were removed and replaced by those of the quail after rotation by 180° so that the lateral edge came into a medial position. Myotome formation and MF cell migration of the grafted quail mesoderm seem to occur as in the host, which suggests a host-specific influence (Fig. 11). Some MF cells, however, leave the somites medially, which indicates an intrinsic control of MF cell migration (Fig. 12). This difference may be explained by assuming that medially migrating cells were already pre-programmed for migration while those migrating laterally would not have formed under normal conditions. The latter seem to have been stimulated for lateral migration after confrontation with the host environment.

CONCLUDING REMARKS

At the limb level of avian embryos, the muscle forming (MF) cells separate from the somites in two ways: they form the myotomes and migrate laterally. Both processes start in the cranial part of each somite, yet in different compartments. The modalities of MF cell separation do not depend on the neural tube, but on environmental influences which might be mediated by ECM components.

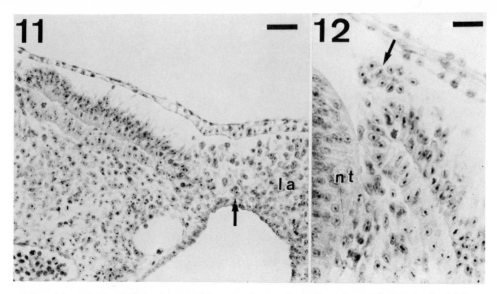

Figs. 11 and 12. Rotated paraxial mesoderm of quail embryos transplanted into chick embryos after turning their medial edge laterally. Reincubation period: 1d. nt, neural tube; la, limb anlage. Normal differentiation of the myotome. Orthotopic migration of MF cells (arrow) into the limb. Fig. 12. Cluster of somite-derived cells (arrow) probably representing MF cells situated at the originally lateral edge of the somite. Bars: Fig. 11, 25 µm; Fig. 12, 15 µm.

REFERENCES

Chevallier, A., Kieny, M. and Mauger, A. 1976, Sur l'origine de la musculature de l'aile chez les oiseaux, C.R.Acad.Sc.Paris, 282:309-311

Chiquet-Ehrismann, R., Mackie, E.J., Pearson, C.A. and Sakakura, T., 1986, Tenascin: an extracellular matrix protein involved in tissue interactions during fetal development and oncogenesis, Cell, 47:131-139.

Christ, B., Jacob, HJ. and Jacob, M., 1974, Über den Ursprung der Flügelmuskulatur. Experimentelle Untersuchungen mit Wachtel- und Hühnerembryonen,Experientia (Basel), 30:1446-1448.

Christ, B., Jacob, HJ. and Jacob, M., 1978, On the formation of the myotomes in embryos. An experimental and scanning electron microscope study, Experientia (Basel), 34:514-516.

Christ, B., Jacob, M., Jacob, HJ., Brand, B. and Wachtler, F., 1986, Myogenesis: a problem pf cell distribution and cell interactions. in: "Somites in developing embryos," DA. Ede, JW. Lash eds., NATO ASI Series, 118: 261-276.

Epperlein, HH., Halfter, W. and Tucker, R., 1988, The distribution of fibronectin and tenascin along migratory pathways of the neural crest in the trunk of amphibian embryos, Development, 103:743-756.

Greenberg, JH., Seppä, S., Seppä, H. and Hewitt AT., 1981, Role of collagen and fibronectin in neural crest cell adhesion and migration, Dev.Biol., 87: 251-266.

Hynes, R.O., 1981, Fibronectin and its relation to cellular structure and behaviour, in: "Cell biology of extracellular matrix, E.D. Hay ed., Plenum, New York, pp. 295-334.

Jacob, HJ., Christ, B. and Jacob, M., 1974, Die Somitogenese beim Hühner-
    embryo, Experimente zur Lageentwicklung des Myotom, Verh.Anat.Ges.,
    68:581-589.
Jacob, M., Christ, B. and Jacob, HJ., 1978, On the migration of myogenic
    stem cells into the prospective wing region of chick embryos, A scanning
    and transmission electron microscope study, Anat.Embryol., 153:179-193.
Jacob, M., Christ, B. and Jacob, HJ., 1979, The migration of myogenic cells
    from the somites into the leg region of avian embryos, An ultrastructural
    study, Anat.Embryol., 157:291-309.
Kaehn, K., Jacob, HJ., Christ, B., Hinrichsen, K. and Poelmann, RE., 1988,
    The onset of myotome formation in the chick, Anat.Embryol., 177:191-201.
Krenn, V., Gorka, P., Wachtler, F., Christ, B. and Jacob, HJ., 1988, On the
    origin of cells determined to form skeletal muscle in avian embryos,
    Anat.Embryol., 179:49-54.

OBSERVATIONS CONCERNING THE CONTROL OF DIRECTED MYOGENIC CELL

MIGRATION

Beate Brand-Saberi and Veit Krenn

Institute of Anatomy, Ruhr-University Bochum, FRG
Institute of Histology and Embryology, University
of Vienna, Austria

INTRODUCTION

Between the third and fifth day of development, the limb
buds of avian embryos are invaded by a cell population originating from the ventrolateral edges of the adjacent dermomyotomes at wing and leg level (Christ et al. 1974, 1977, 1983;
Chevallier et al. 1976; Jacob et al. 1978, 1979). These cells
are known to be the precursor cells of the skeletal limb muscles. Due to their site of origin near the embryonic body axis
and their destination in the limbs, the myogenic cells initially
have to migrate from a medial to a more lateral position across
the space between the somites and the somatopleural mesenchyme.
Once they have reached the limb anlagen, the myogenic cells
migrate towards the tips of the outgrowing buds (Wachtler et al.
1981, 1982; Brand et al. 1985).

The control of this directed proximo-distal migration is an
intriguing problem in developmental biology and there have been
attempts to explore its mechanism from various angles.
On the one hand, there is some evidence that the distally directed cell migration is controlled by the AER, as was pointed out
by Gumpel-Pinot et al. (1984), by Ede et al. (1984) and by Krenn
(1987). On the other hand, in vitro experiments showed that a
molecular gradient of platelet-derived growth-factor (PDGF) or
PDGF-like substances can influence myoblast migration (Venkatasubramanian and Solursh 1984). While the latter finding has not
yet been substantiated by in vivo results, the AER hypothesis
fits into the mosaic of evidence that developmental age and the
state of differentiation of the mesenchyme are the factors controlling myogenic cell migration in the avian limb bud (Brand
1987; Brand-Saberi et al. 1989; Brand-Saberi 1990; Brand-Saberi,
in press a). This concept was introduced by the term "juvenility
hypothesis" (Brand 1987; Brand-Saberi et al., in press b).

Developmental Patterning of the Vertebrate Limb
Edited by J.R. Hinchliffe *et al.*, Plenum Press, New York, 1991

This concept is based on a series of six experimental set-ups involving interspecific transplantations between embryos of the Japanese quail and White Leghorn chicken embryos. In the experiments, limb bud mesenchyme containing myogenic cells from quail embryos or quail somitic tissue was grafted into the dorsal myogenic region of chick embryos. Alternatively, distal parts of quail limb buds were grafted to chick limb bases. The implantation experiments were made with the intact host's AER left in situ as well as after excision of the AER. Different stage combinations were made in all types of experiments. Moreover, the grafts were placed at variable proximo-distal levels, so for instance into the progress zone.
Myogenic donor cells were scored as having migrated when muscle blastemas that were discontinuous with the grafting site were marked with quail nuclei. The distance between the migrated cells and the grafting site which was detectable by the presence of non-migratory quail tissues (cartilage, soft connective tissue) was at least 300 µm. Under the conditions stated above, a distal migration of myoblasts was found to occur independently of the presence or absence of the AER (Fig. 1).

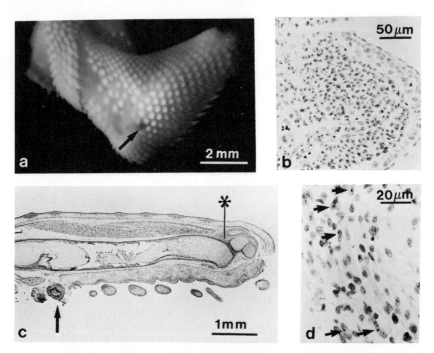

Fig. 1a Macroscopic aspect of a host wing that has received a graft (arrow) at the level of the proximal ulna after AER excision.

1b The dermis in the area of the arrow in 1a and 1c is partially made up of quail cells indicating the grafting site.

1c Longitudinal section through the operated wing with grafting site (arrow) and site of distal-most myotubes containing quail nuclei (asterisk), as shown in 1d.

1d Distal chimeric myotubes at the level of the asterisk in 1c containing quail nuclei (arrows).

Moreover, myogenic cell migration depended on the age relationship between donor and host, i.e. myoblasts only migrated when the host tissue was younger than the donor tissue.
When grafted to the progress zone, myogenic cells were found to migrate distally and proximally, even in older hosts (Fig. 2).

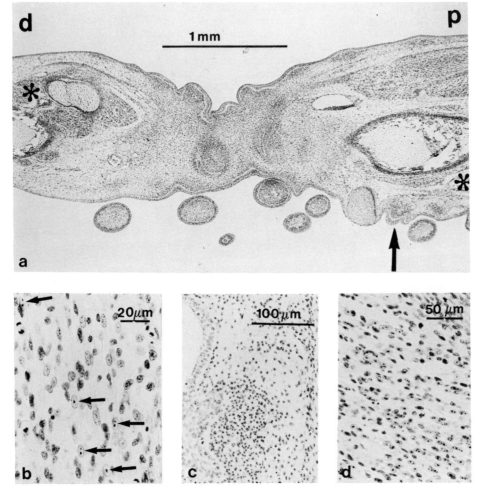

Fig. 2a Longitudinal section of a specimen that has received a graft from a <u>younger</u> quail embryo which was implanted into the host's progress zone. Arrow: grafting site. Asterisks: positions of migrated myogenic quail cells in muscle blastemas; p: proximal; d: distal.

2b Detail from a distal muscle blastema. Arrows point to quail nuclei in chimeric myotubes.

2c Detail of grafting site (arrow in 2a). Dermis and soft connective tissue are heavily marked with quail cells.

2d Detail from a muscle blastema proximal to the grafting site showing the abundance of quail nuclei in chimeric myotubes.

A proximally directed migration also occurred in combination experiments in which older distal parts of donor limb buds were grafted to younger host stumps (Fig. 3). Vice versa, there was no participation of myoblasts from the younger stump in distal muscle blastemas of the older donor. Finally, a proximal migration was induced by grafting a piece of younger chick mesenchyme proximal to a piece of quail mesenchyme in older host mesenchyme.

Summing up, the results indicate that no immediate signalling influence emanates from the AER to the myogenic cells, since migration can occur in the absence of the AER. Moreover, the absence of the AER can be compensated for by the choice of a grafting site that is either located distally or in a younger host mesenchyme. Thus, it could be argued that the AER has an indirect influence upon myogenic cell migration, since it keeps the distal mesenchyme (progress zone) in an undifferentiated state (Toole 1972). Once it has exerted its influence on the

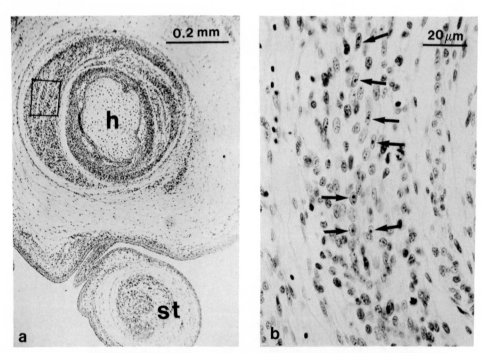

Fig. 3a Transverse section through a specimen that has re-
       ceived a distal limb graft from an older quail em-
       bryo. The graft was initially connected by a "stalk"
       (st) to the stump, but became detached later on.
       Myogenic cells from the graft must have traversed
       the stalk, since quail nuclei are found in chimeric
       muscle blastemas at humerus (h) level.
    3b Detail from the muscle blastema indicated in 3a.
       Arrows point to quail nuclei in myotubes.

distal mesenchyme cells, the AER itself is dispensable, at least for a certain time during which myogenic cells can migrate.

THE SUBSTRATE OF JUVENILITY

This leads us to the question, what is the substrate of "juvenility"? What is it that resides in a "young" and in a more distal mesenchyme and makes it attractive to myogenic cells?
There are several possibilities: "Juvenility" could be a property of the undifferentiated cells themselves, i.e. of the cell membrane that the migrating cells contact.
It could also be a property of the extracellular matrix, or even a combination of both. In any case, these properties could be based on a temporal and spatial gradient of adhesiveness in the limb bud. A developmental increase of adhesiveness leads to a cessation of myogenic cell migration and as a result, migration is always directed towards the younger and less differentiated mesenchyme.
A developmental mutant of the chick, the *talpid³*- mutant, has been shown to affect cell adhesiveness. *Talpid³*- cells are more adhesive to one another as was shown by in vitro experiments by Ede and Agerbak (1968) and Ede and Flint (1975 a, b). In interspecific grafting experiments, the migration of *talpid³*-myogenic cells in normal quail limb buds was found to be much reduced in comparison to normal myogenic cells (Lee and Ede 1989).

The Effect of Hyaluronic Acid on Myogenic Cell Migration

Looking at the extracellular matrix in limb buds, the only substance known to be distributed along a spatial gradient (Kosher et al. 1981) and to occur in early stages at a higher concentration than in later stages (Toole et al. 1984) is hyaluronic acid (HA) or - in its physiological form - hyaluronate. According to Kosher et al. (1981), it is present with a high concentration in the progress zone and with proximally decreasing levels throughout the dorsal and ventral myogenic zone.
Our own electron microscopical findings show that mesenchyme from young (st. 21 HH) donor embryos contains more HA than mesenchyme from old embryos (st. 25 HH; Fig. 4). So distal and "young" mesenchyme have something in common: the high concentration of HA which proximal and "old" mesenchyme do not have. Consequently, it appeared worth while to test the effect that exogenous HA has on the migratory behaviour of myogenic cells. To this purpose, we chose a stage relation between donor and host that does not normally support myogenic cell migration, i.e. the hosts were older than the donors. After having received the grafts, the host wings were micro-injected with 40 μg/μl HA (Merck) in phosphate-buffered saline (Fig. 5).
In order to trace the injected HA in the operated limb bud, its distribution in the mesenchyme was visualized at EM level using the ruthenium red method (Luft 1971 a, b) and tannic acid according to Singley and Solursh (1981) in some specimens (Fig. 6 a). Under these conditions, quail nuclei were found in chimaeric myotubes distal to the grafting site in 13 of 22 cases (Krenn et al., in press). In the cases where migration had taken place, the age difference between donor and host was no more than three HH-stages.

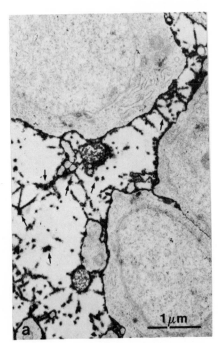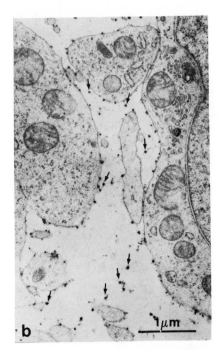

Fig. 4a Electron micrograph showing a detail from the limb
        bud mesenchyme of a stage 21 HH chick embryo. Ruthe-
        nium red-staining according to Luft (1971 a,b). The
        dark extracellular material consists of glycosamino-
        glycans.
   4b Electron micrograph showing a detail from the limb
        bud mesenchyme of a stage 25 HH chick embryo. The
        glycosaminoglycan content of the extracellular ma-
        trix is very much reduced.

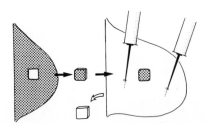

Fig. 5   Diagram showing the injection of hyaluronic acid
          into a chicken limb bud that has received a graft
          from the myogenic zone of a younger quail embryo.

This means that HA can render limb bud mesenchyme that does not support myogenic cell migration in controls (injected with PBS only, Fig. 6 b) again attractive to myoblasts. It has a rejuvenating effect on the mesenchyme, at least to a certain degree.
The adverse effect of HA on cell differentiation has been widely described for myoblasts and chondrocytes (Toole 1973; Elson and Ingwall 1980; Kosher et al. 1981), both cell types that need cell contact in order to differentiate.

Does exogenous HA act by keeping mesenchyme cells less differentiated or does it act via receptor proteins on the myoblasts (Toole et al. 1984; Underhill and Toole 1981) or simply by widening the intercellular spaces (Bellairs 1982) for the migrating cells?
In order to answer this question, we used a synthetic substance with physicochemical properties similar to those of HA: dextran sulfate (Wells 1973).
On injection of dextran sulfate (DS) into wing buds that had received grafts containing myogenic cells, these were again enabled to migrate in older host mesenchyme.
This rules out the possibility that HA acts via specific receptors on the myoblasts, since myoblasts can be assumed not to possess receptors for synthetic substances.

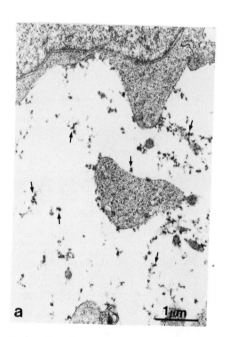 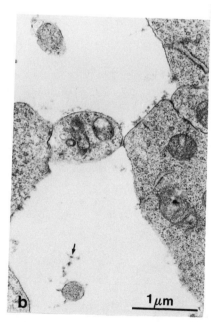

Fig. 6a Hyaluronic acid (HA) visualized within the wing mesenchyme of a chick embryo of stage 22 HH having received a microinjection of HA. Method according to Singley and Solursh (1981). Arrows: HA.
6b Hyaluronic acid (HA) visualized within the wing mesenchyme of a stage 22 HH-control embryo having received an injection of saline. Method according to Singley and Solursh (1981). Arrows: HA

It seems that HA acts by elevating the swelling pressure of the mesenchyme and by opening the intercellular spaces to the migrating myoblasts. Since it is concentrated at the distal tip of the limb bud, myoblasts migrate towards the open intercellular spaces in the distal mesenchyme. So directed migration is the consequence of myogenic cells following a gradient of HA? Or do myogenic cells "read" cell surface properties reflecting the state of differentiation of the mesenchyme cells they encounter? This state of differentiation is kept low in the presence of high levels of HA and possibly by any other substance that keeps cells apart and hinders differentiation in this way. Once differentiation of the mesenchyme cells is too advanced to be held up and/or the cell surface properties of "juvenility" have disappeared, the injection of exogenous HA or DS does not have an effect, as in the cases in which the age of host and donor differed by more than three stages.

The Possible Role of the Cell Membrane in Myogenic Cell Migration

There are indications for a participation of the cell membrane in the control of myogenic cell migration. One of them is the reactivity of the distal limb bud mesenchyme with a monoclonal antibody raised against young chick limb buds (Brand-Saberi, in press a). The strongest fluorescence occurs in the progress zone with a decrease in the dorsal and ventral myogenic area (Fig. 7). The fluorescence does not occur when the tissue is fixed with acetone, or treated with a detergent (triton), suggesting that an antigen of the cell surface is detected rather than an intracellular substance.

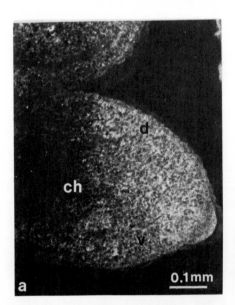
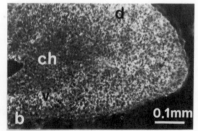

Fig. 7a,b   A monoclonal antibody binding to the distal zone of chick limb buds and to the dorsal and ventral zone. Longitudinal sections through stage 20 HH limb buds. d: dorsal; v: ventral; ch: chondrogenic zone.

Another indication that myogenic cells use cell membrane signals in addition to extracellular matrix components is the presence of focal and close contacts (10 - 20 nm) between myogenic and mesenchyme cells in the limb bud (Fig. 8).

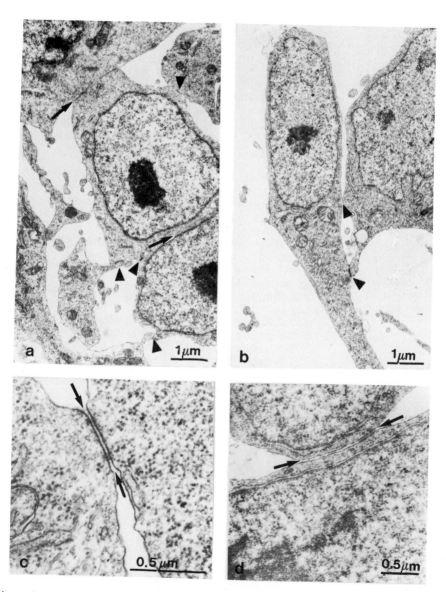

Fig. 8a Cell-contacts in the limb mesenchyme after interspecific replacement of the brachial somites. Arrowheads:
focal contacts. Arrows: close contacts.
8b Two quail cells contacting one another and processes from mesenchyme cells.
8c Quail cell contacting a mesenchyme cell. Note
that the shape of the quail cell suggests that it is
migrating.
8d and 8d Higher magnification of close contacts
between quail and chick cells.

Especially the latter type of cell contact has been described in another population of cells invading the limb bud: the
axon growth cones of the neurons (Lewis et al. 1983). In grafting experiments in which older wing buds were combined with
younger stumps, axons were found to invade the limb buds very
rapidly. In the converse experiments, their growth was held up
by grafting younger wing buds to limb bud stumps to which the
axons had already well advanced (Swanson and Lewis 1982). So
neurons seem to have a preference for the opposite properties of
the mesenchyme: they invade "old" rather than "young" mesenchyme.

Yet another population of cells, the endothelial cells of
the limb bud are known to avoid areas of high HA-concentrations
in the avascular sleeve near the ectoderm (Feinberg and Beebe
1983; Solursh 1984). If myogenic cell migration was merely guided by the concentration of HA, they would end up in the avascular sleeve of the limb bud, and in the progress zone, or at
least they would also colonize these areas.
Since myogenic cells are known not to do so (-the only migratory
populations in the prospective dermis are neural crest-derived
melanoblasts and perhaps lymphatic cells-) we postulate that
myoblast orientation within the limb bud is achieved by a balance between the space available for migration and local cell-
contacts to mesenchymal cells which can only be made when the
concentration of HA is not too high. In this way, they can migrate under normal conditions in a front just proximally to the
progress zone (Brand et al. 1985), but not actually in it.
Once differentiation of the mesenchyme cells has started, the
first changes are likely to occur at the level of the cell membrane. Even if space is now made available (caused by exogenous
HA or DS), the mesenchyme cells are no longer attractive for
transient contacts, maybe because their adhesiveness is too
strong: so myogenic cells stop to migrate.

Acknowledgements

The authors wish to thank Hans Hake and Zhao Bin for technical
assistance, Marion Koehn for photographic work and Brigitte
Scharf for typing the manuscript.

REFERENCES

Bellairs, R., 1982, Gastrulation processes in the chick embryo.
    In: Cell Behaviour (Bellairs, R., Curtis, A., Dunn, G.
    eds.), pp. 395-428, Cambridge University Press, London
Brand, B., 1987, Zur Morphogenese und Musterbildung der Vogel-
    extremität unter besonderer Berücksichtigung der Migra-
    tion der myogenen Zellen. Eine experimentelle, ultra-
    strukturelle und vergleichende Untersuchung.
    Dissertation, Bochum.
Brand-Saberi, B., Krenn, V., Christ, B., 1989, The control of
    directed myogenic cell migration in the avian limb bud.
    Anat. Embryol., 180:555

Brand-Saberi, B., 1990, Interactions in myogenic cell migration. Verh. Anat. Ges., 83 (Anat. Anz. Suppl. 166):93

Brand, B., Christ, B., Jacob, H.J., 1985, An experimental analysis of the developmental capacities of distal parts of avian leg bud. Am. J. Anat., 173:321

Brand-Saberi, B., in press (a) Befunde zur Kontrolle der gerichteten Zellmigration myogener Zellen in der Vogelextremität. Verh. Anat. Ges., 85 (Anat. Anz. Suppl. 168)

Brand-Saberi, B., Krenn, V., Ostermayer, H., in press (b), Die Juvenilitätshypothese: Ein Erklärungsansatz zum proximo-distalen Migrationsverhalten von Myoblasten. Verh. Anat. Ges., 85 (Anat. Anz. Suppl. 168)

Chevallier, A., Kieny, M., Mauger, A., 1976, Sur l'origine de la musculature de l'aile chez les oiseaux. C.R.Acad.Sci.Ser D 282:309

Christ, B., Jacob, H.J., Jacob, M., 1974, Über den Ursprung der Flügelmuskulatur. Experimentelle Untersuchungen an Wachtel- und Hühnerembryonen. Experientia, 30:1446

Christ, B., Jacob, H.J., Jacob, M., 1977, Experimentelle Befunde zur Muskelentwicklung in den Extremitäten von Hühnerembryonen. Verh. Anat. Ges., 71:1231

Christ, B., Jacob, H.J., Jacob, M., Wachtler, F., 1983, On the origin, distribution and determination of avian limb mesenchymal cells. In: Limb Development and Regeneration (Kelley, Goetinck, MacCabe, eds.) Part B, pp. 281-291, A. Liss, New York.

Ede, D.A., Agerbak, G.S., 1968, Cell adhesion and movement in relation to the developing limb pattern in normal and talpid mutant chick embryos. JEEM, 20:81

Ede, D.A., Flint, O.P., 1975 a, Intercellular adhesion and formation of aggregates in normal and talpid mutant chick limb mesenchyme. J. Cell Sci., 18:97

Ede, D.A., Flint, O.P., 1975 b, Cell movement and adhesion in the developing chick wing bud: Studies on cultured mesenchyme cells from normal and talpid mutant embryos. J. Cell Sci., 18:301

Ede, D.A., Gumpel-Pinot, M., Flint, O.P., 1984, Orientated movement of myogenic cells in the avian limb bud and its dependence on presence of the apical ectodermal ridge. In: Matrices and cell differentiation. (Kemp, RB, Hinchliffe, JR, eds.), pp. 427-438, A. Liss, New York

Elson, H.R., Ingwall, J.S., 1980, The cell substratum modulates skeletal muscle differentiation. J. Supramol. Struct., 14:313

Feinberg, R.N., Beebe, D.C., 1983, Hyaluronate in vasculogenesis Science, 220:1177

Gumpel-Pinot, M., 1974, Contribution du mésoderme somitique à la genèse du membre chez l'embryon d'Oiseau. C.R.Acad. Sci. Ser. D 279:1305

Jacob, M,, Christ, B., Jacob, H.J., 1978, On the migration of myogenic stem cells into the prospective wing region of chick embryos. Anat. Embryol., 153:179

Jacob, M., Christ, B., Jacob, H.J., 1979, The migration of myogenic cells from the somites into the leg region of avian embryos. Anat. Embryol., 157:291

Kosher, R.A., Savage, M.P., Walker, K.H., 1981, A gradation of hyaluronate accumulation along the proximodistal axis of the embryonic chick limb bud. JEEM, 63:85

Krenn, V. 1987, Über das Migrationsverhalten von Myoblasten in der Extremitätenanlage. Verh. Anat. Ges., 81:561

Krenn, V., Brand-Saberi, B., Wachtler, F.,(submitted for publication), Hyaluronic acid influences the migration of myoblasts within the avian embryonic wing bud.

Lee, N.K.H., Ede, D.A., 1989, The capacity of normal and talpid mutant fowl myogenic cells to migrate in quail limb buds. Anat. Embryol., 170:395

Lewis, A., Al-Ghaith, L., Swanson, G., Khan, A., 1983, The control of axon outgrowth in the developing chick wing. In: Limb Development and Regeneration (Fallon, JF, Caplan, AI, eds.), Part A, pp. 195-205, A. Liss, New York

Luft, J.H., 1971 a, Ruthenium red and violet. I. Chemistry, purification, methods of use for electron microscopy and mechanism of action. Anat. Rec., 171:347

Luft, J.H., 1971 b, Ruthenium red and violet. II. Fine structural localization in animal tissues. Anat. Rec., 171:369

Singley, C.T., Solursh, M., 1981, The spatial distribution of hyaluronic acid and mesenchymal condensation in the embryonic chick wing. Dev. Biol., 84:102

Solursh, M., 1984, Ectoderm as a determinant of early tissue pattern in the limb bud. Cell Differ., 15:17

Swanson, G.J., Lewis, J., 1982, The timetable of innervation and its control in the chick wing bud. JEEM, 71:121

Toole, B.P., 1972, Hyaluronate turnover during chondrogenesis in the developing chick limb and axial skeleton. Dev. Biol., 29:321

Toole, B.P., 1973, Hyaluronate and hyaluronidase in morphogenesis and differentiation. Am. Zool., 13:1061

Toole, B.P., Goldberg, R.L., Chi-Rosso, G., Underhill, C.B., Orkin, R.W., 1984, Hyaluronate-cell interactions. In: The Role of Extracellular Matrix in Development, pp. 43-66, A. Liss, New York

Underhill, C.B., Toole, B.P., 1981, Receptors for hyaluronate on the surface of parent and virus transformed cell lines. Binding and aggregation studies. Exp. Cell Res., 131:419

Venkatasubramanian, K., Solursh, M., 1984, Chemotactic behavior of myoblasts. Dev. Biol., 104:428

Wachtler, F., Christ, B., Jacob, H.J., 1981, On the determination of mesodermal tissues in the avian embryonic wing bud. Anat. Embryol., 161:283

Wachtler, F., Christ, B., Jacob, H.J., 1982, Grafting experiments on determination and migratory behaviour of presomitic, somitic and somatopleural cells in avian embryos. Anat. Embryol., 164:369

Wells, J.D., 1973, Salt activity and osmotic pressure in connective tissue. I. A study of solutions of dextran sulphate as a model system. Proc. R. Soc. Lond. B., 183:399

EXTRACELLULAR MATRIX AND MUSCLE FORMATION

Arnold I. Caplan

Skeletal Research Center
Department of Biology
Case Western Reserve University
Cleveland, Ohio 44106   USA

Introduction

Something doesn't form from nothing. Thus, for a morphology to
develop, a guide or boundaries must be established; this guide is referred
to as the progenitor structure. Morphologies form from a genomically
coded sequence of events in which both intrinsic and extrinsic factors for
developing and differentiating cells combine to define "positional information" allowing cells to fabricate tissues within specific shapes and
boundaries and, thus, generating specific shapes. For example, how do
individual muscle morphologies form? What are the progenitor boundaries?
Figure 1 shows the morphology of chick embryonic thigh muscles at Hamburger-Hamilton developmental stage 36. What controls the formation of
these individual muscles or, conversely, how do specific muscle masses
segregate from one another? These questions stimulated us to study the
extracellular molecules and cellular aspects of chick embryonic thigh
muscle morphogenesis, with special emphasis on the boundaries of these
events.

There are several cell types found in developing muscle: myoblasts,
myotubes, connective tissue fibroblasts, vascular endothelial cells and
nerve or neural support cells. Importantly , it is the specific interactions of all of these cells with one another which controls, contributes
to and supports morphogenesis. No one cell type alone is capable of
forming a muscle; the myotubes, alone, are able to fabricate the functional aspect of this tissue, the contractile apparatus. However, they cannot
form a discrete morphology, a muscle, without interacting with the surrounding connective tissue fibroblasts. Thus, in order to better understand the individual components of muscle morphogenesis, it is necessary
to dissect out the various cellular and molecular layers and contributors.
For example, several investigators have clearly demonstrated that, although every myotube is eventually surrounded by a basement membrane,
greater than 90% of the molecules in this extracellular matrix structure
are synthesized by surrounding fibroblasts, not by the myotubes[2-6]. This
example clearly demonstrates the complexity of the events involved in the
genesis of muscle substructure. Indeed, in sectioned specimens, it is
difficult to identify the cells which are fabricating the myotube basement
membrane in vivo; it appears that there are few, if any, fibroblasts close
enough to the developing myotubes to fabricate the surrounding basement
membranes.

Developmental Patterning of the Vertebrate Limb
Edited by J.R. Hinchliffe *et al.*, Plenum Press, New York, 1991

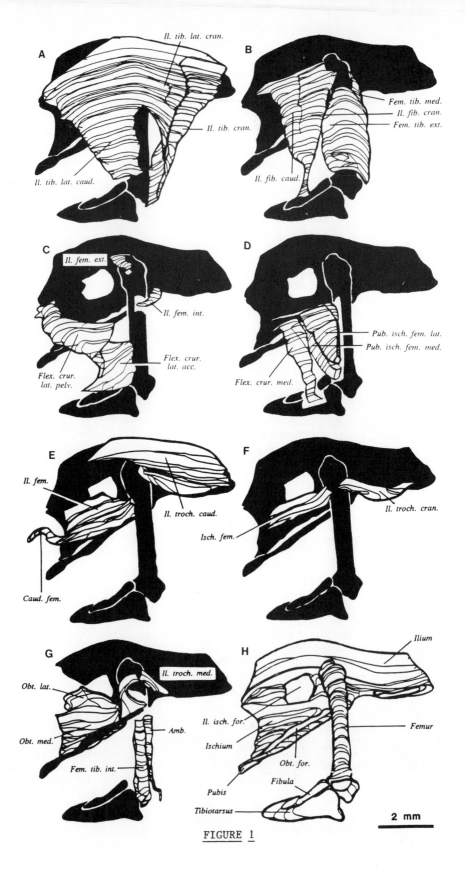

FIGURE 1

Three families of extracellular matrix molecules can be found in muscle and other developing and maturing tissues: collagens, proteoglycans and glycoproteins. The collagens often provide rigid molecules which control the tensile strength of tissue. Proteoglycans interact with these collagens in precise ways and the proteoglycans, by virtue of their large, highly negatively charged saccharides, structure many times their mass of water. And the glycoproteins function to connect cells to each other or to the matrix itself. Our previous studies of the biosynthesis of proteoglycans synthesized by muscle both in vivo and in vitro during development, maturation, regeneration and aging indicate the presence of at least one development- and muscle-specific large chondroitin sulfate proteoglycan (M-CSPG). This molecule is depicted in Figure 2[7] and is synthesized[8-10] by myoblasts and myotubes only during embryonic development and muscle regeneration in adults[11]. We have hypothesized that, because of the large size of this proteoglycan and the unusually large chondroitin-6-sulfate glycosaminoglycan chains, these molecules serve both to reserve space for future myotube expansion and serve as part of the progenitor matrix for replacement by muscle basement membranes[12]. Although we analyzed the newly synthesized proteoglycans temporally, we did not fully analyze the morphological distribution of these important matrix molecules.

Experimentation

To further explore the genesis of extracellular matrix boundaries and morphogenesis, we have used antibodies specific to unique extracellular matrix molecules. The objective is to identify the molecular components of progenitor matrices and to understand the precise replacement of these molecules with the final or adult molecules. To accomplish such a precise and detailed study, a specific, unique locus must be sampled at several different developmental time-points. This requires that distinctive landmarks be used to identify a specific locus at different times. Because no such landmarks are available, Richard Drushel[13] serially sectioned a chick embryo thigh at stages 36, 39 and 43. From these sections, he reconstructed the 3-dimensional arrangement of bone and musculature

---

Figure 1. Three-dimensional reconstruction of St. 36 (day 10) embryonic chick thigh musculature in right lateral view[13]. Graphical reconstruction was performed by orthogonal projection of tracings from serial cross sections. A-G show successively deeper layers of muscles (lateral to medial). H shows the underlying skeleton. In C, D and G some of the skeleton (black) has been cut away to better show medial muscle attachments. Abbreviations: Amb. ambiens, Caud. fem. caudo-femoralis, Fem. tib. ext. femoro-tibialis externus, Fem. tib. int. femoro-tibialis internus, Fem. tib. med. femoro-tibialis medius, Flex. crur. med. flexor cruris medialis, Flex. crur. lat. pelv. flexor cruris lateralis pars pelvica, Flex. crur. lat. acc. flexor cruris lateralis pars accessoria, Il. fem. Ilio-femoralis, Il. fem. ext. ilio-femoralis externus, Il. fem. int. ilio-femoralis internus, Il. isch. for. ilio-ischiadic foramen, Il. fib. cran. ilio-fibularis pars cranialis, Il. fib. caud. ilio-fibularis pars caudalis, Il. tib. cran. ilio-tibialis cranialis, Il. tib. lat. cran. ilio-tibialis lateralis pars cranialis, Il. tib. lat. caud. ilio-tibialis lateralis pars caudalis, Il. troch. caud. ilio-trochantericus caudalis, Il. troch. cran. ilio-trochantericus cranialis, Il. troch. med. ilio-trochantericus medialis, Isch. fem. ischio-femoralis, Obt. for. obturator foramen, Obt. lat. obturator lateralis, Obt. med. obturator medialis, Pub. isch. fem. lat. pubo-ischio-femoralis lateralis, Pub. isch. fem. med. pubo-ischio-femoralis medialis. Reproduced from Drushel and Caplan[13], used by permission of The Journal of Morphology, Alan R. Liss, Inc. (1991).

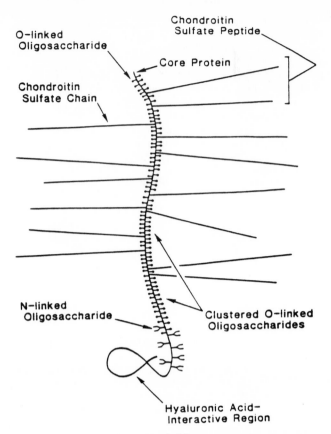

Figure 2. Schematic representation[6-10] of the large chondroitin sulfate proteoglycan synthesized by embryonic chick skeletal muscle (M-CSPG). The core protein is indicated by a heavy line, the chondroitin sulfate chains by lighter lines, the O-linked oligosaccharides by solid circles and the N-linked oligosaccharides by forked lines.

(the exact anatomy) as depicted in Figure 1. Based on examinations of both the serial sections and the reconstructions, we identified the junction of the pars pelvica and pars accessoria of the flexor cruris lateralis muscle as a landmark conserved throughout later thigh development (Figure 3). Using this landmark, we are able to return to a precise locus when we immunostain as a function of development. Because we are sampling at one precise locus, the temporal aspects of cellular and molecular dynamics may be more confidently assessed.

We have selected three families of monoclonal antibody probes: one which binds to various glycosaminoglycans, one against specific proteoglycan core proteins which have these glycosaminoglycans attached and one to specific matrix glycoproteins and collagens[14]. The detailed observations will be presented elsewhere[15] along with the summary provided in Table 1. from Fernandez et al[15].

The observations focus on four regions which undergo distinct patterns of expression and organization of extracellular matrix components during muscle development. An example of the complex immunostaining patterns we observed is provided in Figure 4. Importantly, some of the components exhibit coordinate regulation, while others are independently

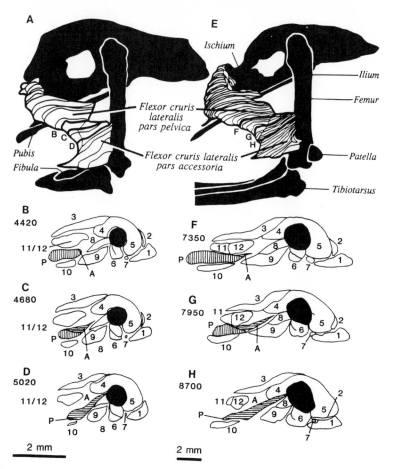

<u>Figure</u> <u>3</u>. Example of a conserved muscular anatomic landmark: junction between the pars pelvica and pars accessoria of the chick flexor cruris lateralis muscle[15]. <u>A</u>. Three-dimensional reconstruction at St. 36 (day 10). <u>B</u>-<u>D</u>. Representative cross sections from <u>A</u> showing the junction at its beginning (<u>B</u>), middle (<u>C</u>) and end (<u>D</u>). P, vertical stripes: pars pelvica. A, horizontal stripes: pars accessoria. <u>E</u>. Three-dimensional reconstruction at St. 39 (day 13). <u>F</u>-<u>H</u>. Representative cross sections from <u>E</u>. Numbers represent the distance in micrometers of each section from the iliac crest along the proximo-distal axis. Number key for muscles: 1 ilio-tibialis cranialis, 2 ilio-tibialis lateralis pars cranialis, 3 ilio-tibialis lateralis pars caudalis, 4 femoro-tibialis externus, 5 femoro-tibialis medius, 6 femoro-tibialis internus, 7 ambiens, 8 pubo-ischio-femoralis lateralis, 9 pubo-ischio-femoralis medialis, 10 flexor cruris medialis, 11 ilio-fibularis pars caudalis, 12 ilio-fibularis pars cranialis. Reproduced from Fernandez <u>et</u> <u>al</u>.[15], submitted.

regulated (<u>Table</u> <u>1</u>). The <u>perimysium</u> contains dermatan·sulfate core protein and dermatan-sulfate glycosaminoglycan epitopes at stage 39, chondroitin-4- and -6-sulfate epitopes by stage 43 and keratan sulfate epitope by stage 46. The <u>epimysium</u> acquires keratan sulfate and chondroitin-4-sulfate epitopes by stage 39 and chondroitin-6-sulfate by stage 43. The <u>endomysium</u> does not stain with dermatan sulfate proteoglycan core protein epitope until stage 43 and dermatan sulfate glycosaminoglycan is not detected in this region until stage 46; keratan and chondroitin-4-sul-

Table 1. MY-174, 383, etc. are monoclonal antibodies to the core protein, glycosaminoglycan epitopes on specific proteoglycans or matrix proteins such as tenascin (M1). Epimysium (Ep), perimysium (P), endomysium (Em), weakly reactive (+) strongly reactive (++) unreactive (−), structure not present (*). Reproduced from Fernandez et al[15], submitted.

DISTRIBUTION OF EXTRACELLULAR MATRIX EPITOPES
DURING CHICK LEG MUSCLE DEVELOPMENT

			MY-174	3B3	CB-1	2B6/ABC	5D4	2B6/ACII	F-31	F-33	M1
St. 36		Ep	−	−	++	++	−	−	−	−	−
		P	*	*	*	*	*	*	*	*	*
		Em	++	−	+	−	−	−	++	+	−
St. 39		Ep	−	−	++	++	++	++	−	−	−
		P	++	++	++	++	−	−	−	−	−
		Em	+	++	−	−	−	−	++	+	++
St. 43		Ep	++	++	++	++	++	++	−	−	−
		P	++	++	++	++	−	+	−	−	−
		Em	++	++	++	−	−	−	++	++	++
St. 46		Ep	++	++	++	++	++	++	−	−	−
		P	++	++	++	++	++	+	−	−	−
		Em	++	++	++	++	−	−	++	++	++

fate are undetected in this region. The myotendinous junction is localized as a specialized region within the epimysium by the binding of both the probes for tenascin and keratan sulfate.

Interpretation

Based on the observations of Mayne, Kieny and others[2-6,14], we assume that muscle fibroblasts are responsible for synthesizing the majority of the matrix molecules associated with the various matrix boundaries. If this is the case, then there are either uniquely different families of fibroblasts responsible for each of these regions or, more likely, the same fibroblasts synthesize uniquely different combinations of matrix molecules in response to different local cues or positional information. This further implies that, since muscle size, shape and insertion (myotendinous junction) pattern are genetically determined, such cues are stringently regulated and somehow specify not only the type of molecules synthesized but the exact location of their deposition, concentration and organization. This level of regulation is not only the case for proteins (i.e., collagens, proteoglycans, core protein, tenascin, etc.) but also for post-translational events such as the sulfation pattern of sugars in glycosaminoglycans or the epimerization of glucuronic to iduronic acid to form dermatan chains. The level of molecular management necessary to establish, modify and eventually replace these boundaries with adult structures is complex in both its precision and its temporal control. Our

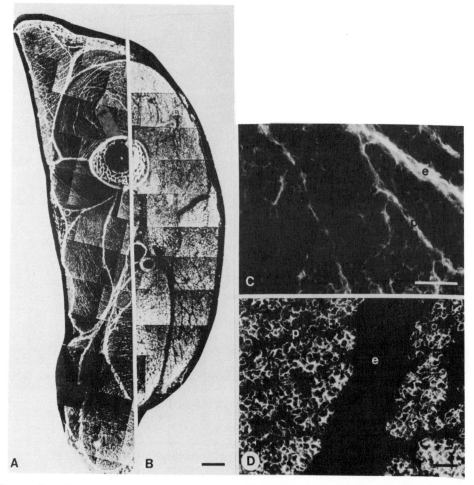

Figure 4. Cross-sections through stage 39 (day 13) thigh. A. shows the distribution of muscle chondroitin sulfate proteoglycan core protein (MY-174) in epimysium and perimysium. B. Chondroitin-6-sulfate glycosaminoglycan (3B3/ABC) shows staining in the region of myotubes and an absence of staining in epimysium and perimysium. Bar represents 500 um. C. High magnification of MY-174 staining showing the positive reaction in epimysium, perimysium and also around groups of myotubes. C. High magnification of 3B3/ABC staining around groups of myotubes but not in the epimysium or perimysium. Perimysium (p), epimysium (e). Bar represents 50 um. Reproduced from Fernandez et al.[15], submitted.

current studies have established the basic outline of these boundaries and their changes during morphogenesis. In the future, we intend to identify the controlling elements and ascertain how these morphology-generating events are regulated.

Acknowledgements

The experimentation reviewed here represents the efforts of D.A. Carrino, J. E. Dennis, R. F. Drushel and M. S. Fernandez, to whom I owe my thanks for their collaborative efforts and thoughtful interaction. Supported, in part, by grants from NIH and Muscular Dystrophy Association of America.

References

1. V. Hamburger and H. L. Hamilton, A series of Normal Steps in the Development of the Chick Embryo, J. Morph. 88:49-92 (1951).

2. R.D. Sanderson, J.M. Fitch, Linsenmayer and R. Mayne, Fibroblasts Promote the Formation of a Continuous Basal Lamina During Myogenesis In Vitro, J. Cell Biol. 102:740-747 (1986).

3. U. Kuhl, M. Ocalan, R. Timpl, R. Mayne, E. Hay and K. von der Mark, Role of Muscle Fibroblasts in the Deposition of Type IV Collagen in the Basal Lamina of Myotubes, Differentiation 28:164-172 (1984).

4. R. Mayne, S. Swasdison, R.D. Sanderson and M.H. Irwin, Extracellular Matrix, Fibroblasts and the Development of Skeletal Muscle, in: "Cellular and Molecular Biology of Muscle Development," Alan R. Liss, Inc., New York (1989).

5. M. Kieny and A. Mauger, Immunofluorescent Localization of Extracellular Matrix Components During Muscle Morphogenesis in Normal Chick Embryos, J. Exp. Zool. 232:327-341 (1984).

6. U. Kuhl, R. Timpl and K. von der Mark, Synthesis of Type IV Collagen and Laminin in Cultures of Skeletal Muscle Cells and Their Assembly on the Surface of Myotubes, Dev. Biol. 93:344-354 (1982).

7. D.A. Carrino and A.I. Caplan, Structural Characterization of Chick Embryonic Skeletal Muscle Chondroitin Sulfate Proteoglycan, Conn. Tiss. Res., 19:35-50 (1989).

8. D.A. Carrino and A.I. Caplan, Isolation and Preliminary Characterization of Proteoglycans Synthesized by Skeletal Muscle, J. Biol. Chem., 257:14145-14154 (1982).

9. D.A. Carrino and A.I. Caplan, Isolation and Partial Characterization of Proteoglycans Synthesized In Vivo by Embryonic Chick Skeletal Muscle and Heart, J. Biol. Chem., 259:12419-12430 (1984).

10. H.E. Young, D.A. Carrino and A.I. Caplan, Change in Synthesis of Sulfated Glycoconjugates During Muscle Development, Maturation and Aging in Embryonic to Senescent CBF-1 Mouse, Mech. Ageing Devel., 53:179-193 (1990).

11. D.A. Carrino, U. Oron, D.G. Pechak and A.I. Caplan, Reinitiation of Chondroitin Sulfate Proteoglycan Synthesis in Regenerating Skeletal Muscle, Develop. 103:641-656 (1988).

12. D.A. Carrino and A.I. Caplan, Proteoglycan Synthesis During Skeletal Muscle Development, in: "Molecular Biology of Muscle Development, UCLA Symposia on Molecular and Cellular Biology," Vol. 29 (C. Emerson, D.A. Fischman, B. Nadal-Ginard and M.A.Q. Siddiqui, eds.), Alan R. Liss, Inc., New York (1986).

13. R.F. Drushel and A.I. Caplan, Three-Dimensional Reconstruction and Cross-Sectional Anatomy of the Thigh Musculature of the Developing Chick Embryo (Gallus gallus), J. Morphol. 208:1-17 (1991).

14. R. Mayne and R.D. Sanderson, The Extracellular Matrix of Skeletal Muscle, Collagen Rel. Res. 5:449-468 (1985).

15. M.S. Fernandez, J.E. Dennis, R.F. Drushel, D.A. Carrino, K. Kimata, X. Yamagata and A.I. Caplan, submitted (1991).

CONTROL OF MUSCLE MORPHOGENESIS AND ENDPLATE PATTERN

IN LIMB MUSCLES OF AVIAN CHIMERAS

Milos Grim

Department of Anatomy
Charles University Prague
Czechoslovakia

INTRODUCTION

Formation of limb muscles involves migration of undifferentiated myogenic cells from somites into the mesenchymal limb rudiment, subsequent differentiation, and in situ pattern formation that eventually gives rise to individual muscles (for review Christ et al., 1986).

Myogenic cells are determined long before their migration starts (Krenn et al., 1988). However, they do not contain any information concerning muscle pattern (Chevallier et al., 1977; Christ et al., 1978; Wachtler et al., 1981; Jacob et al., 1983; Grim et al., 1986). Therefore it can be assumed that determination of muscular pattern is the result of interactions between myogenic and non-myogenic cells. The mode of this interaction remains to be studied.

The quail-chick marker technique (Le Douarin, 1969) and the species-specific differences either in form and pennation or in the distribution of muscle fibre types between muscles of the chick and quail (Grim et al., 1983; 1985) allows to analyse the development of motoneuron-muscle interactions and muscle pattern formation in chimeric embryos. In interspecific grafting experiments the source of myogenic material of limb muscles was heterotopically replaced or the nerves were forced to innervate an inappropriate limb.

RESULTS AND CONCLUSIONS

The development of four muscles was analysed: 1) the extensor medius brevis, which shows in the chick wing a well formed muscle belly, while in the quail its anlage degenerates after the 12th day of development; 2) the flexor digitorum superficialis muscle, which is a thin long muscle in the chicken wing. In the quail this muscle is shorter and more vigorous; 3) the ulnimetacarpalis muscle, which consists of a focally and a multiply innervated portion in the chicken as well as in the quail wing; 4) the plantaris muscle, which in the chicken leg consists mainly of multiply innervated slow fibres, while in the quail focally innervated fast muscle fibres predominate. This difference provides a possibility to force appropriate (i.e. originating from the same motoneurone pool) or inappropriate motoneurons of one species to innervate the plantaris muscle of the other species.

Developmental Patterning of the Vertebrate Limb
Edited by J.R. Hinchliffe *et al.*, Plenum Press, New York, 1991

I. MUSCLE MORPHOGENESIS

The brachial somites or the brachial neural tube were replaced by somites or by neural tube of the quail. Orthotopic or heterotopic grafts were used (Fig. 1). The analysis of the flexor digitorum superficialis and the extensor medius brevis muscles shows that the development of the individual muscle form has not been disturbed by the exchange of source of myogenic material or the replacement of motoneurons (Jacob et al., 1983; Grim et al.; 1986). The factors which control muscle pattern formation cannot be suggested to be situated within the myogenic cells themselves or the ingrowing nerves. Thus muscle morphogenesis must be determined by influences coming from the limb mesenchyme.

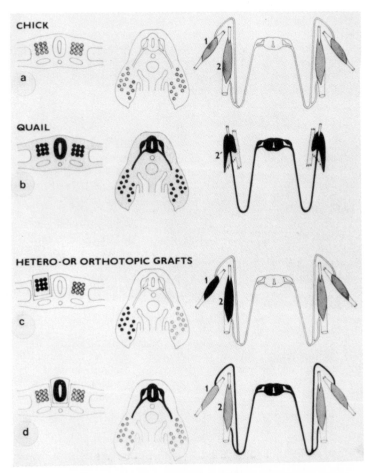

Fig. 1. Schematic representation of species-specific muscle morphogenesis in chick (a), in quail (b) and in chick-quail chimeras after replacement of brachial somites (c) or brachial neural tube (d) of the chick by ortho- or heterotopic quail grafts. Left: embryos at 2 day of incubation: the type of grafting procedure depicted in c and d. Centre: embryos at the wing level on embryonic day 4. Circles - myogenic cells of chick (gray) and quail (black) origin. Right: species-specific muscle forms. 1 - extensor medius brevis muscle of the chick (in the quail not developed); 2,2' - flexor digitorum superficialis muscle.

This concept has recently been supported by the observation that muscle pattern may be established in the absence of myogenic cells (Grim and Wachtler, 1990). Cells from the lateral portion of the epiblast of quail embryos (HH-stage 3) were grafted into the premuscular mass of the right wing bud of a chicken embryo (HH-stage 23). These cells are not determined as myogenic cells (Krenn et al., 1988). The grafted quail cells developed into mononucleate, fibroblast-like cells that formed the belly of the extensor medius longus muscle (Fig. 2) which is characterised by a normal topography and form as revealed by computer-aided 3D-reconstruction.

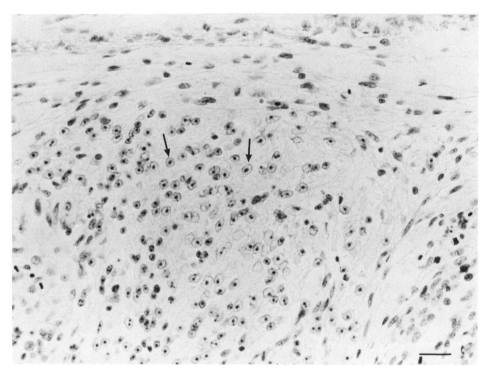

Fig. 2. Transversal section of the extensor medius longus muscle from the wing anlage that recieved a quail graft. The muscle is formed by grafted fibroblast-like quail cells (arrows) and of chicken cells forming endomysium and perimysium. Bar: 20 μm, Feulgen reaction.

This finding corroborates the observation that the morphogenesis of muscle does not depend on the presence of muscle cells (Christ et al., 1978; Lanser & Fallon, 1987). Furthermore it shows that the morphogenetic signal ultimately resulting in the formation of a muscle can also be interpreted by non-muscle cells.
Therefore muscle morphogenesis and differentiation of muscle cells seem to be independent processes.

II. ENDPLATE PATTERN AND MUSCLE FIBRE TYPE DIFFERENTIATION

Differentiation and distribution of muscle fibre types and development of motor endplate pattern are controlled by hierarchically acting neurogenic and myogenic factors.
When the plantaris or the ulnimetacarpalis muscle are innervated by inappropriate motoneurons, the endplate pattern inherent in the muscle anlage itself becomes realized. On the other hand if appropriate motoneurons of one species innervate the plantaris muscle of the other species, the endplate pattern and muscle fibre type differentiation are dictated by the motoneurons (Fig. 3, 4, 5).

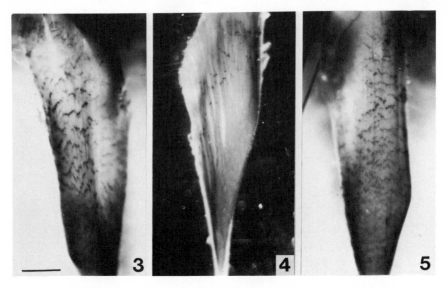

Fig. 3 and 4. Dorsal surface of the plantaris muscle of chick (3) and
quail (4) dissected out from prehatching embryos.

Fig. 5. Chick-like endplate pattern of the quail plantaris
muscle from a quail leg grown in place of the chick
leg. Bar: 1 mm.

A sequence of hierarchically acting factors has been proposed to bring these different results in line (Grim et al., 1989). There are two identical programmes of which the one under neuronal control has priority over that one localized in the muscle. This is true for the normal development and might generate the high neuro-muscular specifity. If under experimental conditions the neuronal programme and the peripheral programme differ, the axons and muscle fibres interact selectively with respect to their inherent characteristics and the muscle specific programme becomes expressed. If there is a lack of a functionally corresponding axon type, muscle fibres might become innervated by non-corresponding motoneurons which alter the muscle fibre type. This concept might help to reconcile the different experimental results and controversial theories dealing with this subject.

REFERENCES

Chevallier, A., Kieny, M., Mauger, A., 1977, Limb somite relationship: origin
of the limb musculature, J.Embryol.exp.Morphol., 41:245-258
Christ, B., Jacob, HJ., Jacob, M., 1978, Zur Frage der regionalen Determination
der frühembryonalen Muskelanlagen. Experimentelle Untersuchungen an
Wachtel- und Hühnerembryonen, Verh.Anat.Ges., 72:353-357.
Christ, B., Jacob, M., Jacob, HJ., Brand, B., Wachtler, F., 1986, Myogenesis:
A problem of cell distribution and cell interactions. in: "Somites in
developing embryos," R. Bellairs, DA. Ede, JW. Lash eds., Plenum
press, New York & London, pp. 261-275.
Grim, M., Christ, B., Klepacek, I., Vrabcova, M., 1983, A comparison of motor endplate distribution and the morphology of some wing muscles of
the chick and quail, J.Histochem., 15:289-291.
Grim, M., Christ, B., Jacob, HJ., Vrabcova, M., 1985, Chick and quail specific
pattern of endplates and fibre types in the plantaris muscle, Physiol.
Bohemoslov., 34:247-249.

Grim, M., Christ, B., Jacob, HJ., 1986, Über die Entstehung des Muskel- und Endplattenmusters. Experimentelle Untersuchungen an Extremitätenanlagen von Huhn-Wachtelchimären, Verh.Anat.Ges., 80:127-135.

Grim, M., Nensa, K., Christ, B., Jacob, HJ., Tosney, KW., 1989, A hierarchy of determining factors controls motoneuron innervation. Experimental studies on the development of the plantaris muscle (Pl) in avian chimeras, Anat.Embryol., 180:179-189.

Grim, M., Wachtler, F., (1990), Muscle morphogenesis in the absence of myogenic cells, Anat.Embryol., (in press).

Jacob, HJ., Christ, B., Grim, M., 1983, Problems of muscle pattern formation and of neuromuscular relations in avian limb development, in: "Limb development and regeneration, Part B, RO. Kelly, PF. Goetinck, JA. MacCabe eds., A.R. Liss, New York, pp. 333-341.

Krenn, V., Gorka, P., Wachtler, F., Christ, B., Jacob, HJ., 1988, On the origin of cells determined to form skeletal muscle in avian embryos, Anat. Embryol., 179:49-54.

Lanser, HE., Fallon, JF., 1987, Development of wing-bud-derived muscles in normal and wingless chick embryos: A computer assisted three-dimensional reconstruction study of muscle pattern formation in the absence of skeletal elements, Anat.Rec., 217:61-78.

Le Douarin, N., 1969, Particularites du noyau interphasique chez la caille japonaise (Coturnix coturnix japonica). Utilisation de ces particularites comme marquage biologique dans les recherches sur les interactions tissulaires et les migrations cellulaires au cours de l'ontogenese, Bull.Biol.Fr.Belg.,103:435-452.

Wachtler, F., Christ, B., Jacob, HJ., 1981, On the determination of mesodermal tissues in the avian embryonic wing bud, Anat.Embryol., 161:283-289.

REVIEW OF THE EXTRACELLULAR MATRIX IN RELATION TO LIMB DEVELOPMENT

John F. Fallon

Department of Anatomy
University of Wisconsin
Madison, WI 53706, USA

The cells that eventually make up the limb originate from the somite and somatopleure. A major area of investigation in limb development research is how the cells become organized with a core of bone-cartilage, surrounded by muscle, which is wrapped in connective tissue, and finally the dermis and epidermis. Each of these components is served by blood vessels and nerves also arranged in predictable patterns. It has been obvious for some time that there must be cues or signals in the environment which the limb bud cells generate and then use to permit the patterned differentiation just outlined. This can be investigated on a number of levels. For example: how is the pattern of signalling controlled; what are the signals; how are the signals presented; how are the signals received by target cells; what are the changes in the target cells, and so forth?

These were among the topics which developed in the day-long presentations on the extracellular matrix (ECM) in limb development. M Solursh began with a clear definition of the current understanding of the extracellular matrix components, their receptors and possible functions during limb development. He reported that large, mesenchymal proteoglycan (PG-M) was first detected by immunohistochemical means at Hamburger and Hamilton stage 17 in the chick limb bud distal mesoderm and subsequently found in the prechondrogenic areas while being maintained in the sub-ridge mesoderm. This specific distribution is intriguing because of the ability of PG-M to interact with other more widely distributed ECM molecules eg fibronectin and type I collagen. He speculated PG-M could be a key regulatory ECM molecule. Along the same line of thought, Solursh made mention of the differential localization of the cell surface heparan sulfate proteoglycan, syndecan, in the ectoderm and mesoderm of the mouse limb bud and of integrins in the chick limb bud mesoderm.

Solursh's presentation laid the foundation for the major theme of the ECM session, viz. the patterned distribution of ECM molecules and cell surface components. Using immunohistochemical methods M Critchlow presented data indicating that staining for chondroitin sulfate proteoglycan was absent from the diffuse basal lamina of the chick apical ridge. Further, and most interesting, was the observation that staining for the avian integrin complex was found only in the membranes of the apical ridge and not the dorsal and ventral epithelia. Integrin expression would appear to be a molecular cell surface marker of the differentiated apical ridge cell.

Developmental Patterning of the Vertebrate Limb
Edited by J.R. Hinchliffe *et al.*, Plenum Press, New York, 1991

The limb bud mesoderm cells are surrounded by a hyaluronan (HA) -rich matrix. The ability of this molecule to bind water results in the separation of the mesodermal cells. Recently, Toole and co-workers have demonstrated HA-dependent pericellular coats. Toole reported on a series of observations and experiments relating to a specific expression of an HA receptor on the cell surfaces of prospective myogenic and chondrogenic cells. He proposed that in the presence of excess HA these cells proliferate and are able to migrate. The reduction of HA synthesis concurrent with receptor-mediated endocytosis of HA would account for the observed condensation of both precartilage and pre-muscle cells that is a prelude to their differentiation.

Toole's observations and hypothesis fit very well with the report of B Brand-Saberi. She showed that there was a proximal to distal migration of the myogenic lineage within the limb. Experimental manipulation showed that myogenic cells always migrated to regions of excess HA. Furthermore, dextrane sulfate substituted for HA indicating these cells seek a water-rich environment. When such an environment was removed, myogenic cells stopped migrating. B Christ demonstrated that the routes muscle-forming cells take from the somite to the limb are rich in fibronectin. When a somite was grafted after 180° rotation the muscle-forming cells took the proper route to the limb bud indicating a specific muscle forming cell-ECM interaction.

Why HA synthesis is reduced and it is actively removed by endocytosis is a critical question. Both Toole and R Kosher presented evidence that growth factors such as fibroblast growth factor and transforming growth factor play a role. Moreover, cell contact and cell communication as demonstrated by Brand-Saberi and Kosher respectively may also be important. TGF-beta 1 and -beta2 have been implicated in chondrogenic differentiation. R Kosher reported that TGF-beta3 also promoted chondrogenesis in culture. However, he also reported that TGF-beta3 messenger RNA is expressed at high levels in pre-chondrogenic cells in vitro and declined as condensation occurred. Messenger RNA for TGF-beta1 was not detectable in these cultures. An exciting observation was that an anti-sense oligonucleotide against TGF-beta3 inhibited chondrogenesis in culture. Kosher also reported that messenger RNA for syndecan began to accumulate during chondrogenesis in culture. The possibility that there is a relationship between the rise and fall of TGF-beta3 gene expression and syndecan gene expression is provocative. It will be critical to devise in vivo approaches which will determine whether or not this hypothesis is true.

Another major theme of the day was the relationship of patterns within the ECM to specific patterns of the vasculature, cartilage, muscle and tendons. D Wilson presented a review of the exclusion of blood vessels from cartilage and their subsequent invasion into the forming bone. He argued effectively that these are not passive phenomena but rather directed in some way. While mention was made of possible mechanisms, it was clear that this is a fertile area for research efforts.

The possibility that the connective tissue that wraps muscle is responsible for the form of particular muscles is becoming established. A Caplan presented an overview of an extensive body of work on the ECM changes which occur over developmental time in the epimysium, perimysium and endomysium of chick thigh muscles. Using a battery of monoclonal antibodies to EM molecules and serial reconstructions he made an elegant case for the establishment and modification of ECM boundaries within and between muscles over developmental time. The challenge is to understand how the detailed developmental program is controlled. Caplan's work lays out the developmental anatomy and is an excellent foundation for experimental work.

The problem of how different parts of the forming limb are assembled was addressed in the analysis of tendon formation. N Holder presented very interesting studies on muscleless, regenerating amphibian limbs. He demonstrated that tendons, muscle connective tissue elements and cartilage elements form relatively normally in regenerating limbs without muscle. It would be of great interest to determine whether the ECM progression described by Caplan can be demonstrated in the forming muscles of the regenerating limb and whether the same progression occurs without muscle cells being present.

Holder's work fits nicely with a report by M Ros indicating an ECM prepattern for the long autopodium tendons of the chick foot. The "mesenchyme lamina" arises at stage 27 and stains with antibodies to type I collagen, and tenascin but not laminin or type IV collage. Pretendon fibroblasts condense on the mesenchyme lamina and use it as a scaffolding to form the dorsal and ventral tendons. There was evidence that the ectodermal basal lamina may be a key to setting up the scaffolding-like pattern.

The use of ECM substrate as a scaffolding for fibroblasts was explored by A Harris. He reported on a variety of experiments which indicate that fibroblasts use collagen to orient themselves. He also presented data indicating that fibroblasts were capable of generating enough force to rearrange collagen into particular orientations. The presentations by Holder, Ros and Harris were at the tissue level of organization. How individual cells orient in the area they find themselves - what the cues are to set the stage for scaffolding formation now challenges us.

All the preceding work has dealt with the ECM in normal development. J. Hurle has developed an interesting experimental system. He has demonstrated that removal of the surface ectoderm from the chick foot interdigit 3-4 either surgically or with trypsin results in supernumerary cartilage formation. In this situation, cells which normally die continue to live. After ectoderm removal, either surgically or after trypsin treatment, a prechondrogenic condensation forms and develops in many cases into a digit with easily recognizable phalanges with joints. These results are fascinating. Is there a removal of ectodermal inhibition of chondrogenesis in subjacent mesenchyme? How is the supernumerary structure determined? What does all of this mean for the theories of limb patterning? Hurle makes an interesting discussion of many of these points. It will be important to integrate his finding into how we think about limb pattern formation.

A clear picture of the developmental anatomy of the ECM in the limb is beginning to emerge. The data on ECM, growth factors and specific gene expression by limb cells in culture are provocative. These approaches point the way to experimental manipulations in vivo that may lead to fundamental insights into how limb bud cells achieve a limb. The sessions on ECM at the next limb meeting should be very exciting.

THE EVOLUTION OF CONNECTIVE AND SKELETAL TISSUES

Brian K. Hall

Department of Biology
Dalhousie University
Halifax, Nova Scotia
Canada B3H 4J1

INTRODUCTION

I was charged with providing an overview and background paper for the third day of the Workshop devoted to "The Developmental Basis of Limb Evolution." As the Workshop has concentrated on skeletal tissues, I too will emphasize skeletal tissues and will attempt to place fin and limb skeletogenesis in the context of other vertebrate skeletal elements, both developmentally and evolutionarily. This picks-up a theme developed by Lewis Wolpert in his introduction to the Workshop, viz. that limb skeletogenesis should not be considered in isolation and that we should not expect the mechanisms of limb skeletogenesis to be unique.

This overview is intended to provide a bridge between the developmental papers presented on days one and two and the evolutionary papers presented on day three. As such it provides the background for those papers presented on the third day dealing with the embryology and paleontology of the fin to limb transition (Coates, Thorogood, Vorobeyeva), the development of limb musculature and the pectoral girdle (Cihak, Burke) and limb patterning (Hinchliffe, Muller).

To relate fin and limb skeletogenesis to skeletogenesis in other regions of the embryos necessitates both a comparative approach and a discussion of various embryonic skeletogenic regions. As I will document, the skeleton can be mesodermal or neural crest in origin, exo- or endoskeletal, bony or cartilaginous, and bone can develop by intramembranous, perichondral or endochondral ossification. Both the development and the evolutionary origin of the fin and limb skeleton have to be seen against this diverse background.

Developmental Patterning of the Vertebrate Limb
Edited by J.R. Hinchliffe *et al.*, Plenum Press, New York, 1991

I will also evaluate theories of the origin of the paired vertebrate appendages. A combination of developmental diversity of the skeleton and a lack of critical data for key developmental events, such as the origin of the fin rays, makes acceptance of any of the theories difficult. Either none fit all the developmental data, or there is insufficient, or often no, developmental data base from which to choose between alternate theories.

I treat five topics in this overview.

1. An overview of the embryonic sources of skeletal tissues is provided.
2. The common features of the development of skeletal tissues, irrespective of their embryological origin are discussed.
3. I then examine the distinction between the vertebrate exoskeleton and endoskeleton
4. The time of evolutionary appearance of skeletal tissues is discussed.
5. Lastly, I examine theories of the origin of the paired appendages in the context of developmental and evolutionary data on skeletal tissues presented in sections 1-4.

1. AN OVERVIEW OF THE EMBRYONIC SOURCES OF SKELETAL TISSUES

Trunk (postcranial) mesoderm is a major source of skeletal and connective tissues. Mesoderm (the embryonic germ layer) breaks-up into mesenchyme (a loose meshwork of mesenchymal cells) which then migrates to the site of skeletogenesis.

Somitic mesoderm adjacent to the spinal cord subdivides into dermatome, myotome and sclerotome which give rise to connective, muscular and skeletal tissues respectively (Hall,1977). All the skeletal cells of the vertebrae and ribs arise from sclerotomal mesenchyme. Lateral extensions from the dermomyotome in fishes, reptiles and birds (and possibly also in other vertebrates for which experimental data is lacking) migrate into the developing limb buds to produce the muscle of the limb.

Lateral plate mesenchyme provides skeletal and connective tissue of the paired appendages (fins, limbs) and the skeleton of the pelvic and pectoral girdles with the exception of varying amounts of the scapula (see Burke, this volume).

Fins and limbs are therefore chimaeric; appendicular muscles arise from somitic mesoderm, while skeleton and connective tissue arise from lateral plate mesoderm.

Although mesoderm contributes some skeletal tissues to the head (the roof of the skull), the bulk of the cranial and all facial skeletal and connective tissues arise from mesenchyme derived from the embryonic neural crest (Hall and Hörstadius, 1988). As neural crest cells arise in the neural tube, this is essentially an ectodermal origin of these skeletal tissues.

Cranial neural crest-derived mesenchyme forms cartilage, bone, dentine of teeth and mesenchyme (and cartilage where present) of the heart. Trunk neural crest-derived mesenchyme produces connective tissue for the median fins in amphibians and fishes, may provide dentine and bone of dermal scales (Smith and Hall, 1990) but has not been reliably demonstrated to contribute to the limb skeleton. (As we will see below, the rays of the fin skeleton, which consist of dermal bone, sometimes associated with dentine, may be of neural crest origin).

Thus, the vertebrate skeleton is both ectodermally and mesodermally-derived.

2. THE COMMON FEATURES OF THE DEVELOPMENT OF SKELETAL TISSUES

I identify six features as common and the minimal developmental events required for skeletogenesis wherever it occurs in the embryo. Hall (1988) and Smith and Hall (1990) provide recent evaluations of these features.

(a). Mesenchyme

The basic cellular organization from which skeletal and connective tissues arise is mesenchyme, sometimes called ectomesenchyme when it arises from neural crest cells. Mesenchyme is not unique to vertebrates but is found in many but not all invertebrates where it can arise from ecto- meso- or endoderm (Willmer, 1990). The distinctiveness of mesenchymal populations and how early subpopulations of mesenchymal cells segregate are open questions (Hall, 1978).

(b) Cell migration

Some migration of mesenchymal cells precedes the development of most if not all skeletal elements. Thorogood (this volume) discusses such movement as visualized in real time in the fish fin.

Cell migration may be very extensive as in the migration of neural crest or somitic mesenchymal cells or very limited as in the translocation of lateral plate mesenchyme into the limb field. The important points to stress are (1) that preskeletogenic cells experience different embryonic environments as they migrate and (2) that cellular environments exert epigenetic control over differentiation and sometimes over morphogenesis (see d below and Maclean and Hall, 1987).

(c) Condensation

Mesenchymal cells aggregate into condensations at the site of skeletogenesis. Solursh, Kosher and Toole (this volume) discuss the condensation process. The process of condensation triggers specific gene activity, such as amplification of mRNA for the synthesis of specific products associated with the onset of cell diferentiation. Such products are type I collagen and osteonectin for osteogenesis and type II collagen and cartilage-specific proteoglycan core protein for chondrogenesis. Condensation represents the first important stage initiating skeletogenic cell differentiation through differential gene expression.

(d) Epigenetic activation : epithelial-mesenchymal interactions

In all sites of skeletogenesis studied, interaction of preskeletal mesenchymal cells with an associated epithelium is required for skeletal differentiation to be initiated. Such interactions trigger the onset of condensation formation (Fyfe and Hall, 1983; Hall, 1988). Typical interactions are the limb ectoderm and subsequently the apical ectodermal ridge with limb mesenchyme (later in development the epithelium inhibits chondrogenesis; Hurle, this volume), sclerotomal mesenchyme with notochordal and spinal cord extracellular matrix products (Hall, 1977) and craniofacial mesenchyme with craniofacial epithelia (Hall, 1988).

(e) Regionalization of skeletogenic mesenchyme

There is an anterior-posterior (rostro-caudal) regionalization of skeletogenic mesenchyme. This is seen most obviously in the skeletogenic neural crest which extends along the neural tube from mid-prosencephalon caudal to somites 4 or 5 in those representative lampreys, teleosts, anurans, urodeles and birds that have been examined (Hall and Hörstadius, 1988). There is experimental verification of this regionalization; more anterior mesenchyme contributes more anterior skeletal elements, more posterior mesenchyme more posterior elements (Langille and Hall, 1988a,b; Hall and Hörstadius, 1988).

Regionalization is also seen, although less well understood, in the rostro-caudal organization of somitic mesoderm into cervical, thoracic, abdominal and caudal vertebrae, and in the organization of limb mesenchyme into anterior and posterior limb fields, corresponding to anterior and posterior appendages.

(f) Common cell types

Irrespective of embryonic origin, with one exception, the same cell types arise from neural crest and mesodermally-derived skeletogenic

306

mesenchyme. Chondrocytes and osteocytes, cartilage and bone, intramembranous, perichondral and endochondral osteogenesis occur in both. However dentine and the associated basal bone which attaches teeth into the jaws is exclusively a neural crest-derivative (Smith and Hall,1990).

3. THE DISTINCTION BETWEEN THE EXOSKELETON AND THE ENDOSKELETON

Patterson (1977) in a seminal paper argued for the independence of the two vertebrate skeletons; the exoskeleton and the endoskeleton, for in reality vertebrates possess two skeletal systems. Recently, Smith and Hall (1990) reviewed the developmental and fossil evidence for the evolutionary separation and developmental origins of these two skeletons.

The exoskeleton consists of dermal scales or armour composed of dentir underlain by bone and overlain by enamel. Bone is the primitive skeletal tissue of the exoskeleton, although cartilage can appear secondarily in dermal bone in fishes, birds and mammals (Smith and Hall,1990). The fundamental tissue in the exoskeleton is therefore odontogenic consisting of dentine and associated basal bone, initially in isolated scales or odontodes. Given the argument outlined above that this odontogenic population of cells is of neural crest origin, and other evidence available (Smith and Hall,1990) it is argued that the exoskeleton is of neural crest origin.

The endoskeleton consists of cartilage which is often, although not always replaced by bone. Cartilage is therefore the primitive skeletal tissue of the endoskeleton. Dentine is never associated with the endoskeleton which can be mesodermal as in the vertebrae, ribs, limb and dorsal skull bones, or of neural crest origin, as in the chondro- and viscerocranium. As will be discussed in section 5 below, the fin skeleton may have a composite origin, the fin rays being derived from the dermal exoskeleton and the basal elements (pterygiophores) from the endoskeleton.

I summarize what has been said so far by paraphrasing the ninth of twelve postulates proposed by Smith and Hall (1990) viz, that initiation of differentiation of skeletogenic and odontogenic tissues, whether neural crest or mesodermally derived, cranial or trunk, exo- or endoskeletal, is controlled epigenetically by one or more epithelial-mesenchymal interactions in an epigenetic cascade.

4. THE TIME OF EVOLUTIONARY APPEARANCE OF SKELETAL TISSUES

As discussed by Hall (1975) all skeletal tissues with the exception of endochondral bone appear equally early in the fossil record, i.e. cartilage, bone (cellular and acellular), dentine, and enamel can all be recognized with confidence.

Maisey (1988) performed a cladistic analysis on the appearance of skeletal tissues (see his Fig. 3). The sequence of appearance of mesenchymal derivatives is as follows (with appropriate "earliest" groups in parentheses); somatic tail with lateral muscle bands (urochordates), somitic and lateral plate mesoderm (amphioxus), neural crest and neurocranial endoskeleton, trunk somites subdivided into dermatome, myotome and sclerotome, and a cartilaginous axial skeleton (*Myxine*), dermal exoskeleton of neural crest origin (Heterostraci), paired appendages and perichondrial ossification in the head (Osteostraci), postcranial somitic skeleton of perichondral bone, and postcranial appendicular skeleton (Anaspida).

It is clear that at the time of the origin of the paired appendages vertebrates possessed somites subdivided into the three regions, a skeletogenic exoskeleton and endoskeleton derived from neural crest, and a cartilaginous axial skeleton of somitic mesodermal origin. These were the mesenchymal and skeletogenic cell populations available to be co-opted for the formation of a fin skeleton.

5. THEORIES OF THE ORIGIN OF THE PAIRED APPENDAGES

Three major theories have been proposed for the origin of the paired appendages. These theories have been reviewed, primarily from the morphological point of view by Goodrich (1958), Westoll (1958), Jarvik (1965), Maderson (1967) and Schaeffer (1977).

(a) The gill arch theory

The paired fins and pectoral, but not pelvic girdles, are postulated to have arisen from skeletal elements of the posterior gill arches. However, the cartilaginous gill arch skeleton is of neural crest origin and the cartilaginous limb skeleton is mesodermal. Extensive migration would be required to derive the posterior fins from a gill arch and this theory takes no account of the origin of the pelvic girdle. As summarized by Goodrich (1958):
"indeed the (gill arch) theory could hardly have been conceived except at a time when the ontogeny of fins was scarcely known at all (p. 126).

(b) The body spine theory

According to this theory, ventrolateral body spines coalesced and acquired a membranous cover. Loss of intervening spines left two centres as paired appendages. As body spines are dermal this theory can only apply to the fin rays and not to the cartilaginous pterygiophores and it says nothing of the origin of the girdles.

(c). The fin fold theory

It is known that the median, unpaired fins of larval amphibians are induced by trunk neural crest and contain neural crest-derived mesenchyme (Du Shane, 1935) and it is assumed that the same holds for fishes. Paired fins can be induced bu neural crest or by lateral plate mesoderm (Schaeffer,1977). but do not contain neural crest-derived mesenchyme (Thomson,1987). The induction of fin folds by lateral plate mesoderm would bring a chondrogenic mesenchyme into association with ectoderm, establishing the physical proximity required for an epithelial-mesenchymal interaction. Intermediate regions of a continuous fold could be lost because of loss of the somitic extensions which are known, at least in reptiles, to maintain the apical ectodermal ridge. Accommodating dermal fin rays into this theory is a difficulty.

However, if fin rays are of neural crest origin, as the association with dentine in some fishes would support, then a combination of the fin fold and body spine theories would fit the currently available developmental data. The following is presented as a model.

(i) Fin folds are induced by lateral plate mesoderm or trunk neural crest and are associated with dermal body spines.

(ii) The basal cartilaginous elements of the fin (pterygiophores) develop from lateral plate mesoderm because of epigenetic association between lateral plate mesoderm and the ectoderm.

(iii) The fin rays either develop from dermal exoskeletal elements (if fin rays are neural crest) or from superficial lateral plate mesoderm (if bone associated with dentine can be shown to form from mesoderm).

Whichever of the alternatives in (iii) is correct, the fin is viewed as a composite structure, with basally developing cartilages and apically developing bony fin rays.

The transition from fin to limb involves loss of any trace of the dermal (neural crest or mesodermal) component. Considerable variability would be expected in the fin-limb transition in a situation where the basal cartilaginous elements were expanding to replace apical elements, a situation that is seen in the fossil record.

Developmentally, a search for the embryonic origin of all mesenchymal elements of the fish fin would greatly advance our knowledge of the evolution of conective and skeletal tissues.

ACKNOWLEDGEMENTS

I thank Drs A. Burke, A. Graveson, T. Miyake, P. Mabee and S. Smith for helpful comments. Research support was provided by the Natural Sciences and Engineering Research Council of Canada (grant A5056).

REFERENCES

Du Shane, G. P., 1935. An experimental study of the origin of pigment cells in amphibia. J. exp. Zool. 72:1.

Fyfe, D. M., and Hall, B. K., 1983. The origin of the ectomesenchymal condensations which precede the development of the bony scleral ossicles in the eyes of embryonic chicks. J. Embryol. exp. Morph. 73:69.

Goodrich, E. S., 1958. "Studies on the Structure and Development of Vertebrates," Dover Publications, Inc., New York.

Hall, B. K., 1975. Evolutionary consequences of skeletal development. Amer. Zool. 15:329.

Hall, B. K., 1977. Chondrogenesis of the somitic mesoderm. Adv. Anat. Embryol. Cell Biol. 53(4): 1.

Hall, B. K., 1978. "Developmental and Cellular Skeletal Biology," Academic Pres, New York.

Hall, B. K., 1988. The embryonic development of bone. Amer. Sci. 76(2):174.

Hall, B. K., and Hörstadius, S., 1978. "The Neural Crest." Oxford University Press, Oxford.

Jarvik, E., 1965. On the origin of girdles and paired fins. Israel J. Zool. 14:141.

Langille, R. M., and Hall, B. K., 1988a. The role of the neural crest in the development of the cartilaginous cranial and visceral skeleton of the medaka, Oyyzias latipes (Teleostei). Anat. Embyol. 177:297.

Langille, R. M., and Hall, B. K., 1988b. Role of the neural crest in development of the trabecular and branchial arches in embryonic sea lamprey, Petromyzon marinus (L). Development 102:301.

Maclean, N., and Hall, B. K. 1987. "Cell Commitment and Differentiation," Cambridge University Press, Cambridge.

Maderson, P. F. A., 1967. A comment on the evolutionary origin of vertebrate appendages. Amer. Nat. 101:71.

Maisey, J. G., 1988. Phylogeny of early vertebrate skeletal induction and ossification patterns. Evol. Biol. 22:1.

Patterson, C., 1977. Cartilage bones, dermal bones and membrane bones, or the exoskeleton versus the endoskeleton. in "Problems in Vertebrate Evolution," S. M. Andrews, R. S. Miles and A. D. Walker, eds., Academic Press, London.

Schaeffer, B., 1977. The dermal skeleton in fishes. in "Problems in Vertebrate Evolution," S. M. Andrews, R. S. Miles and A. D. Walker, eds., Academic Press, London.

Smith, M. M., and Hall, B. K., 1990. Development and evolutionary origins of vertebrate skeletogenic and odontogenic tissues. Biol. Rev. 65:277.

Thomson, K. S., 1987. Speculations concerning the role of the neural crest in the morphogenesis and evolution of the vertebrate skeleton. in "Developmental and Evolutionary Aspects of the Neural Crest," P. F. A.

Maderson, ed, John Wiley & Sons, New York.

Westoll, T. S., 1943. The origin of the primitive tetrapod limb. _Proc. R. Soc. London_ 131:373.

DEVELOPMENTAL APPROACHES TO THE PROBLEM OF TRANSFORMATION OF LIMB STRUCTURE

IN EVOLUTION

Richard Hinchliffe

Department of Biological Sciences
University College of Wales
ABERYSTWYTH, Wales, UK

1   INTRODUCTION

This paper is intended as an introduction to the subsequent
contributions which are concerned with the developmental basis of limb
evolution at a number of levels, whether genetical, embryological (both
experimental and comparative) or palaeontological. Even to put this topic
on the agenda of a development meeting reflects changing attitudes,
particularly a degree of dissatisfaction with the neo-Darwinian "new
evolutionary synthesis" which has long dominated evolutionary theory(Horder,
1989). The "synthesis" has been largely concerned with identifying
selection pressures and their effects on gene frequencies within
populations, to the exclusion of developmental biology.

An important aspect of vertebrate evolution is the origin and the
structural transformation of the limb for many different adaptive
"purposes". The overall similarity in limb structure, regardless of its
specialisation, was recognised by Owen (1849), who used the limb as an
illustrative example of homology. Fundamental similarities include the
stylopod (containing a single proximal element, eg humerus), zeugopod (two
middle elements, eg radius and ulna) and autopod (carpus or tarsus and
radiating digits.). One obvious "variation on the theme" is in the number
of digits which are frequently reduced from the 5-digit pentadactyl
"archetype" – or, as Coates (these proceedings) demonstrates, from the 7- or
8- digits of early tetrapods. Other variation is found in carpus or tarsus
structure, in the phalangeal formula of different digits, or in the
specialised morphology of individual elements, as in the digits of birds.
My question is – how far can we explain in developmental terms, the basis of
this combination of similarity and variation?

2   SIMILARITY

Deconstructing the archetype

One answer to this question is the attempt to construct a common
archetype, in which the general precedes the specific. This approach
(Holmgren, 1933) still characterises recent accounts of vertebrate palaeontology and evolution (Jarvik, 1980). According to these accounts, a
generalised archetype representing the adult limb skeleton of the first
tetrapods, appears as the pattern of prechondrogenic mesenchymal
condensations in the limb bud. It is claimed there is a one-to-one

Developmental Patterning of the Vertebrate Limb
Edited by J.R. Hinchliffe *et al.*, Plenum Press, New York, 1991

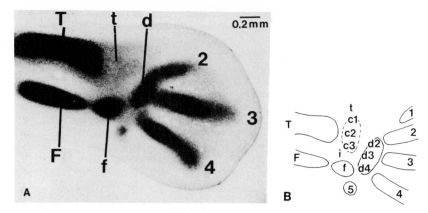

FIG 1. A: Sulphate-labelled autoradiograph of stage 26/7 chick hind limb.
Three tarsus pre-chondrogenic elements are present (or will form), f,
fibulare: t, tibiale position; d, distal tarsal 2/3.
B: The archetypal theory claims 10 tarsal elements (shown here
superimposed on the pattern in A): centrale 1-4 (c 1-4); distal
tarsals 2-4 (d 2-4); i, intermedium; f, fibulare, t, tibiale, F,
fibula; T, tibia; 1-5 metatarsals (Holmgren, 1955).

correspondence between each ancestral adult skeletal element and the
corresponding pre-chondrogenic condensation. Holmgren in fact identifies a
"canon" of the fundamental elements comprising the limb skeletal archetype.
Then the general archetype is modified later in development by specialised
secondary fusion or regression of elements (Montagna, 1945, Hamilton, 1952,
Romanoff, 1960). This approach is recapitulationary, assuming that the
adult forms of ancestors are repeated in the embryonic stages of their
descendents.

Recent reassessment of this theory fails to confirm it (Muller and
Alberch, 1990, Hinchliffe and Griffiths, 1983). A more precise
visualisation of the early developing skeletal pattern is provided by
autoradiographic techniques, using radiosulphur to label the sulphated
glycosaminoglycans (which include chondroitin sulphate) of the earliest
extracellular matrix. This shows very clearly that, at least in more
specialised limbs such as those of birds, there is no one-to-one
correspondence between condensation and archetype elements. In the bird leg
bud, where 10 tarsus prechondrogenic condensations should represent the
archetype (Holmgren, 1955), only 3 can be found (Fig 1), precursors of the 3
elements of the definitive adult (Hinchliffe, 1977). The pattern of
prechondrogenic elements is clearly species-specific: it does not pass
through an archetypal pattern of elements.

However, it is worth noting that the bird leg does share common pattern
features with other (non-urodele) tetrapod limbs, as pointed out by Shubin
and Alberch (1986). These include a 4,(5), 3, 2, 1 sequence in digit
appearance, and an asymmetrical branching pattern which sweeps in a
posterior proximal to anterior distal direction. Part of this pattern is
the "digital arch" which lies at the base of the metatarsals (or
metacarpals) of the digits. At this level developing tetrapod limbs do have
a common pattern, but not as a recapitulation of each tetrapod ancestral
skeletal element. It is also worth pointing out, that with the discovery by
Coates (these proceedings) of early tetrapods with 7 or 8 digits, strict
adherence to a recapitulated archetype would demand the appearance in

development of <u>all</u> these digits in descendents, which has never been claimed.

Although the limb archetype may be dismissed as a <u>description</u> of developing pattern, it also provides no developmental <u>mechanism</u>, and it is to the modern techniques of developmental biology, which include experimental embryology and cell culture, that one must turn in attempting to establish the common mechanisms likely to underlie the homology of the limb

The hierarchy of processes controlling tetrapod limb development

The limb may be regarded as an example of a "condition generated form" (Waddington, 1962), arising not from the units from which limb buds are built, but from "the interactions of a number of initial, spatially distributed conditions".

Mainly through experiments on chick limb buds but also on other tetrapod species, 3 essential processes can be identified controlling limb outgrowth and skeletal patterning. These are (Fig 2)

1) Induction of mesenchymal outgrowth by the AER (the apical ectodermal ridge, a distal thickening of the ectoderm)

2) Generation of spaced prechondrogenic condensations by the mesenchyme.

3) The role of the ZPA (zone of polarising activity) in controlling pattern across the antero-posterior axis. Specifically, the ZPA appears to impose asymmetry on the prechondrogenic branching process of the mesenchyme.

Modern techniques have enabled these processes to be dissected out. Numerous experiments on the AER of birds and other tetrapod classes demonstrate its role in limb bud outgrowth (Summerbell, 1974, Reiter and

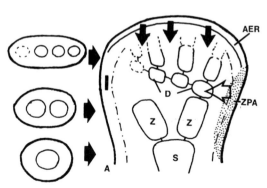

Fig 2. Schematic account of the principal interacting processes in tetrapod limb buds (except urodeles), representing the branching and segmenting of the prechondrogenic condensations. Cross sectional shape is shown, reduced in size (left). A, anterior; AER, apical ectodermal ridge, inducing mesenchyme outgrowth; D, digital arch; I, ectodermal inhibition of chondrogenesis in adjacent mesenchyme; S, stylopod; Z, zeugopod; ZPA, zone of polarising activity, controlling antero-posterior polarity and generating skeletal asymmetry.

Solursh, 1982, Raynaud, 1985'. Limb bud mesenchyme is able to generate spaced condensations, the in ovo process being to some extent mimicked under in vitro conditions (Ede, 1982, Ide, 1989). The inhibition of chondrogenesis imposed by the ectoderm on adjacent mesenchyme (Solursh, 1984) is an important factor in positioning condensations centrally rather than peripherally.

The role of the ZPA in control of antero-posterior polarity in chick limb buds (Tickle et al, 1982, Saunders, 1972), and interpreted as a "positional information" source is well known. Numerous experiments show that when limb bud mesoderm is disaggregated and rejacketed in an ectodermal cap (so that the recombinant lacks a ZPA reference point) it forms a simple branching system of chondrogenic elements (Pautou, 1973, MacCabe et al, 1973, reviewed Hinchliffe 1985, 1990). For asymmetry and recognisable individual digits, a ZPA must be added. The talpid[3] polydactylous mutant of the chick demonstrates the same principle. Lacking a ZPA activity, talpid[3] limbs form digits which lack individuality (Hinchliffe and Ede, 1967, Ede, 1982 and these proceedings). (Note that the less extreme talpid[2] has a weak ZPA with some anterior to posterior distinction between digits (Fallon, these proceedings)).

Table 1 Mechanisms of Limb (or Fin) Development Common to the Vertebrate Classes.

| | Fish | Amphibians | | Amniotes | | |
	TELEOSTS	URODELE	ANURA	REPTILES	BIRDS	MAMMALS
AER PRESENT	X	X	X	X	X	X
AER INDUCES OUTGROWTH	X [1]	X [2]	X [3]	X [4]	X [5]	X [6]
ZPA GRAFT PRODUCES DUPLICATION		X [7*]	X [8]	X [9]	X [10]	NOT TESTED
ZPA GRAFT PRODUCES DUPLICATION IN AVIAN WING-BUD		— [11]	— [11]	X [12]	X [12]	X [12]

Similar AER and ZPA mechanisms are found throughout tetrapod limb buds and also, in the case of the AER, in fish fin buds.

References   1 Bouvet, 1970: 2 Thornton, 1957: 3 Tschumi, 1957: 4 Raynaud, 1985:   5 Saunders, 1948: 6 Milaire, 1973:   7 Slack, 1976 (*posterior flank on limb disc): 8 Cameron and Fallon, 1978: 9 Honig, 1984: 10 Saunders, 1972: 11 Fallon, (personal communication): 12 Fallon and Crosby, 1977:

While the work just quoted is on chick limb bud development, there is evidence from other avian species, and from all the other tetrapod classes of AER and ZPA developmental mechanisms (Table 1). Urodeles, anurans, reptiles and mammals all provide good evidence of AER induction of outgrowth and of ZPA control of polarity. Regarding the ZPA, for technical reasons preaxial grafting has not always been possible within the tetrapod class (as in mammals), but in these cases duplications have been obtained using the chick wing bud as an assay system (Table 1).*

There is also evidence of trans-class communication between ectodermal jackets complete with their AERs (or grafted AERs), and mesodermal cores. This has been demonstrated by the successful outgrowth and skeletal differentiation of mammal-avian (Jorquera and Pugin, 1971) and reptile-avian chimeric limb buds (Fallon, Muller, these proceedings).

It is too early to add retinoic acid effects to this list, since too few species have been examined. There is indeed evidence for duplication, though this is across the antero-posterior axis in avian limb buds, and along the proximo-distal axis in urodele amphibians and along both axes in anurans (Maden, 1985). Cross-class comparison of homeobox gene expression patterns (eg Duboule, Hill, these proceedings) should also soon be possible.

In conclusion, it is argued here that the basis of limb structural similarity is to be found in common shared developmental processes involving AER and ZPA and demonstrated across the tetrapod classes. The next section will examine the question of the developmental basis of structural transformation of the limb in evolution.

3 VARIATION

Variation in definitive limb stucture in different classes and species takes many forms, whether at the level of the overall pattern of individual elements (eg of carpus or tarsus) or in their specific growth and morphogenetic patterns (Shubin, these proceedings). Here, I will focus on the mechanisms which could generate digit loss or gain.

Quite small boundary changes in the interacting system described in Fig 2 would have this effect. The length and period of inductive action of the AER is critical since both should affect the number and distribution of the distal mesoderm cells available for skeleton formation. Alternatively, a direct increase or decrease in the number of these cells would have similar effects. Depending on their distribution, increased cell numbers would generate additional digits preaxially, or increase the number of phalangeal elements in a digit. Quantitative changes in cell number would become qualitative change in the skeletal pattern.

Such an interpretation is not merely theoretical; it may be demonstrated by "real" limb buds. Thus chick wing and leg buds, which both have localised preaxial and distal areas of mesenchymal cell death (Hinchliffe, 1982, 1985) have reduced or lost the anterior no 1 digit. They have relatively short AERs compared with the mouse and rat limb buds which lack the cell death zones and generate skeletons with 5 digits (Milaire, 1977). Talpid[3] mutant limb buds also lack the normal avian cell death zones, have excessively elongated AERs, and generate polydactylous limb skeletons with 7 or 8 digits (Hinchliffe and Ede, 1967).

A theoretical model (Fig 3) clarifies the "boundary change" interpretation of structural change. It incorporates processes already demonstrated (Fig 2) such as inhibition of chondrogenesis by the ectoderm

---

* Negative results have been obtained when "ZPA tissue" (posterior border) from urodele and anuran limb buds has been grafted preaxially into avian wing buds. However, in view of the very different avian/amphibian blood composition, it may be questioned whether these ZPAs function normally. Within amniotes, such trans-class ZPA grafts give duplications.

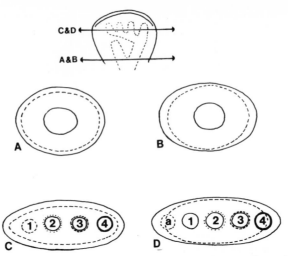

Fig 3. A 'boundary change' model of structural change (see text). A 10% increment in mesenchyme cell number changes the limb bud shape symmetrically (A, C) or asymmetrically (B, D). New condensations are placed at 3 x radius distance from the centre of their predecessors. Only in D will enough additional space be created to permit the formation of an additional condensation (a), located distally and preaxially. A and B, proximal sections, C and D distal sections through a limb bud (p, posterior). Solid line post-, and dotted line pre-increment mesenchyme-ectoderm boundary.

and imposition of asymmetry by the ZPA, so that the sequence of digits formation is 4, (5), 3, 2, 1. It assumes that each condensation creates a zone of inhibition so that each new condensation is placed approximately at a 3 x radius distance from the centre of its predecessor. This will produce a proximal to distal branching, as in Shubin and Alberch's model. Assuming a 10% (approximately) increment in the number of mesenchyme cells, consider the consequences of placing these additional cells either symmetrically around the edge, or asymmetrically, increasing the long axis. The symmetrical increment will have no effect on skeletal pattern either proximally or distally. But the asymmetrical increment will create enough additional space distally for an additional digital element preaxially, while having no effect proximally. Given small changes in cell number, our model suggests that any increases or decreases in digital number should be preaxial, while the proximal pattern should be more stable.

The model can be tested against 2 sets of evidence; variation in definitive limb skeletal structure, and modification of limb form through alterations in limb bud size, whether in mutants or by experiment. In amphibian species Alberch (1990) has reviewed evidence supporting this model, but here I will discuss some less familiar examples.

"Crossopterygians" (see Coates, Vorobyeva, these proceedings) may be regarded as a Devonian group close to the first tetrapods, and their fossilised paired fin skeletons have frequently been regarded as "pre-limbs", suitable structures for transformation into tetrapod limbs and demonstrating homology at the proximal level (humerus, radius, ulna) with the pentadactyl limb (Gregory and Raven, 1941, Holmgren, 1933, Hinchliffe and Johnson, 1980). Four Crossopterygian species have had their fin paddle skeletal structure described, and they demonstrate clearly the principle of

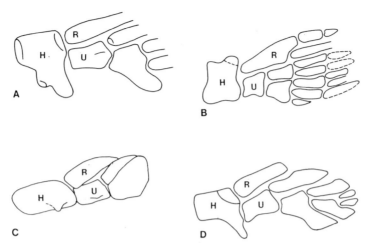

Fig 4. Pectoral lobed fin skeletons from 4 species of fossil Rhipidistian Crossopterygians, close relatives of the first tetrapods. Note the similarity of structure proximally, while the number of branchings varies distally. See text for interpretation. A, Sterropterygion; B, Sauripterus; C, Panderichthys; D, Eusthenopteron (From Vorobyeva, these proceedings; Vorobyeva and Hinchliffe, 1991).

distal variability and proximal stability (Fig 4 – Hinchliffe. 1989, Vorobyeva, 1975, Vorobyeva and Hinchliffe, 1991). Moreover, the new evidence regarding the earliest tetrapod limbs (Coates and Clack, 1990, Coates, these proceedings) demonstrated that although having the standard structure at the proximal level these limbs had 6, 7 or 8 digits, with the "additional" digits localised preaxially, compared with the later tetrapod limb forms in which 5 is the maximum number.

Table 2   Digit Reduction in Lizards

Lacerta viridis	1 2 3 4	5
Ophiomurus tridactylus	2 3 4	(5)
Chalcides chalcides	2 3 4	(5)
Ophiodes striatus	3 4	
Scelotes gronovii	4	

The digits which survive in the fore and hind limbs of various lizard species showing different degrees of natural digital reduction (Brackets represent survival of metacarpal or tarsal). Data from Raynaud and Clergue-Gazeau, 1986.

Evidence of digital reduction in lizards provides another test of the theory. Amongst the lizards, skinks demonstrate a particularly wide range of limb reduction patterns. Greer (1990) has demonstrated within species of a single genus (Lerista: Australian sand-swimmers) a range of limb forms from pentadactyl to complete absence of digits. The digits disappear in the following sequence: 1,(5), 2, 3, 4, so that the order of loss of anterior elements in evolution precisely reverses the order in development where this has been studied in other amniotes. Although developmental evidence is lacking in Lerista species, it is provided in the 3-toed skink, Chalcides chalcides, demonstrating that the retained digits are 2, 3 and 4 (Raynaud et al, 1987).

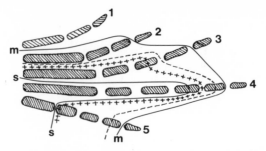

Fig 5. Digit reduction patterns following Ara—C (a mitotic inhibitor)
treatment of <u>Lacerta viridis</u> (the Green Lizard) embryos. Different
lines show the effects ranging from mild (m) to severe (s). (Data
from Raynaud and Clergue—Gazeau, 1986).

Evidence from experimental embryology also supports the theory.
Raynaud and Clergue—Gazeau, (1986) used a mitotic inhibitor, ara—C, to
reduce the quantity of limb bud mesenchyme and to examine the effect on
digital pattern in the green lizard, <u>Lacerta viridis</u>, which normally has 5
digits. Raynaud obtained digit loss ranging from mild to severe, with a 1,
2, 3 order of loss of anterior digits, and digit 4 as the last to be
eliminated (Fig 5). Raynaud emphasises that such experimental modification
of developmental mechanisms provides an insight into the mechanisms of the
evolutionary reduction of limbs.

Supporting evidence is also provided by mutant genes which increase the
antero—posterior axis of the limb bud and affect digital number, eg <u>Luxate</u>
in the mouse, and <u>polydactyly</u> in the chick. A survey of such polydactylous
mutants showed that in most cases the supernumerary digits were localised
preaxially (Hinchliffe and Ede, 1967).

4 CONCLUSION

The homology of the limb is best regarded as a consequence of a
hierarchy of interacting processes common to limb development, rather than
as the possession in development of a precisely similar set of ancestral
skeletal elements as a prechondrogenesis pattern. There is considerable
evidence in limb evolution for:-

i)   distal variability/proximal stability and for

ii)   preaxial localisation of the loss or gain of digits in evolution.

Such changes may be interpreted as the consequences of boundary changes
to the developing limb system with its interacting processes.

Experimental evidence suggests that differences in limb form of the
type just described are due to "developmental constraints" rather than
random variation.

Where do we go from here? For evolutionary morphologists,
developmental criteria are likely to be useful in the "naming of parts" eg
in assigning identity in digit reduction, or, in the absence of a single
ancestral archetype, in recognising that carpal and tarsal bones are not
identical throughout the tetrapods. In addition, the new data which is
being provided by studies on homeobox expression patterns and retinoic acid
effects, promises fresh insight into the question of limb homology, both at
the level of similarity and difference. But the complexity of a regulative
and interactive developmental system, such as the limb, with its cell
migrations and inductions, makes it rather unlikely that we will find

Drosophila-style pattern genes "for polydactyly" or "for the radius" other than genes acting indirectly on morphology for example through control of cell division. Meanwhile, there is a legitimate alternative approach (represented by Muller and Alberch, 1990, Muller, these proceedings) which assumes a similar set of processes in developing limbs in tetrapods, but identifies changes in their timing (truncation or prolongation) as leading to major structural transformations (as in crocodile and bird limbs). We need both approaches.

REFERENCES

ALBERCH. P., (1990). The logic of monsters: evidence for internal constraint in development and evolution. In "Ontogenese et Evolution", Colloque International du CNRS, Geobiol, suppl. 12: 21-57.

BOUVET, J., (970). Etablissement de la carte des territoires presomptifs du bourgeon de la nageoire pectoral chez la Truite indigene (Salmo trutta fario L) a l'aide d'excisions et de marques colorees. Anns. Embryol. Morph. 3: 315-328.

CAMERON, J.-A., and FALLON, J. F., (1977). Evidence for polarising zone in the limb buds of Xenopus laevis. Devel. Biol. 55: 320-330.

COATES, M. I. and CLACK, J. A., (1990). Polydactyly in the earliest known tetrapod limbs. Nature. 347: 66-69.

EDE, D. A., (1982). Levels of complexity in limb-mesoderm cell culture systems in "Differentiation in vitro", eds M. M. Yeoman and D. E. S. Truman. Cambridge: Cambridge University Press, pp 207-229.

FALLON, J. F., and CROSBY, G. M., (1977). Polarising zone activity in limb buds of amniotes. In "Vertebrate Limb and Somite Morphogenesis", eds. D. A. Ede, J. R. Hinchliffe and M. Balls. Cambridge: Cambridge University Press, pp 55-69.

GREER, A. E., (1990). Limb reduction in the Scincid lizard genus, Lerista. 2. Variations in the bone complements of the front and rear limbs and the number of post-sacral vertebrae. J. Herpetol. 24: 142-150.

GREGORY, W. K., and RAVEN, H. C., (1941). Studies on the origin and early evolution of paired fins and limbs. Ann. N. Y. Acad. Sci. 42: 273-360.

HAMILTON, H. L., (1952). "Lillie's Development of the Chick". New York: Henry Holt.

HINCHLIFFE, J. R., (1977). The chondrogenic pattern in chick limb morphogenesis: a problem of development and evolution. In "Vertebrate Limb and Somite Morphogenesis", eds. D. A. Ede, J. R. Hinchliffe and M. Balls. Cambridge: Cambridge University Press, pp 293-390.

HINCHLIFFE, J. R., (1982). Cell death in vertebrate limb morphogenesis. Progress in Anatomy. 2: 1-17.

HINCHLIFFE, J. R., (1985). 'One, two, three' or 'Two, three, four': an embryologist's view of the homologies of the digits and carpus of modern birds. "Beginnings of birds" (eds Hecht, M. K., Ostrom, J. H., Viohl, G., Wellnhofer, P.) Jura-Museums, Eichstatt.

HINCHLIFFE, J. R., (1990). An evolutionary perspective of the developmental mechanisms underlying the patterning of the limb skeleton in birds and other tetrapods. In "Ontogenese et Evolution", Colloque International du CNRS, Geobios, suppl. 12: 217-225.

HINCHLIFFE, J. R., and EDE, D. A., (1967). Limb development in the polydactylous talpid³ mutant of the fowl. J. Embryol. Exp. Morph. 17: 385-404.

HINCHLIFFE, J. R. and GRIFFITHS, P.J., (1983). The prechondrogenic patterns in tetrapod limb development and their phylogenetic significance. In "Development and Evolution", eds. B. C. Goodwin, N. H. Holder, and C. C. Wylie. Cambridge: Cambridge University Press, pp 99-121.

HINCHLIFFE, J. R. and JOHNSON, D. R., (1980). "The Development of the Vertebrate Limb." Oxford University Press.

HOLMGREN, N., (1933). On the origin of the tetrapod limb. _Acta Zool_. 14: 184-295.

HOLMGREN, N., (1955). Studies on the phylogeny of birds. _Acta Zool_. 36: 243-328.

HORDER, T. J., (1989). Syllabus for an embryological synthesis. In "Complex Organismal Functions: Integration and Evolution in Vertebrates", eds D. B. Wake, and G. Roth. Dahlem Konferenzen, J. Wiley, pp 315-348.

IDE, H. and AONO, H. (1988). Retinoic acid promotes proliferation and chondrogenesis in the distal mesodermal cells of chick limb bud. _Dev. Biol_. 130: 767-773.

JARVIK, E., (1980). "Basic Structure and Evolution of Vertebrates", Vol 2, London, Academic Press.

JORQUERA, B., and PUGIN, E., (1971). Sur le comportement du mesoderme et de l'ectoderm du bourgeon de membre dans les eschanges entre le poulet et le rat. _C.R. Hebd Seanc. Acad. Sci_. Paris. D, 272: 1522-1525.

MacCABE, J. A., SAUNDERS, J. W.(Jnr), and PICKETT, M. (1973). The control of the antero-posterior and dorso-ventral axes in embryonic chick limbs constructed of dissociated and reaggregated limb bud mesoderm. _Devel. Biol_. 31: 323-325.

MADEN, M. (1985). Retinoids and the control of pattern in limb development and regeneration. _Trends Genet_. 1: 103-107.

MILAIRE, J. (1973). Indices d'une participation morphogene de l'epiblaste au cours du developpement des membres chez les mammiferes. _Arch. Biol_. 84: 87-114.

MILAIRE, J. (1977. Rudimentation digitale au cours du developpement normale de l'autopode chez les mammiferes. In "Mecanismes de la rudimentation des organes chez les embryones de vertebres". Colloque CNRS, Paris. 266: 221-233.

MONTAGNA, W. (1945). A re-investigation of the development of the wing of the fowl. _J. Morphol_. 76: 87-113.

MULLER, G., and ALBERCH, P. (1990). Ontogeny of the limb skeleton in _Alligator mississippiensis_: developmental invariance and change in the evolution of archosaur limbs. _J. Morphol_. 203: 151-164.

OWEN, R. (1849). "On the Nature of Limbs". London: J. Van Voorst.

PAUTOU, M.-P. (1973). Analyse de la morphogenese du pied des oiseaux a l'aide de melanges cellulaires interspecifiques. I. Etude morphologique. _J. Embryo. Exp. Morph_. 29: 175-196.

RAYNAUD, A. (1985). Development of limbs and embryonic limb reduction. In "Biology of the Reptilia" vol. 15B, ed. C. Gans, New York, Wiley, pp 59-148.

RAYNAUD, A. (1990). Developmental mechanisms involved in the embryonic reduction of limbs in reptiles. _Int. J. Dev. Biol_. 34: 233-243.

RAYNAUD, A., and CLERGUE-GAZEAU, M. (1986). Identification des doigts reduits ou manquants dans les pattes des embryons de lezard vert (_Lacerta viridis_) traites par la cytosine-arabinofuranoside. Comparaison avec les reductions digitales naturelles des especes de reptiles serpentiformes. _Arch. Biol. (Bruxelles)_. 97: 279-299.

RAYNAUD, A., CLERGUE-GAZEAU, M., BONS, J., and BRABET, J. (1987). Nouvelles observations fondee sur les caracteres du metatarsien lateral et sur la structure du tarse, relatives a la formula digitale du Seps tridactyle (_Chalcides chalcides_, L). _Bull. Soc. Hist. Nat. Toulouse_. 123:127-132.

REITER, R. S., and SOLURSH, M. (1982). Mitogenic property of the apical ectodermal ridge. _Dev. Biol_. 93: 28-35.

ROMANOFF, A. L. (1960). "The Avian Embryo." New York: Macmillan.

SAUNDERS, J. W(Jr). (1948). The proximo-distal sequence of origin of the parts of the chick wing and the role of the ectoderm. _J. exp. Zool_. 108: 363-404.

SAUNDERS, J. W.(Jnr). (1972). Developmental control of three-dimensional polarity in the avian limb. Ann. N. Y. Acad. Sci. 193: 29-42.

SHUBIN, N. H., and ALBERCH, P. (1986). A morphogenetic approach to the origin and basic organization of the tetrapod limb. Evol. Biol. 20: 319-387.

SLACK, J. M. W. (1976). Determination of polarity in the limb. Nature (Lond). 261: 44-46.

SOLURSH, M. (1984). Ectoderm as a determinant of early tissue pattern in the limb bud. Cell Differ. 15: 17-24.

SUMMERBELL, D. (1974). A quantitative analysis of the effect of excision of the AER from the chick limb-bud. J. Embryol. exp. Morph. 32: 651-660.

THORNTON, C. S. (1957). The effect of apical cap removal on limb regeneration in Amblystoma larvae. J. exp. Zool. 134: 357-381.

TICKLE, C., ALBERTS, B., WOLPERT, L., and LEE, J. (1982). Local application of retinoic acid to the limb bud mimics the action of the polarising region. Nature (Lond). 296: 564-565.

TSCHUMI, P. A. (1957). The growth of the hind limb bud of Xenopus laevis and its dependence upon the epidermis. J. Anat. 91: 149-173.

VOROBYEVA, E. I. (1975). Notice on Panderichthys rhombolepis (Gross) from Lode in Latvia (Gauja-beds, Upper Devonian). N. Jahrb. Geol. Palaont. 5: 315-320.

VOROBYEVA, E. I., and HINCHLIFFE, J. R. (1991). The fin-limb transformation in tetrapod evolution: new palaeontological and developmental data. (In Russian). J. Common Biology. (In press).

WADDINGTON, C. H. (1962). "New Patterns in Genetics and Development". Columbia University Press.

NEW PALAEONTOLOGICAL CONTRIBUTIONS TO LIMB ONTOGENY AND PHYLOGENY

Michael Coates

University Museum of Zoology
University of Cambridge
Downing Street
Cambridge, CB2 3EJ, UK

INTRODUCTION

Recently, together with my colleague Dr Jennifer Clack, (Coates and Clack,1990) I reported the discovery of polydactylous, Devonian tetrapod limbs: the forelimb of *Acanthostega* (Jarvik, 1952), and the hindlimb of *Ichthyostega* (Jarvik, 1952,1980). These data , and other recent discoveries of early tetrapod material, combine to produce an almost completely new picture of the emergence of terrestrial vertebrate life. The first aim of this paper is to provide an updated palaeontological data-base for the study of vertebrate limb development.

The fossil record remains a unique source of long-term evolutionary data despite its many imperfections. Fossils have the potential to surprise us with anatomical phenomena which cannot be predicted by extrapolation from extant taxa. But, like all data, fossil interpretation is loaded with theoretical preconceptions, from identification to insertion within a broader context. The second aim of this paper is to illustrate the importance of comprehending the relative quality of the original data, and any subsequent interpretive process. Only after this may palaeontology be able to contribute usefully to multidisciplinary research areas such as tetrapod limb development.

EARLY LIMB DATA

Estimates of the age of tetrapod origin depend upon which data are considered acceptable. The earliest known skeletal remains (Jarvik, 1952; Lebedev, 1984; Coates and Clack,1990) date from the Famennian, the most recent stage of the Devonian, ~360 Myr BP (Cowie and Bassett, 1990). Earlier, indirect evidence of tetrapods exists in the form of fossil trackways. Trackways are of highly variable quality, and tell us nothing conclusive about the morphology of the feet from which they originated. The best known (and preserved) of these is of Frasnian age (Upper Devonian, ~375 Myr BP)(Warren and Wakefield, 1972), from Eastern Australia. Other, isolated foot-prints are known from the Mid-Devonian of Russia (Martyanov, 1960) and Brazil (Leonardo,

Developmental Patterning of the Vertebrate Limb
Edited by J.R. Hinchliffe *et al.*, Plenum Press, New York, 1991

1983). The earliest known track-way has been dated as potentially Basal Devonian (Warren et al, 1986), ~410 Myr BP, again from Eastern Australia. However, isolated imprints may be artifactual, and the basal Devonian trackway can only be indirectly attributed to a geological formation of the aforementioned age. Thus there could be ~50 Myr of unknown tetrapod evolution, which precedes the diverse morphologies of the earliest known skeletal remains.

The forms of the limbs of *Acanthostega* and *Ichthyostega* (Coates and Clack,1990), and *Tulerpeton* (Lebedev, 1984), differ considerably although they are all polydactyl. The acanthostegid forelimb is preserved as part of a remarkably intact, mostly articulated specimen, Geological Museum of Copenhagen (MGUH) field number 1227. The bones of the limb (humerus, radius, ulna, carpal fragments and digits) lie almost as they must have in life, and are exposed from the dorsal surface (figs 1a & b). The humerus is L-shaped, and resembles closely the humeri of other early tetrapods, such as *Eoherpeton* (Smithson, 1985), *Proterogyrinus* (Holmes, 1984), and *Greererpeton* (Godfrey, 1989). The large entepicondylar, ectepicondylar, and deltoid processes suggest that the limb musculature must have undergone considerable differentiation relative to the simple dorsal and ventral muscle blocks of the pectoral fin insertion of most fish. Andrews and Westoll (1970a) have proposed a provisional scheme of the musculature associated with the humeri of the osteolepiform fish, *Eusthenopteron*. Osteolepiforms are generally considered to be the sister-group of tetrapods (Panchen and Smithson, 1987); refer to Rackoff (1980) for a further consideration of the mechanics of their paired fins. It would seem equally appropriate to apply this scheme to the forelimb of *Acanthostega*. The articulatory facets for the radius and ulna are situated terminally. The spatulate radius is similar to the first pre-axial radial of the pectoral fin of *Eusthenopteron* (Andrews and Westoll, 1970a). The most significant difference is that the flattened distal region bears no grooves for the insertion of lepidotrichia. The radius is considerably longer than the broad, squat ulna. This is a characteristic unique to *Acanthostega* amongst the tetrapods, but shared with all known osteolepiform pectoral fins.

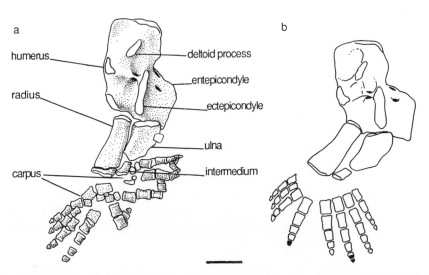

Fig. 1. Forelimb of *Acanthostega*. (a) Drawing from specimen MGUH 1227. (b) Reconstruction of forelimb; restored parts blocked-in. Scale bar=10mm. Anterior edge to left of figure.

The proportions of the radius and ulna are therefore considered to be primitive relative to all other known limbs. The ulna lacks any form of olecranon process, normally associated with the ability to effectively flex the distal part of the limb.

The carpus is poorly preserved, with only a single well ossified subcylindrical bone, interpreted as the intermedium. The digits lie in a broad arch which spans the width of the distal end of the limb. There are eight in total, with the phalangeal formula of 3,3,3,4,4,5,5,4 as listed from the anterior to posterior edges of the limb. None of the digits is duplicated across the span, and no contributory structures have been found during preparation of the fossil which might suggest that the octodactylous condition is a taphonomic artifact.

The hindlimb of *Ichthyostega* (specimen field number MGUH 1394) (figs 2a,b, & c) is essentially similar to Jarvik's (1952, 1980) published description. The ichthyostegid limb depicted in Carroll's volume (1988, fig.9-10a, p.164) should be disregarded (it has aquired five additional bones; proportions of the remainder are inaccurate). The important new data yielded by MGUH 1394 concern the tarsus and digits. First, there is a more extensive series of distal tarsals. Second, the digits conventionally numbered as III and IV contain an additional phalanx. Third, the tibiale and prehallux, plus digit I (as reconstructed by Jarvik), are replaced by a cluster of three small digits. The limb therefore bears seven digits, with the phalangeal formula of 3,4,2,3,4,4,3 (listed from anterior to posterior). The anterior-most three digits are separate entities, and not in any sense fused. Examination of ichthyostegid specimens in Stockholm have confirmed this observation.

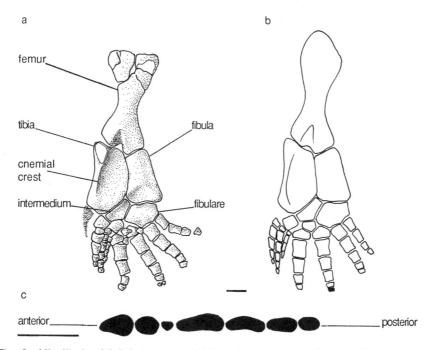

Fig. 2. Hindlimb of *Ichthyostega*. (a) Drawing from specimen MGUH 1394. (b) Reconstruction of hindlimb; restored parts blocked-in. (c) Transverse sections of proximal phalanges, from specimen A166b. Scale bars=10mm. Anterior edge to left of figure.

Stockholm specimen A166b contains the proximal parts of a full digital complement, preserved "in-the-round"; the bone has been etched away to leave a detailed natural mould. Latex casts of this specimen enabled a detailed examination to be made of the digits in transverse section (fig. 2c). The third digit from the anterior edge is not only shorter, but also much narrower than flanking digits.

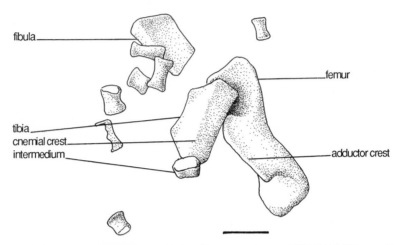

Fig. 3. Hindlimb of *Acanthostega,* drawn from specimen MGUH1227, mostly viewed from ventral surface. Scale bar=10mm.

The forelimb of *Tulerpeton* (Lebedev, 1984) has not yet been described formally. However, the published reconstruction shows it to have a digital formula of 2,3,4,5,4,2, (listed from anterior to posterior). The radius and ulna are of equal length, and the carpus appears to consist of broad subrectangular bones similar to those figured, but unreconstructed, in the carpal region of *Acanthostega* (fig. 1a).

It is currently unclear whether the hindlimb of *Tulerpeton* bears five or six digits. Superficially, it resembles the hindlimbs of more recent tetrapods, in contrast to the paddle-like structure of *Ichthyostega* (Bjerring, 1985; Coates and Clack, 1990). Preliminary results suggest that *Acanthostega* has a similarly formed hindlimb (fig. 3). The femur, tibia, fibula, intermedium and assorted phalanges (specimen MGUH 1227) are well preserved but disarticulated. The phalanges are much larger, relative to the span of the tibia and fibula, than those of *Ichthyostega.* The tibia and fibula, however, are ichthyostegid-like: both dorsoventrally flattened, tibias bearing an elongate cnemial crest.

The femora of all three Devonian tetrapods are similarly proportioned, and all three bear a huge ventral ridge (50%+ of total length). This ridge is interpreted as an adductor crest, a neomorph (and therefore autapomorphous) feature of tetrapods (none is found on osteolepiform femora, Andrews and Westoll, 1970a&b).

INTERPRETATION OF FOSSIL DATA

The three earliest known tetrapods have an anatomically diverse range of limbs, none of which corresponds to established concepts of primitive limb morphology (cf. any current text-book which covers vertebrate anatomy).

Patterson (1981) has identified four ways by which ontogeny and phylogeny can be analysed with the assistance of palaeo-data:

1. fossils can reveal non-homologies,
2. fossils can refute certain synapomorphies or ontogenetic sequences,
3. fossils can suggest the sequence of acquisition of derived characters (and can therefore be used to define the minimal age of taxa),
4. fossils can supply biogeographic data.

The way in which the Devonian limb data corroborates Shubin and Alberch's (1986) model of limb morphogenesis (Coates and Clack, 1990) is essentially a manifestation of the second point on Patterson's list. But, if fossil data is to shed further light on the evolution of limbs, then the quality of the data, and the justification for any phylogenetic hypothesis drawn from it, must be made explicit. By these means it may be possible to avoid the propagation of spurious evolutionary scenarios, such as the remote origination of limbs from the archipterygium-like fins of the chondrichthyan *Xenacanthus* (Horder, 1989, fig.1). Unsurprisingly, this transformational series includes a major homological discontinuity, from which an abrupt evolutionary transition could be inferred (illustrating Patterson's third point). Clearly, the validity of any inference of evolutionary tempo and mode drawn from this transformation must be proportional to the validity of the phylogeny upon which it is based. In this example, *Xenacanthus* is known to be a highly derived, Permian, fresh water shark (Zangerl, 1981), and its superficially archipterygial fins are derived from a metapterygial pattern (Dick, 1981). Xenacanth fins, therefore, post-date the earliest tetrapods by ~80 Myr, and are taxonomically far removed. In comparison to tetrapod limbs, they can illustrate nothing more than the derivation of different skeletal arrangements from what may be a common, but chronologically distant origin.

The ancestry of tetrapod limbs is usually traced only as far as their closest sarcopterygian (lobe-finned) relative. The closest known limb-like fins, stratigraphically and morphologically, are known from Middle and Upper Devonian fishes: *Eusthenopteron* (Jarvik, 1980; Andrews and Westoll,1970a), *Sauripterus* (Gregory and Raven, 1941; Andrews and Westoll, 1970b), and *Panderichthys* (Vorobjeva, 1975). All three of these are too recent to represent more than archaic hang-overs of an earlier form from which tetrapods arose. All three are frequently referred to as "crossopterygians" (eg. Hinchliffe, 1989b) although rarely is it made clear which of several meanings of the term is being used. The Sarcopterygii were designated by Romer (1966) to include the Dipnoi (lungfishes) and the Crossopterygii, the latter group containing all other sarcopterygian taxa which were assumed to be of common ancestry. The Crossopterygii in this sense is a paraphyletic grade-group, and must be defined partly by the absence of tetrapod characters. The group includes porolepiforms (=holoptychiids), osteolepiforms, rhizodonts, actinistians (=coelacanths), and onychodonts (=struniiforms). The interrelationships of these groups are undefined, although a selection, including the osteolepiforms, may be termed "rhipidistian" because of its close proximity to the origin of tetrapods. Examples of this imprecise usage can be found in Romer (1966), Moy-Thomas and Miles (1971), and Carroll (1988) (fig. 4a). The Crossopterygii as a natural, monophyletic, subdivision of the Sarcopterygii has been formally characterised by Schultze (1987) and Long (1989), and includes tetrapods (fig. 4c). However, as a result of the struggle to refute the proposed lungfish-tetrapod sister-grouping (Rosen et al., 1981) (fig. 4b), there have been several rigorous reexaminations of sarcopterygian interrelationships. A new consensus is emerging, partly based upon new data obtained from primitive, Lower Devonian sarcopterygians, in which the constituent members of the Crossopterygii have become dissociated (Maisey, 1986, fig. 4d; Panchen and Smithson, 1987).

Crossopterygians have thus become a polyphyletic assemblage, and the term should be disregarded except for use within a historical context. Selection of fin data on the basis of it being crossopterygian is therefore probably misinformative with regard to the long-term canalisation of fin development into limbs.

The term "Rhipidistian" (Romer, 1966; Carroll, 1988), refers to the crossopterygian subset of Porolepiforms+Osteolepiforms. This set is either paraphyletic (Schultze,1987), or polyphyletic (Maisey,1986; Panchen and Smithson, 1987), and should be treated with similar caution.

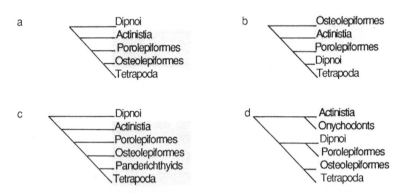

Fig. 4. The distribution of crossopterygians (=shaded taxa) within alternative cladograms depicting interrelationships of fish-groups with tetrapods. (a) Adapted from Carroll (1988). (b) Adapted from Rosen et al. (1981). (c) Adapted from Schultze (1987) and Long (1989). (d) Adapted from Maisey (1986).

The quality of actual palaeontological data, as distinct from its phylogenetic interpretation, should be given similarly careful consideration. Certain examples of early limbs and fins which are figured frequently in texts concerning the onto- and phylogenetic aspects of limb development contain large amounts of speculative restoration. The pectoral fin of *Sauripterus* illustrates this problem. Ever since it was first described (Hall, 1843, in Andrews and Westoll, 1970b), homologies have been suggested between the proximal elements and tetrapod limb bones. Yet the quality of the material (Upper Devonian of Pennsylvania, USA; incomplete pectoral girdle and fin; isolated scales) has been criticized for almost as long. Andrews and Westoll noted that Newberry, in 1889, stated "The pectoral fin... is too imperfect for study". Nevertheless, Gregory (1913) and Broom (1913) established the place of this specimen in limb-evolution literature, where it has remained and recurred ever since. Rarely is any hint of the poor quality of the original indicated (Andrews and Westoll, 1970b, text fig.15, pl.VIIIc; Pers.Obs. cast taken from original specimen). Andrews and Westoll concluded that *Sauripterus* is probably the earliest and most specialised known member of the rhizodonts, a group of sarcopterygians with close affinities to the osteolepiforms (fig.4). A rather better preserved rhizodont pectoral fin has been recently described from what is considered to be the most primitive known form, *Barameda* (Long, 1989), which originates from the Lower Carboniferous of Australia. Neither of these fins displays closer affinities than the pectoral fin of the osteolepiform *Eusthenopteron* to the pectoral limbs of early tetrapods (Andrews and Westoll, 1970a

Jarvik, 1980; Coates and Clack, 1990). Other aspects of rhizodont morphology preclude them further from a sister-group relationship with tetrapods (Schultze, 1987; Long, 1989). *Sauripterus* is, therefore, too specialized, poorly preserved, and taxonomically remote to provide useful information about the transition from sarcopterygian fin to tetrapod limb.

Gregory and Raven's (1941) pentadactyl restoration of the forelimb of *Eryops* is a similarly inappropriate choice of fossil data to illustrate primitive tetrapod limb structure. *Eryops* is a Lower Permian temnospondyl, and therefore a member of a group thought to be closely associated with the origin of extant amphibians (Milner et al,1986;Panchen and Smithson, 1988) (fig. 5). Temnospondyls are characterised partly by the posession of a four digit manus; *Eryops* specimens show the incomplete remains of no more than four digits on the forelimb. In fact, this early limb now appears to be relatively derived, and occurs some ~60Myr after the earliest known limbs. Significantly, the pentadactyl restoration of this forelimb shares with Jarvik's (1980) original reconstruction of the ichthyostegid hindlimb several hypothetical stuctures thought to be serially homologous: a prepollex/ prehallux, and postminimi. These elements were restored presumably to correspond to Holmgren's (1933,1939) influential, theoretical scheme of "canonical" limb elements. The contrived consistency of many restored fossil limbs with Holmgren's and other's views has led to the presentation of a great deal of apparently unambiguous, but in fact erroneous, text-book data. Uncritical use of this form of fossil data is unlikely to provide an informative, evolutionary perspective on the development of tetrapod limb patterning.

A PHYLOGENY OF EARLY TETRAPODS: PATTERNS OF LIMB EVOLUTION

The phylogenetic tree of early tetrapods shown in figure 5 incorporates the most complete early limb-data currently available. This tree is based upon Panchen and Smithson's (1988) recent cladistic analysis of early tetrapod interrelationships (no further character analyisis has been undertaken), plotted against a geological time-scale (Cowie and Basset,1990). The minimum age of taxa can be estimated from the earliest occurence of constituent members (Patterson's third point, see above), or the consequent points of divergence for the major tetrapod groups. Certain additions have been made to Panchen and Smithson's data: *Acanthostega* (Coates and Clack, 1990), *Tulerpeton* (Lebedev, 1984), the earliest known amniote (Smithson, 1989), and an elaborated microsaur phylogeny (Carroll and Gaskill, 1978).

Acanthostega is not yet known to share synapomorphies with any groups other than tetrapods as a whole, and has therefore been placed, *incertae sedis* as the sister-group of all other tetrapods. *Tulerpeton* has been placed *incertae sedis* at the base of the reptiliomorph ramus (discussed below), on the basis of Lebedev's (1985) comments concerning its anthracosaur-like features. The earliest known amniote has been placed similarly *incertae sedis* at the base of the amniote ramus because it is not yet fully described; not because of stratophenetic assumptions (cf. previously discussed rhizodont anatomy).

Panchen and Smithson have emphasized the "very tentative nature" (1988,p.25) of their cladogram, and note that it is based upon relatively few synapomorphies. While a cladistic analysis of interrelationships is at least explicit in its reasoning, they acknowledge that the cladistic use of parsimony (Panchen, 1982) is not a reliable detector of convergence (homoplasy).

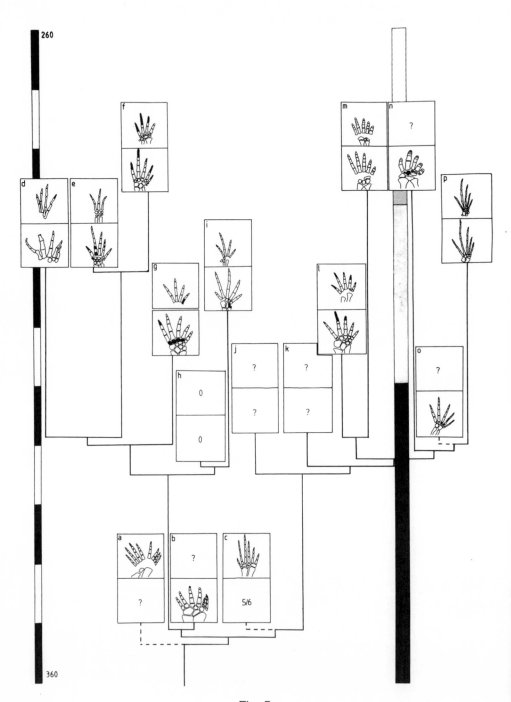

Fig. 5

Fig. 5.   A phylogenetic tree of the interrelationships of early tetrapods, based upon Panchen and Smithson's (1988) cladistic analysis. A time scale is on the left side; each division represents 10Myr. The scale on the right represents 4 geological systems: Upper Devonian, Lower Carboniferous, Upper Carboniferous, Lower Permian (geo-chronological correlation from Cowie and Bassett,1990). Each box contains the earliest, relatively complete limb-data for each major tetrapod group, the base of which is aligned with its estimated age. The upper division of each box contains forelimb data, and the lower division hindlimb data.All restored skeletal parts are blocked-in; all anterior edges are on the right. Broken lines connect with taxa placed *incertae sedis*. Note that age of limb data does not always coincide with minimum age of divergence of taxonomic group. (a) *Acanthostega gunnari,* (Jarvik,1952; Coates and Clack, 1990). Hindlimb insufficiently known. (b) *Ichthyostega* sp. (Jarvik, 1952; Coates and Clack, 1990). Forelimb insufficiently known. (c) *Tulerpeton curtum,* (Lebedev, 1984) Hindlimb uncertain: 5 or 6 digits. (d) Temnospondyli, considered to be sister-group of extant amphibia, represented by *Amphibamus lyelli,* forelimb after Carroll (1964), hindlimb after Hook and Baird (1964). Earliest temnospondyl dates from Lower Carboniferous (Milner *et al*, 1986). (e) Microsauria, *Hyloplesion longicostatum,* (Carroll and Gaskill,1978), a microbrachomorph. (f) Microsauria, *Pantylus cordatus,* (Carroll and Gaskill, 1978), a tuditanomorph. (g) Colosteidae, forelimb from *Colosteus scutellus,* (Hook, 1983), hindlimb from *Greererpeton burkemorani,* (Godfrey, 1989). (h) Aistopoda, *Lethiscus stocki,* (Wellstead, 1982). (i) Nectridea, *Sauropleura scalaris,* (Bossey, Ph.D thesis, unpublished). (J) Loxommatoidea, post-cranial skeleton currently unknown, cranial morphology in Beaumont (1977). (K) *Crassigyrinus scoticus,* limbs incompletely known (Panchen and Smithson, 1990). (l) Anthracosauroideae, *Proterogyrinus scheeli,* (Holmes, 1984), earliest anthracosaur dates from Lower Carboniferous (Milner *et al ,* 1986). (m) Seymouriamorpha, *Seymouria baylorensis ,* (White, 1939). (n) Diadectomorpha *Diadectes ,* sp. (Romer and Byrne, 1931), forelimb insufficiently known. (o) Amniota, stem amniote from Smithson (1989), forelimb insufficiently known. (p) Amniota, *Palaeothyris acadianum,* (Heaton and Reisz, 1985).

This phylogenetic tree (fig.5) is only one of several equally valid alternatives which may be derived from the same cladogram (Wiley, 1981): it covers a time span of 110 Myr, from the Upper Devonian to the Lower Permian; almost a third of known tetrapod evolution. Most major groups have diverged before the end of the Lower Carboniferous, ~330 Myr Bp. The first, and major branching event may have occured before the date of earliest known skeletal material. *Ichthyostega* is considered to be a member of the batrachomorph branch (which leads ultimately to extant amphibia), while *Tulerpeton* is assigned to the reptiliomorphs. If the limb patterns of these two branches are compared, then it appears that the reptiliomorphs rapidly fix upon a pentadactyl arrangement, while the batrachomorphs exhibit greater variability (with no evidence of ever having had a pentadactyl manus). The most dramatic change in limb morphology found in this phylogeny occurs between the Devonian tetrapods and the earliest of the Lower Carboiferous forms, *Lethiscus* (Wellstead, 1982). *Lethiscus* is the earliest known aistopod, an entirely limb-less and girdle-less group of snake-like forms (with up to two hundred vertebrae and short tails). The aistopods are usually grouped with the nectrideans because of similarities in vertebral structure, although this has been recently questioned (Milner et al.1986). Panchen and Smithson even consider the question of whether they are characterisable as tetrapods at all. This early and relatively rapid specialisation has been remarked upon before (Baird, 1964; Wellstead, 1982), although "rapid" in this case encompasses approximately 25 Myr.

Apart from *Acanthostega* (see previous discussion), it is probably too soon to categorise Devonian tetrapod limbs as either primitive or generalised with regard to subsequent evolution. In contrast, the hindlimbs of *Ichthyostega* and *Acanthostega* could be considered as derived or specialised, relative to a hindlimb such as that of *Proterogyrinus* (Holmes, 1984) (fig.5). However, currently there is no adequate out-group data for comparison. It would, therefore, be premature to pair-off *Ichthyostega* and *Acanthostega* as members of an ichthyostegalian sub-group (eg. Long, 1990), on the basis of these limbs.

CONCLUSION: FOSSIL LIMBS AND EVOLUTIONARY PROCESSES

Fossils can suggest possible evolutionary patterns, but can they reveal anything about evolutionary processes? Palaeontological data is frequently criticised because of the incomplete nature of both the fossil record and individual specimens. Fossils, especially vertebrates, are also incomplete relative to their own ontogeny: only one stage may be preserved; usually a mature adult. However, developmental processes are not necessarily concealed; fossil morphologies can only be interpreted as results of the same developmental processes that we observe today. Therefore, early fossil limbs may not be entirely uninformative about the developmental changes which occured during the fin to limb transition. As previously discussed (Coates and Clack, 1990), the acanthostegid and ichthyostegid limb morphologies corroborate the limb-patterning morphogenetic scheme proposed by Shubin and Alberch (1986). In this it is suggested that the metapterygial (main) axis of the ancestral fin now runs across the digital arch. The acanthostegid forelimb may therefore be interpreted as a "peramorphic" (Mc Namara, 1986) osteolepiform pectoral fin: the pectoral limb is an anteriorly curled fin which has developed through additional ontogenetic stages relative to an ancestral condition. It is impossible to ascribe this change to any one of the more precisely defined peramorphic processes: hypermorphosis, acceleration, or predisplacement. However, these heterochronic terms are essentially descriptive, referring to the effects of currently unknown developmental causes.

Polydactyly in the earliest known limbs can be interpreted, alternatively, as a product of known developmental processes, such as a broad "tissue domain" (Oster *et al*,1988) in the distal region of the limb bud, or an expansive apical ectodermal ridge (Saunders,1977). Hinchliffe (1989a), recognising the assymmetry of osteolepiform fins, has previously suggested the potentially long term, evolutionary influence of the zone of polarising activity. Similarly, now that the development of tetrapod limbs appears to be initiated and to an unknown extent controlled by a homoeotic selector gene complex (Dolle *et al*,1989), we may infer that the development of the earliest limbs was also regulated in this way. Intriguingly, Lewis and Martin (1989) have noted that the expression of this homoeotic selector gene complex fits neither a simple proximo-distal or antero-posterior gradient: it is skewed relative to both axes. Can this phenomenon be explained by the derived distal-anterior curvature of limbs relative to their proximodistally arranged ancestral fins? It is known that, like the variable pattern of early tetrapod limbs, there is considerable variability in the body/ground-plan of the fossil ancestry of the myriapod/hexapod lineage (Anderson,1973), also apparently regulated by homoeotic genes. Are these variable early morphologies the result of simple, unelaborated homoeotic selector gene complexes?Trilobites were described recently as having the influence of homoeotic genes inscribed upon their carapaces. Perhaps the hidden morphological repertoire of the fossil record can be used further to complement advances in developmental research.

ACKNOWLEDGEMENTS

I am grateful to Drs T.R. Smithson, J.A. Clack, and A.C. Milner for assistance and constructive criticism during the preparation of this manuscript. The Geological Survey of Greenland, Denmark; the Carlsberg Foundation (Copenhagen); the Natural Environment Research Council, and the University Museum of Zoology, Cambridge, provided support for the 1987 Greenland expedition, led by Dr S.E. Bendix-Almgreen, during which the acanthostegid and ichthyostegid limbs were collected.

REFERENCES

Anderson, D.T., 1973, "Embryology and phylogeny in annelids and arthropods" O.U.P., Oxford.
Andrews, S.M., and Westoll, T.S., 1970a, The postcranial skeleton of *Eusthenopteron foordi* Whiteaves, Trans. R. Soc. Edinb. , 68:207-329.
Andrews, S.M., and Westoll, T.S., 1970b, The postcranial skeleton of rhipidistian fishes excluding *Eusthenopteron*, Trans. R. Soc. Edinb., 68:391-486.
Baird, D., 1964, The aistopod amphibians surveyed, Breviora , 206:1-17.
Beaumont, E.H., 1977, Cranial morphology of the Loxommatidae (Amphibia: Labyrinthodontia), Phil. Trans. R. Soc., B280:29-101.
Bjerring, H.C., 1985, Facts and thoughts on piscine phylogeny, in: "Evolutionary Biology of Primitive Fishes," R.E. Foreman, A. Gorbman, J.M. Dodd, and R. Olsson eds., Plenum, New York.
Broom, R., 1913, Origin of the cheiropterygium, Bull. Am. Mus. N. H., 32:453-464.
Carroll, R.L., 1964, Early evolution of the dissorophid amphibians, Bull. Mus. Comp. Zool. Harvard., 131:163-250.
Carroll, R.L., 1988, "Vertebrate palaeontology and evolution," W.H. Freeman and Company, New York.
Carroll, R.L. and Gaskill, P., 1978, The Order Microsauria, Mem. Am. Philos. Soc.,126:1-128.
Coates, M.I. and Clack, J.A., 1990, Polydactyly in the earliest known tetrapod limbs, Nature,347:66-69.
Cowie, J.W. and Bassett, M.G., 1990, I.C.S.:I.U.G.S. 1989 Global stratigraphic chart, in:"Palaeobiology, a Synthesis," D.E.G. Briggs and P.R. Crowther, eds., Blackwell, London.
Dick, J.R.F., 1981, *Diplodoselachi woodi* gen. et sp. nov., an early Carboniferous shark from the Midland Valley of Scotland, Trans. R. Soc. Edinb. Earth Sci.,72:99-113.
Dolle, P., Izpisua-Belmonte, J.C., Falkenstein, H., Renucci, A., and Duboule, D., 1989, Coordinate expression of the murine Hox-5 complex homeobox-containing genes during limb pattern formation, Nature, 342:767-772.
Godfrey, S.J., 1989, The postcranial skeletal anatomy of the Carboniferous tetrapod *Greererpeton burkemorani* Romer 1969, Phil. Trans. R. Soc., B323:75-133.

Gregory, W.K., 1913, Crossopterygian ancestry of the amphibia, <u>Science</u>, 37:806.

Gregory, W.K. and Raven, H.C., 1941, Studies on the origin and early evolution of paired fins and limbs, <u>Ann. N.Y. Acad. Sci.</u>, 17: 273-360.

Heaton, M.J. and Reisz, P.R., 1986, Phylogenetic relationships of captorhinomorph reptiles, <u>Can. J. Earth Sci.</u>, 23:402-418.

Hinchliffe, J.R., 1989a, Evolutionary aspects of the developmental mechanisms underlying the patterning of the pentadactyl limb skeleton in birds and other tetrapods, <u>in</u>:"Fortschritte der Zoologie",35:226-229, H. Splechtna and C.H. Hilgers eds., Gustav Fischer Verlag, Stuttgart.

Hinchliffe, J.R., 1989b, Reconstructing the archetype:innovation and conservatism in the evolution and development of the pentadactyl Limb, <u>in</u>: "Complex organismal functions: integration and evolution in vertebrates," D.B. Wake and G. Roth eds., John Wiley and Sons Ltd, Chichester.

Holmes, R., 1984, The Carboniferous amphibian *Proterogyrinus scheeli* Romer, and the early evolution of tetrapods, <u>Phil. Trans. R. Soc.</u>, B306:431-527.

Holmgren, N., 1933, On the origin of the tetrapod limb, <u>Acta Zool. (Stockholm)</u>., 30:459-484.

Holmgren, N., 1939, Contribution to the question of the origin of the tetrapod limb, <u>Acta Zool. (Stockholm)</u>., 20:89-124.

Hook, R.W., 1983, *Colosteus scutellus* (Newberry), a primitive temnospondyl amphibian from the Middle Pennsylvanian of Linton, Ohio, <u>Am. Mus. Novit.</u>, 2770:1-41.

Hook, R.W. and Baird,D., 1984, *Ichthycanthus platypus* Cope,1877, reidentified as the Dissorophoid Amphibian *Amphibamus Lyelli*, <u>J. Paleont.</u>, 58:697-702.

Horder, T.J., 1989, Syllabus for an embryological synthesis, <u>in</u>: "Compex organismal functions: integration and evolution in vertebrates," D.B. Wake and G. Roth eds., John Wiley and Sons Ltd, Chichester.

Jarvik, E., 1952, On the fish-like tail in the ichthyostegid stegocephalians, <u>Meddr. Gronland</u>, 114:1-90.

Jarvik, E., 1980, "Basic structure and evolution of vertebrates" vol.1, Academic, London.

Lebedev, O.A., 1984, First discovery of a Devonian tetrapod vertebrate in U.S.S.R., <u>Dokl. Akad. Nauk. SSR.</u>, 278:1470-1473.

Lebedev, O.A., 1985, The first tetrapod: description and occurrence, <u>Priroda</u>, 1985 (11):26-36.

Leonardi, G., 1983, *Notopus petri* nov. gen., nov. sp., : Une empreinte d'amphibien du Devonien au Parana (Bresil), <u>Geobios</u>, 16:233-239.

Lewis, J. and Martin, P., 1989, Limbs: a pattern emerges, <u>Nature</u>, 342:734-735.

Long, J.A., 1989, A new rhizodontiform fish from the Early Carboniferous of Victoria, Australia, with remarks on the phylogenetic position of the group, <u>J. Vert. Paleo.</u>, 9:1-17.

Long, J.A., 1990, Heterochrony and the origin of tetrapods, <u>Lethaia</u>, 23:157-167.

Mc Namara, K.J., 1986, A guide to the nomenclature of heterochrony, <u>J. Paleont.</u>, 60:4-13.

Martyanov, H.E., 1960, Otpechatok pyatepalogo sleda, <u>Priroda</u>, 1960 (9):115.

Maisey, J.G., 1986, Heads and tails: a chordate phylogeny, <u>Cladistics</u>, 2:201-256.

Milner, A.R., Smithson, T.R., Milner, A.C., Coates, M.I., and Rolfe, W.D.I., 1986, The search for early tetrapods, <u>Mod. Geol.</u>10:1-28.

Moy-Thomas, J.A. and Miles, R.S., 1971, "Palaeozoic fishes," Chapman and Hall, London.

Oster, G.F., Shubin, N., Murray, J.D. and Alberch,P., 1988, Evolution and morphogenetic rules: The shape of the vertebrate limb in ontogeny and phylogeny, Evolution, 42:862-884.

Panchen, A.L., 1982, The use of parsimony in testing phylogenetic hypotheses, Zool. J. Linn. Soc., 74: 305-328.

Panchen, A.L. and Smithson, T.R., 1987, Character diagnosis, fossils and the origin of tetrapods, Biol. Rev., 62:341-438.

Panchen, A.L. and Smithson, T.R., 1988, The relationships of the earliest tetrapods, in: "The phylogeny and classification of the tetrapods, volume 1: Amphibians, Reptiles, Birds," M.J. Benton, ed., Clarendon Press, Oxford.

Panchen, A.L. and Smithson, T.R., 1990, The pelvic girdle and hindlimb of Crassigyrinus scoticus (Lydekker) from the Scottish Carboniferous and the origin of the tetrapod pelvic skeleton, Trans. R. Soc. Edinb. Earth Sci., 81:31-44.

Patterson, C.P., 1981, Significance of fossils in determining evolutionary relationships, Ann. Rev. Ecol. Syst., 12:195-223.

Rackoff, J.S., 1980, The origin of the tetrapod limb and the ancestry of tetrapods, in:"The terrestrial environment and the origin of land vertebrates," A.L. Panchen ed., Academic Press, London.

Romer, A.S., 1966, "Vertebrate Paleontology," University Press, Chicago.

Romer, A.S. and Byrne, F., 1931, The pes of Diadectes: notes on the primitive tetrapod limb, Palaeobiologica, 1931(IV Band):25-48.

Rosen, D.E., Forey, P.L., Gardiner, B.G., and Patterson, C.P., 1981, Lungfishes, tetrapods, palaeontology, and plesiomorphy, Bull. Am. Mus. Nat. Hist., 167:159-276.

Saunders, J.W., 1977, The experimental analysis of chick limb-bud development, in: "Vertebrate limb and somite morphogenesis," D.A.Ede, J.R. Hinchliffe, and M. Balls, eds., University Press, Cambridge.

Schultze, H.P., 1987, Dipnoans as sarcopterygians, J. Morph. sup.1:39-47.

Shubin, N.H. and Alberch, P., 1986, A morphogenetic approach to the origin and basic organisation of the tetrapod limb, Evol. Biol. , 20:319-387.

Smithson, T.R., 1985, The morphology and relationships of the Carboniferous amphibian Eoherpeton watsoni Panchen, Zool. J. Linn. Soc., 85:317-410.

Smithson, T.R., 1989, The earliest known reptile, Nature, 342:676-678.

Worobjewa (Vorobyeva), E.I., 1975, Bermerkungen zu Panderichthys rhombolepis (Gross) aus Lode in Lettland (Gaujashchichten, Oberdevon), N. Jarbuch. Geol. Palaont. , Monatschafte 1975: 315-320.

Warren, J.W. and Wakefield, N.A., 1972, Trackways of tetrapod vertebrates from the Upper Devonian of Victoria, Australia, Nature, 238:469-470.

Warren, A., Jupp, R. and Bolton, B., 1986, Earliest tetrapod trackway, Alcheringa, 10:183-186.

Wellstead, C.F., 1982, A Lower Carboniferous aistopod amphibian from Scotland, Palaeontology,25:193-208.

White, T.E., 1939, Osteology of Seymouria baylorensis Broili, Bull. Mus. Comp. Zool. Harvard,85:325-409.

Wiley, E.O., 1981, "Phlogenetics: the theory and practice of phylogenetic systematics," John Wiley and Sons Inc., New York.

Zangerl, R., 1981, "Chondrichthyes 1. (Paleozoic Elasmobranchii). Handbook of Palaeoichthyology; vol.3," Fischer, Stuttgart.

THE FIN-LIMB TRANSFORMATION: PALAEONTOLOGICAL AND

EMBRYOLOGICAL EVIDENCE

Emilia I. Vorobyeva

Institute of Evolutionary Morphology and Ecology
MOSCOW, USSR

The problem of the protetrapod pentadactyl limb still remains under discussion. Different forms among sarcopterygians have been claimed to possess the archetype of such a limb (Jarvik, 1980; Rosen et al, 1981). Sometimes the possibility of reconstruction of this prototype on the evidence from fish fins has been rejected (Rackoff, 1980).

The discovery of Palaeozoic crossopterygians and of the earliest tetrapods during the 1970s and 1980s makes a fresh contribution to this problem. Together with data on the development of recent forms and functional analysis, these palaeontological finds permit the understanding of some of the mechanisms of the evolutionary process. Among these finds, Panderichthys rhombolepis from the Upper Devonian of Latvia (Vorobyeva, 1975) provides considerable interest for the problem. Together with Elpistostege this genus has been included into a new order Panderichthyidae (Vorobyeva, 1989). In the structure of the skull, teeth, and some elements of the postcranial skeleton, these rhipidistians display tetrapod features, in particular, in common with the ichthyostegids (Acanthostega, Ichthyostega). To a certain extent they can be treated as a missing morphological link between fishes and labyrinthodonts and from the cladistic position as a sister group of Tetrapoda (Vorobyeva and Schultze, 1991). At the same time the panderichthyids display a distinct mosaic pattern of morphological evolution, carrying many plesiomorphic features (rhombical scales, the simplified polyplocodont teeth, protocercal caudal fin etc). There are the following unique features of the locomotory apparatus; absence of the dorsal and anal fins, ventrally displaced paired fins, with the pelvic being smaller than the pectoral, and situated close to the caudal fin. Probably the vertebrae are without pleurocentra. Comparatively short intercentra are fused ventrally forming a hemi-ring structure which is connected with densely spaced neural arches by a horizontal suture.   There are large dorsal ribs (about 40) connecting with the neural arch and with the intercentra by sutures. The scapulo-coracoid has a dual attachment to the cleithrum not far from the contact of the latter with the clavicle. There is a well developed coracoidal plate, comparable with that in Ichthyostega, but differing in the presence of a large glenoid fossa and in the more anterior position of the shoulder girdle. The clavicle is small and surrounds the coracoid plate anter-dorsally and anter-ventrally, forming the anterior margin of the glenoid fossa. The ventro-medial margin of the concave glenoid fossa has a small process, which limits the lateral mobility of the humerus. It is the proximal division (stylopod and zeugopod) which reveals the greatest similarity among the known rhipidistians with the

Developmental Patterning of the Vertebrate Limb
Edited by J.R. Hinchliffe *et al.*, Plenum Press, New York, 1991

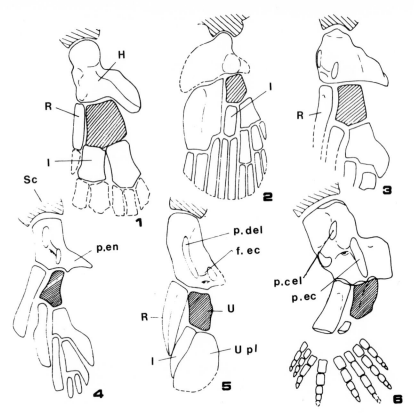

Figure 1. The diversity in pectoral fin structure in rhipidistians

1 - Barameda, after Long, (1989)
2 - Sauripterus, after Andrews and Westoll, (1970)
3 - Sterropterygion, after Rackoff, (1980)
4 - Eusthenopteron, after Jarvik, (1980
5 - Panderichthys, after Vorobyeva, (1975)
6 - Acanthostega, after Coates and Clack, (1990)

(H, humerus; I, intermedium; p del, deltoid process;
p ec, ectepicondylar process; p en, entepicondylar process;
R, radius; U, ulna (shaded); U pl, ulnar plate).

tetrapod limb. This has become more evident after the discovery of
Acanthostega, (Coates and Clack, 1990). The resemblance of Panderichthys
with the latter is observed in the general topography and proportions of
proximal elements (the elongated humerus and radius), and in the shape of
the articulations for radius, ulna, radiale, intermedium and ulnare. On the
ventral surface of the stemlike and dorso-ventrally flattened humerus there
can be discerned entepicondylar, ectepicondylar, and deltoid processes and a
posteriorly placed ectepicondylar foramen (Fig 1). On the dorsal surface
there is a low diagonal crest, probably for the pectoral muscle. The distal
part of the fin skeleton is unusual for rhipidistians. It is represented by

an undivided ulnar plate, to which lepidotrichia are attached. This plate represents one of the components forming the single paddle-shaped fin lobe together with an equal sized intermedium and very elongated radius.

The discovery of Panderichthys supports the suggestion that the archetype of the tetrapod limb must be organized according to the pattern of the uniserial archipterygium: the scapulo-coracoid end of the joint is the concave glenoid fossa ("the screw-shaped joint") for an unpaired basal element (humerus/femur), there are paired sub-basal elements (ulna/tibia, radius/fibula) which are not joined together proximally and distally but have two ball and socket joints with the basal element. All these features are common for Osteolepididae, Rhizodontidae and Panderichthyidae and can be treated as synapomorphic for this taxon.

As for the pectoral fin, this taxon features a long radius, and postaxially positioned ulna, connected with the intermedium and ulnare. To the latter two bones, and less often to the radius the distal radialia are attached.

The other taxon is represented by porolepid rhipidistians, lungfishes and coelacanths. It is characterized by a multisegmented biserial archipterygium and a ball-socket joint between scapulocoracoid and humerus. The last feature can be treated (Druzinin, 1933) as primitive (plesiomorphic for the osteichthyans.

Apart these differences both groups have some features of similarity, which can hardly be taken just as convergences. Firstly, it applies to the presence in the pectoral fin of an unpaired basal element with specific processes, which are partially comparable in rhipidistians and lungfishes (Fig 1). Another common feature is the double composition of the second mesomere, which in the first group is found in the adult condition (ulna/radius), while in the second is present only as an embryonic feature.

The investigations of Druzinin (1933) revealed that in the Neoceratodus embryo (54 day-old specimen) there is a centre of an early prochondral tissue distal to the rudiment of the first mesomere and beside the second mesomere (Fig 2). This centre is continued by a strand of mesenchyme. On the whole this rudiment has a biaxial shape. In the later stages it can be seen that its preaxial branch is independent. The "biaxiality" of the second mesomere can be also traced in the embryogenesis of Protopterus, which, unlike Neoceratodus, features a very short preaxial axis. In an adult Neoceratodus the second mesomere appears as a single structure, although a groove can be traced on its dorsal surface. This groove divides it into the greater postaxial part with the process for attachment of muscles and the smaller preaxial part, bearing the first preaxial ray. In the second mesomere of the pelvic fin there can be traced a long fissure, running about half the length between the analogous elements (Fig 2.1). However, in older specimens the fissure can be missing. It is evident that the second mesomere in the dipnoan fin corresponds to the second element in the archipterygium of the porolepidids. We can also surmise a bifold origin of the latter in embryogenesis. At least, if the first mesomere is compared to the stylopod (humerus/femur), the second undoubtedly represents the zeugopod. Then its greater postaxial part can be homologous with the ulna (fibula), and the smaller preaxial part with the radius. Such an assumption is also confirmed by the morphology of the second mesomere. In porolepidids (Glyptolepis) as in dipnoans it carries a dorsal outgrowth at its postaxial margin (Fig 2.3). Topographically this process is comparable to the entepicondylar process of rhipidistians and tetrapods. In porolepidids the analogue of the dorsal process can be traced in all of the following mesomeres, forming a specific dorsal keel along the postaxial axis of the pectoral fin (Ahlberg, 1989). According to Druzinin (1933) all these processes, as well as the processes on the basal element, could represent the fused and modified radial rays.

341

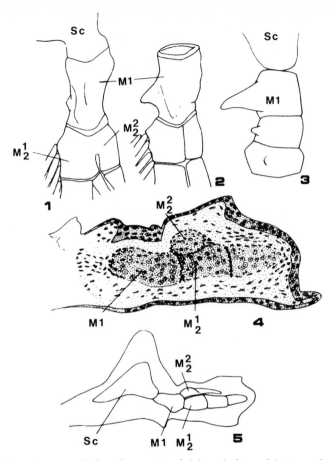

Figure 2. Structure and development of biserial archipterygium in
Neoceratodus (1, 2, 4, 5) (after Druzinin, 1933) and Porolepididae (3 -
after Ahlberg, 1989). 1-2 proximal parts of pelvic (1) and pectoral (2)
fins in dorsal view; 3 - proximal part of the pectoral fin in Glyptolepis in
dorsal view; Sc - scapulocoracoid; M_1 - first mesomere M^1_2, M^2_2 - mesomere
primordia and the result of their fusion in second mesomere; 4 - sagittal
section through the fin of young (20 mm) Neoceratodus; 5 - reconstruction of
the same fin.

The traditional view exemplified by Druzinin (1933) and by Holmgren
(1933) is that the main axis in the dipnoan fin is post axial thus comparing
it with the ulna. After carrying out anatomical, embryological, and
functional analysis Druzinin demonstrated that the Neoceratodus pectoral
fins have three different positions (Fig 3):

1 Embryonic. The fin base lies in the vertical plane, its ventral side
turned laterally. 2 Definitive Normal. The fin in the horizontal
(frontal) plane, its ventral side faces downwards. The preaxial margin is
turned outwards and the postaxial is turned inwards. From this position tbe
fin can be periodically transferred into the vertical state. 3 Vertical
State. If the horizontal position of the first mesomere is retained due to
the twist in the articulation between the first and the second mesomeres,
the fin appears in a "twisted" condition, observed when the fish supports
itself on the bottom via the fins. In this case the pectoral fins acquire

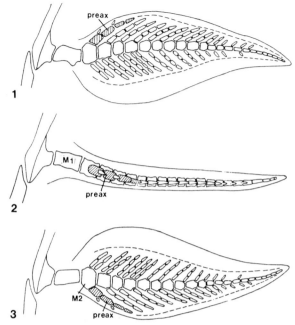

Figure 3. Various positions of pectoral fin in <u>Neoceratodus</u> (after
Druzinin, 1933): 1 - vertical position in embryo (ventral surface faces
laterally); 2 - normal horizontal position; 3 - vertical ("twisted")
position (dorsal surface faces laterally) (M1, M2, mesomeres 1 and 2; preax,
preaxial position).

the position characteristic of the pelvic fins; their morphologically dorsal
side is turned laterally and the preaxial margins face ventrally. Druzinin
noted that such a position recalls the position of the skeletal elements in
the limbs of the lower Tetrapods, when the animal is swimming with its legs
close to the body or is lying on the bottom.

At the same time the normal horizontal position of the <u>Neoceratodus</u> fin
corresponds to that of <u>Panderichthys</u>. This comparison means that the main
axis of the <u>Neoceratodus</u> fin represents the postaxial axis.

If the biaxiality of the second mesomere was embryologically
characteristic of all the Sarcopterygii, two evolutionary trends can be
presumed. 1. Biaxiality was a feature of the common ancestors of
Sarcopterygians, but the lungfishes lose it due to developmental
modification - the fusion of the second mesomere rudiment. This was
followed by the appearance of serial segmentation in the distal part of the
fin. 2. The biaxiality represented a variant in the sarcopterygian
ancestors in the embryonic devlopment of the fin, which was stabilised and
fixed in some adult rhipidistians. In Sarcopterygians (and subsequently
also in tetrapods) the development of the distal (autopod) part of the limb
shows a greater degree of variability (Hinchliffe and Vorobyeva, 1989;
Hinchliffe, these proceedings) expressed in a different succession of serial
segmentation and bifurcation patterns and sometimes fusions (Shubin and
Alberch, 1986).

Plasticity of morphogenesis (see Hinchliffe, these proceedings) was
expressed in diversity of the distal fin morphology, where the number of

343

rays on the postaxial axis varied from 2 (Panderichthys) to 7 Sauripterus. In some cases the processes of bifurcation also affected the preaxial axis, as seen in Sauripterus and Baramedia (Fig 1.1-2). At the same time the formation of the single ulnar plate in Panderichthys (Fig 1.5) probably reflects a retardation of fin development as it takes place also in the limb of aquatic reptiles, where reduction of distal carpal elements is explained by paedomorphic processes (Rieppel, 1989). This structure could have originated from the fusion of ulnare with the analogues of central and carpalia distalia, as seen in the aquatic Urodela in typical cases of limb simplification. The latter are usually regarded as due to an early cessation of morphological differentiation and therefore as cases of neoteny (Schmalhausen, 1915). However in Panderichthys the fin rays attach directly to the ulnare plate, giving an impression that the digits and the carpal zones were neomorphic in aquatic tetrapods and served primarily for attachment of the web, forming the distal lobe, used for swimming and substrate locomotion. Here we can note a functional analogy with Panderichthys, where lepidotrichia provided for lobe elongation. Biomechanical analysis of the locomotory apparatus in Panderichthys leads us (Vorobyeva, Kuznecov, in press) to the conclusion that it was able to move about the substratum (in water and on the beach zone). The movements could have corresponded to the "concertina" mode of locomotion, found in snakes ie bending and dragging of the body with consecutive fixation by means of pectoral and pelvic fins. This mode of locomotion in our opinion presents one of the possible transitions on the way to tetrapod locomotion. Morpho-functional preadaptations to the latter evolved probably independently among fishes and aquatic polydactylous forms of tetrapods, in the first place as an adaptation for "underwater transport" (Edwards, 1989). Probably the appearance of the pentadactyl limb was connected with the transition of tetrapods to territorial locomotion.

REFERENCES

AHLBERG, P. E. (1989), Paired fin skeletons and relationships of the fossil group Porolepiformes (Osteichthyes: Sarcopterygii). Zool. J. Linn. Soc., 96: 119.

ANDREWS, S. M. and WESTROLL, T. S., (970). The postcranial skeleton of Eusthenopteron foordi Whiteaves. Trans R Soc Edin. 68: 207-329.

COATES, M. I. and CLACK, J. A., (1990). Polydactyly in the earliest known tetrapod limbs. Nature. 347: 66-69.

DRUZININ, A., 1933. Ahnlichkeit im Bau der Extremitaten der Dipnoi und Quadrupeda. In "Trav. Lab Morphol. evol., Acad. Sci. USSR". 1: 1. (In Russian with German Abstract).

EDWARDS, I. L. 1989. Two perspectives on the evolution of the tetrapod limb. Amer Zool. 29: 235.

HINCHLIFFE, J. R. and VOROBYEVA, E. I., (1989). A reassessment of the fin-limb transition in tetrapod evolution: new palaeontological and developmental data. Annal. Soc. Roy. Zool. Belg. 119: 42.

HOLMGREN, N. (1933). On the origin of the tetrapod limb. Acta Zool., Stockholm. 14: 185.

JARVIK, E., 1980. Basic structure and evolution of vertebrates. London, Acad. Press.

LONG, I. A., 1989. A new rhizodontiform fish from the early carboniferous of Victoria, Australia, with remarks on the phylogenetic position of the group. J. Vert. Pal. 9: 1.

RACKOFF, J. S., (1980). The origin of the tetrapod limb and the ancestry of tetrapods. In "The terrestrial environment and the origin of land vertebrates". Ed A. L. Panchen. New York - Acad. Press.

REIPPEL, O., (1989). <u>Helveticosaurus zollingeri</u> Peyer (Reptilia, Diapsida) skeletal paedomorphosis, functional anatomy and systematic affinities. <u>Palaeontographica.</u> 208: 123.

ROSEN, D. E., FOREY, P. I., GARDINER, B. G. and PATTERSON, C., (1981). Lungfishes, tetrapods, palaeontology and plesiomorphy. <u>Bull. Amer. Mus. Natur. Hist</u>. 167: 159.

SCHMALHAUSEN, I. I., (1915). The development of amphibian limbs. Publ. <u>Mosc. Univ., Dept. Natur.-Histor</u>., 37. Moscow (in Russian).

SHUBIN, N. H., and ALBERCH, P., (1986). A morphogenetic approach to the origin and basic organization of the tetrapod limb. <u>Evol. Biol</u>. 20: 319.

VOROBYEVA, E. I., 1975. Notice on <u>Panderichthys rhombolepis</u> (Gross) from Lode in Latvia. <u>N. Jarb. Geol. Pataeont</u>. 5: 625.

VOROBYEVA, E. I. 1989. Panderichthyida - a new order of palaeozoic crossopterygians (Rhipidistia). <u>Dokladi Adac. Sci. USSR</u>. 306: 188.

VOROBYEVA, E. I., and SCHULTZE, H.-P., (1971). Panderichthyida and tetrapod origin. In "Origin of the Higher Categories of Tetrapods". Ed. H. P Schultze and L. Trueb). <u>Cornell Univ</u>., Ithaca, New York.

THE DEVELOPMENT OF THE TELEOST FIN AND IMPLICATIONS

FOR OUR UNDERSTANDING OF TETRAPOD LIMB EVOLUTION

Peter Thorogood

Department of Oral Biology
Institute of Dental Surgery
(London University)
256, Gray's Inn Road
LONDON WC1X 8LD, Britain

Elsewhere in this volume are accounts of how the study of adult fins and limbs of extinct, putatively ancestral, vertebrates might reveal the inherent constraints of developmental programmes underlying vertebrate fin and limb formation. Moreover, evolutionary morphologists postulate that phenotypic range within these vertebrate forms reveals 'permissive' variation within such developmental programmes (see papers by Coates and Vorobyeva, this volume). Here I wish to complement that approach by proposing that study of embryonic development of the paired fins in extant fishes can elucidate how variation of adult phenotype in extinct and ancestral forms might have been generated. It is also worth stating at this point that the development of fin buds, unlike that of tetrapod limb buds, has been a grossly neglected area of developmental biology, which is somewhat surprising given the consistent interest in origin and evolution of the tetrapod limb by developmentalists and evolutionists alike. Of the limited work published, much of it concerns the development of the paired fins of actinopterygian teleosts and the brief account which follows is confined to the general and common features described for the small number of species studied to date (namely, _Salmo_ species, _Aphyosemion_ _scheeli_, _Barbus_ _conchonius_ and _Brachydanio_ _rerio_).

FIN BUD DEVELOPMENT

Initially the teleost fin bud resembles a tetrapod limb bud but is far smaller, measuring some 250um in anterio-posterior dimension as compared to 1-1.5mm for the avian embryo at comparable stages of appendage development. It comprises a bulge of packed mesenchyme cells overlayen by an ectoderm, the margin of which is initially thickened and pseudostratified, and closely resembles a typical tetrapod apical ectodermal ridge or 'AER'[1,2,3]. However, unlike the limb bud of tetrapods, the teleost homologue undergoes a transition to a flattened paddle structure which lacks the more familiar stylopod and zeugopod morphology. Internally, the organisational differences are even more apparent. The most profound difference is that the ridge transforms from a thickened,

Developmental Patterning of the Vertebrate Limb
Edited by J.R. Hinchliffe _et al._, Plenum Press, New York, 1991

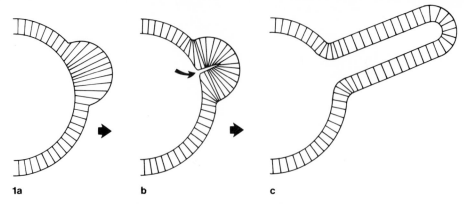

Fig. 1. The ridge-to-fold transition illustrated
diagrammatically in vertical section along the
proximodistal axis. The sequence starts with a
normal, pseudostratified, apical ectodermal
ridge (a), progresses to a ridge with a notch
or groove (b) and ends with a fully developed
apical ectodermal fold (c). (mesenchyme and
periderm are not depicted).

marginal, pseudostratified epithelium into an 'apical ectodermal
fold', so called because the marginal ectoderm appears folded back
on itself with the two basal surfaces apposed and enclosing an
extracellular space[3,4] (see fig.1). Initially this arises from a
notch or narrow groove, running along the ventral aspect of the
marginal ectoderm and oriented along the anterio-posterior axis
of the ridge. This notch subsequently deepens and the marginal
ectoderm grows or extends in a coordinated manner, to produce the
apical ectodermal fold. As far as is known, this structure is
found only in fishes where it occurs in both paired and median fin
primordia (although a small basal notch appears transiently in
some tetrapod limb buds[5]). In vertical section it can be seen that
the apex of the notch ends as an invagination into a single cell[4].
This is an extremely unusual feature for a folded epithelium since
folding morphogenesis virtually always arises as a result of
coordinated shape changes in individual cells (from cuboidal/
columnar to pyramidal) as a result of the 'purse string' effect[6].
The deepening of the notch or fold seems to be the consequence of
unsealing of basal-to-basal adhesions between cells in the initial
notch ('b' in fig.1) and the extension or lengthening of the
apical ectoderm due to successive waves of mitosis which sweep
proximodistally along the fin bud[2].

Mesenchymal cells are initially excluded from this space and
confined to the base of the fin bud. Here the endoskeletal
elements are already starting to form, initially as cartilage
which subsequently undergoes endochondral ossification. Early
stages of chondrogenesis, -condensation and matrix accumulation,
can be observed very early in fin extension (see fig.10 in [3]). In
fact, as far as one can extrapolate from the literature,

endoskeletal elements are either actually forming, or at least developmentally specified, by this period of ridge transformation into a fold.

Meanwhile, within the extracellular space of the fin fold ('c' in fig.1), two parallel arrays of large diameter collagenous fibrils, termed 'actinotrichia', are deposited. Each array is subjacent to one of the apposed basal laminae and the individual fibrils are oriented approximately parallel to one another, at 0.2-0.5um spacings, along the proximodistal axis of the fin bud (see fig.2 in [7]). Each fibril extends most, if not all, of the proximodistal length of the fin fold and grows lengthwise in a fashion which is coordinated with the growth of the fold, which the actinotrichia themselves apparently support. The manner of their growth has not been elucidated but it is generally assumed that they originate from the epithelium and grow lengthways by distal extension.

The very small size of the fin bud has meant that experimental investigation of causal mechanisms is technically difficult. Consequently for the teleost embryo there are no equivalents to such classic experimental strategies as ZPA grafts and AER ablations, which have been used to such effect with avian, mammalian and even amphibian systems. However, certain teleost species produce embryos which initially are optically transparent at early stages of development. We serendipitously discovered that even at organogenetic stages the fin primordia of such species are still transparent, affording unique opportunities to study morphogenetic cell behaviour in vivo using Nomarski microscopy coupled with time-lapse video or cinematography and, for some applications, computer-based image analysis.

From studying cultured fins and intact embryos in this way it can be observed that shortly after maximal fin fold extension is attained, the most distal of the mesenchyme cells at the base of the fin bud start to migrate distalward into the extracellular space of the fold. This coincides with an increase in the volume of extracellular space and the emergence of a number of aligned filopodia on the distal aspect of the individual mesenchymal cells. Quantitative analyses of changes in cell behaviour at this time, including such parameters as rates of filopodial extension and of whole cell translocation into the fold, have been made and reported elsewhere[8]. Ultrastructural and morphometric analysis reveals that both filopodia and cell bodies use the actinotrichia almost exclusively as a migration substratum and avoid, or are excluded from, the exposed basal lamina between adjacent actinotrichia[9]. Thus, the cells display a marked contact-guided migratory response to the highly-structured actinotrichial environment within the fold and this system constitutes one of the most dramatic examples of in vivo contact guidance known[7]. The roles that differential adhesiveness and/or physical factors might play is not clearly understood. However, it can be shown that microfabricated substrata, consisting of parallel ridges and grooves of apparent uniform adhesiveness, can elicit a contact guided locomotory response from these same cells in vitro, indicating that physical factors alone can play a primary role[10]. The vectorial influence driving the aligned and oriented mesenchyme cells distalwards into the fin fold almost certainly arises from a high cell population density proximally, as contact inhibition events with neighbouring cells can be observed at the proximal and lateral aspects of the migrating cells[8].

These mesenchymal cells will subsequently form the definitive fin rays or 'lepidotrichia', which are generally classified as 'dermal' skeleton. Thus, the actinotrichia, -of putative ectodermal origin, not only support the embryonic fin paddle but elicit a guided response from the invading mesenchymal cells and thereby set up the cell organisation and geometry for the formation of the definitive dermal skeleton of the functional fin. The lepidotrichia are bony, segmented structures consisting of serially repeating units along their proximodistal length[11], - each unit being formed in two complementary and adjacent halves known as 'demi-rays'. In larval and adult fins the terminal unit of each lepidotrichium is co-axial with the vestiges of the actinotrichia which persist as tufts of collagen fibrils at the distal ends of the fin rays[12]. Although this spatial relationship between the largely embryonic actinotrichia and the lepidotrichial fin rays has long been known to exist, the mode of developmental transition from one set of structures to the other, and the manner in which actinotrichia determine lepidotrichial pattern, has not been adequately investigated. Lastly, as a number of authors have shown, the paired and medial fins of teleosts are capable of quite considerable regeneration[13,14]. However, it should be noted that regenerative ability is confined to the dermal, fin ray skeleton and is not possessed by the proximal, endoskeletal elements[14,15].

A MORPHOGENETIC MODEL

It is clear that during fin development there is a temporal correlation between the 'ridge' and 'fold' phases, and the formation of the endoskeleton and dermal skeleton respectively. I propose that underlying this temporal correlation is a causal relationship. Whatever endoskeleton subsequently forms is specified during the 'ridge' phase, -a not unjustified extrapolation from studies of the AER of the tetrapod limb bud, which has been shown to have a fundamental role not only in the outgrowth of the limb mesenchyme but also in the 'induction' of the limb skeleton (which in tetrapods is entirely endoskeleton). After the transition to a fin fold, the dermal skeleton (actinotrichia and lepidotrichia) commences to form, reflecting what is apparently a unique property of the fin fold, i.e. the ability to specify or induce dermal skeletal structures. If this is indeed the case then the relative proportions of endo- and dermal skeleton in the fin will have been determined by the duration of the ridge and fold phases. Thus, a prolonged AER will elicit the formation of a proportionately more complex endoskeleton than a short-lived AER. Correspondingly, a short-lived fin fold will produce a limited dermal skeleton whereas a prolonged fin fold would result in a more extensive dermal skeleton. Should such causal relationships exist, and for the tetrapod AER and endoskeleton it has already been demonstrated, then the timing of ridge-to-fold transition during fin development will be critical in terms of determining fin phenotype.

Among the fishes, major fin phenotypes are usually categorised as 'ray-finned' (exemplified by the Actinopterygii) and 'lobe-finned' (characteristic of the Crossopterygii and seen in two dipnoan genera, -Neoceratodus and Lepidosiren, the Australian and South American lungfishes respectively). In the former, the endoskeleton is largely confined to the base of the fin, the bulk of which is supported by dermal skeleton fin rays.

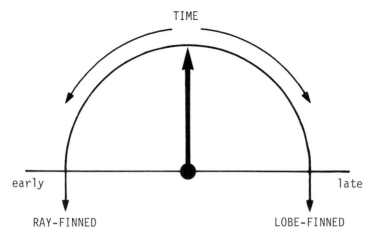

TIME

early | | late

RAY-FINNED | LOBE-FINNED

Fig. 2. A clockface diagram illustrating the phenotypic
consequences of heterochrony. The margin of the
clockface denotes developmental time and the
position to which the hand points signifies the
actual time at which a particular developmental
event occurs. In this case the event in question
is the transition from an AER to an apical ectodermal fold. Shifts in the timing of the event,
causing early or late transitions, will result in
different phenotypes.

In the latter, the endoskeleton can be quite considerable,
extending well away from the trunk and often to the distal tip of
the appendage, with the dermal fin rays confined to the anterior
and posterior marginal fringes of the fin.

If the causal relationships proposed earlier are valid then
netrochronic shifts in the developmental timing of the ridge-to-
fold transition will determine the proportionate amounts of endo-
and dermal skeleton formed in the developing fin bud. Consequently
the timimg of this event will be pivotal in that an early or late
transition will generate ray fins or lobed fins respectively. This
can be visualised on a Gould-type clockface diagram[16] (see fig.2)
where the position of the hand on the clockface denotes the timing
of the developmental event which displays potential heterochrony.

PREDICTIONS ARISING FROM THE MODEL

New phenotypes emerge through the genetic modification of the
underlying mechanism; in other words, the genes involved in
phenotypic change during evolution are also those involved in
normal ontogeny. Given the generally acknowledged importance of
ancestral crossopterygians in tetrapod evolution, any
heterochronic shift towards a later ridge-to-fold transition and

351

the genesis of a lobe-finned phenotype, would have constituted a crucial step in the evolution of a limb-like appendage. However, whereas such evolutionary scenarios cannot be tested directly, it is possible to pursue certain predictions by examination of fin development in extant fishes.

Firstly, do lobe-finned fishes display a late transition of ridge to fold during development of their paired fins? Ideally a longitudinal study of Latimeria chalumnae development should be carried out but given that the chances of obtaining embryos of the correct stages from this rare, viviparous species are extremely remote, then attention must be focussed on the development of the dipnoan Neoceratodus forsteri, the lungfish species with the best developed lobed fins. Reference to the classic monograph on Neoceratodus by Semon[17] reveals a very rudimentary fin fold which not only arises later in fin development than its actinopterygian counterpart, but is also a transient structure of minimal size (see plates XI and XII for illustration of the pectoral and pelvic fins respectively). Interestingly, the sections illustrated are in a plane vertical and parallel to the proximodistal axis of each fin bud, i.e. along the fullest extent of the endoskeleton. Given that most of the dermal skeleton is adjacent to the anterior and posterior margins of the fin, examination of the fin bud in the anterio-posterior plane might reveal a slightly more developed fold and this in itself would prove a further interesting correlation.

Secondly, in extant actinopterygians with pectoral and pelvic fins of disparate size and endoskeletal complexity, are there differences in the timing of the ridge-to-fold transition? Although there are clearly more dramatic examples of such disparity, the relevant information is readily available for salmonids, and in particular, for Salmo trutta. The pectoral fin develops a true AER which subsequently transforms to a typical fold and the pectoral endoskeleton consists, in a proximal to distal sequence, of a coracoid, mesocoracoid, scapula, four basiradials and thirteen radials[18,19]. By contrast, the pelvic fin has a ridge which is so short-lived that it has been termed a 'pseudo apical cap', displays an early transition to a fold and has a limited endoskeleton comprising merely a basipterygium, a metapterygium, two basiradials and three radials[14,18]. Thus, comparison of the pectoral and pelvic fins reveals a clear and direct correlation between the duration of the ridge and the complexity of the endoskeleton, as predicted.

AGENDA FOR FUTURE WORK

A morphogenetic model has been proposed which might explain how a quantitative shift (in terms of the timing) of a developmental event can generate ray-finned or lobe-finned phenotypes. Two emergent predictions are confirmed, at least in a preliminary sense, but far more investigation of each of these predictions is now necessary. If the model is substantiated, then this heterochronic change is likely to be one component in a complex of developmental changes underlying the evolution of the early tetrapod limb. Another component is the apparent disappearance of the dermal skeleton. Cells which form the dermal skeleton are generally assumed, though rarely proven, to originate from the neural crest[20] (and see paper by Hall, this volume). In fact, crest-derived cells are capable of colonizing

352

the early fin bud and both Schwann cells[21] and pigment cells[22] have been reported to do so. That crest-derived ectomesenchyme cells form the lepidotrichia (and their homologues in non-actinopterygian teleosts) remains to be demonstrated. The general absence of dermal skeleton in the tetrapod limb is likely to reflect either the absence of such skeletogenic cells in the limb bud mesenchyme or the absence of an apical ectodermal fold (or both). Clearly, the existence and distribution of crest-derived ectomesenchyme cells in developing fin and limb buds provides a focus for future investigation in this area.

ACKNOWLEDGEMENTS

Some of the work described has been supported by the Wellcome Trust. I am particularly grateful for discussions with Colin Patterson, Peter Forey and Moya Smith prior to writing this paper.

REFERENCES

1. R. G. Harrison, Die Entwicklung der unpaaren und paarigen Flossen der Telostier, Archives Mikroscopische Anatomie, 46:500 (1895).
2. J. Bouvet, Cell proliferation and morphogenesis of the apical ectodermal ridge in the pectoral fin bud of the trout embryo (Salmo trutta fario L.), Wilhelm Roux's Archives, 185:137 (1978).
3. A. T. Wood, Early pectoral fin development and morphogenesis of the apical ectodermal ridge in the killifish, Aphyosemion scheeli, Anat.Rec., 204:349 (1982).
4. P. J. Dane and Tucker, J. B., Modulation of epidermal cell shaping and extracellular matrix during caudal fin morphogenesis in the zebra fish Brachydanio rerio, J.Embryol.Exp.Morph., 87:145 (1985).
5. W. L. Todt and Fallon, J. F., Development of the apical ectodermal ridge in the chick wing bud, J.Embryol.exp. Morph., 80:21 (1984).
6. B. Burnside, Microtubules and microfilaments in amphibian neurulation, Am.Zool., 13:989 (1973).
7. P. Thorogood and Wood, A., Analysis of in vivo cell movement using transparent tissue systems, J.Cell Sci., Supplement 8:395 (1987).
8. A. Wood and Thorogood, P., An analysis of in vivo cell migration during teleost fin morphogenesis, J.Cell Sci., 66:205 (1984).
9. A. Wood and Thorogood, P., An ultrastructural and morphometric analysis of an in vivo contact guidance system, Development, 101:363 (1987).
10. A. Wood, Contact guidance on microfabricated substrata: the response of teleost fin mesenchyme cells to repeating topographical patterns, J.Cell Sci., 90:667 (1988).
11. J. Geraudie and Landis, W. J., The fine structure of the developing pelvic fin dermal skeleton in the trout Salmo gairdneri, Am.J.Anat., 163:141 (1982).
12. J. Geraudie, Fine structural peculairities of the pectoral fin dermoskeleton of two Brachiopterygii, Polypterus senegalus and Calamoichthys calabaricus (Pisces, Osteichthyes), Anat.Rec., 221:455 (1988).
13. J. Bouvet, Establissement de la carte des territoires presomptifs du bourgeon de la nageoire pectorale chez la

truite indigene (_Salmo_ _trutte_ _fario_ L.) a l'aide
d'excisions et de marques colorees, Anns.Embryol.Morph.,
3:315 (1970).

14. J. Geraudie, Les premiers stades de la formation de l'ebauche
    de nageoire pelvienne de Truite (_Salmo_ _fario_ et _Salmo_
    gairdneri) III-capacites de regulation, J.Embryol.exp.
    Morph., 34:407 (1975).

15. R. J. Goss, "Principles of Regeneration," Academic Press, New
    York (1969).

16. S. J. Gould, "Ontogeny and Phylogeny," Belknap Press of
    Harvard University Press, Cambridge, Mass. (1977).

17. R. Semon, Die entwicklung der paarigen flossen des _Ceratodus_
    forsterii, Denkschriften der Medizinisch naturwissenschaf-
    tlichen Gesselschaft zy. Jena, 4:59 (1893).

18. J. Bouvet, Histogenese precoce et morphogenese du squelette
    cartilagineux des ceintures primaires et des nageoires
    paires chez la Truite (_Salmo_ _trutta_ _fario_ L.),
    Arch.Anat.microsc. et Morphol.exper., 57:79 (1968).

19. J. Bouvet, Differenciation et ultrastructure du squelette
    distal de la nageoire pectorale chez la Truite indigene
    (_Salmo_ _trutta_ _fario_ L.) II -differenciation et
    ultrastructure des lepidotriches, Arch.Anat.microsc. et
    Morphol.exper., 63:323 (1974).

20. M. M. Smith and Hall, B. K., Development and evolutionary
    origins of vertbrate skeletogenic and odontogenic tissues,
    Biol.Rev., 65:277 (1990).

21. J. Geraudie, Innervation of the early pelvic fin bud of the
    trout embryo, _Salmo_ _gairdneri_, J.Morph., 184:61 (1988).

22. J. P. Trinkaus, Directional cell movement during early
    development of the teleost, _Blennius_ _pholis_, I. Formation
    of epithelial cell clusters and their pattern and mechanism
    of movement, J.Exp.Zool., 245:157 (1988).

PHYLOGENETICALLY ANCIENT PATTERN IN ONTOGENESIS OF LIMB MUSCLES

Radomír Čihák

Charles University First Medical Faculty
Prague
Czechoslovakia

From the beginning of this century when Lewis (1901/02, 1910), Bardeen
and Lewis (1901) and Bardeen (1906/07) published their impressive illustrations of embryonal muscles, obtained by plastic reconstructions from serial
sections and when Gräfenberg (1905/06) brought the descriptions of stages
of muscle development, it has generally been accepted that limb muscles undergo a "direct" or a "straight up" differentiation in their ontogenesis.
This means that every one muscle primordium separates from the larger blastema, i.e. from the premuscular mass for flexors or extensors, already in
its form and position corresponding to the definitive muscle.

According to this idea the whole ontogenetic process of muscle development thus from the moment of primordium individualization onward goes
only through proportional growth changes - up to its definitive form. This
view, however, does let aside the phylogenetic past of many muscles - their
phylogenetic changes - tacitly supposing that if there were such changes
in the phylogenetic past they are finished and fixed and the muscle in question directly realizes its final form during its ontogenesis. The polemics
against a "non direct" development of muscles during ontogenesis practically appear until a short time ago (e.g. O'Rahilly and Gardner 1975).

Since our research group in the Institute of Anatomy of the Charles
University First Medical Faculty in Prague can - in its results of ontogenetic studies of limb muscles - offer many examples of a "non direct" muscle development, a part of these results is presented here to demonstrate
that the changes of some muscles during their ontogenesis are rather more
complicated than the idea of the "direct differentiation" could concede or
even admit.

In fact, the above quoted direct differentiation was found in all muscles which persist in their phylogenetically ancient form up to human limbs
and where in the muscle primordium - after splitting off from the premuscular mass - only proportional growth changes of muscle belly and tendon and
other minor form changes can be observed. Such muscles are e.g. the deep
flexor of fingers and the common extensor of fingers on the forearm, the
lumbrical muscles of the hand, etc. The variability of e.g. the lumbricales
in comparative anatomy does not concern their form but their number, corresponding with the number of digits of the species in question. On the other side there are considerable changes in some other muscles. Some examples
will be given in the following text.

Developmental Patterning of the Vertebrate Limb
Edited by J.R. Hinchliffe *et al.*, Plenum Press, New York, 1991

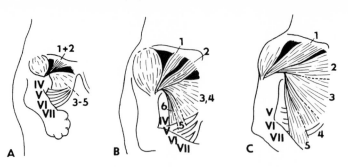

Fig. 1. Three developmental stages of the human pectoralis major muscle, according to microdissections; A - embryo 16,5 mm in Crown-Rump-Length where only two premuscular masses with its own anterior pectoral nerve each are formed; B - 37 mm C-R-Length - inside of two original masses independent primordia are formed, all of them being not yet separated; C - 100 mm C-R-Length - only the clavicular part remains separated, other four primordia rejoin to form the macroscopic sternocostal and abdominal muscle part; IV - VII - the ribs; 1 - the clavicular part; 2 - the manubrial part (both supplied by the 1st /lateral/ anterior pectoral nerve); 3 - the sternocostal part; 4 - the costal part (originally from the 4th, later from the 5th rib); 5 - the abdominal part; 6 - primordium of the pectoralis minor muscle (parts 3 - 6 are supplied by branches of the 2nd /medial/ anterior pectoral nerve).

Almost all results demonstrated here were published in journals. They were obtained by studies of histological serial sections of embryonal human limbs between 13,5 and 100 mm C-R-Length of embryos and fetuses, by microdissections of embryonal and fetal limbs and dissections of muscles of various mammals. The numbers of histological series, dimensions of employed embryos and fetuses and other technical details are given in the cited publications.

The pectoralis major muscle

develops from its premuscular mass by separating two parts of it, one on the clavicle, the second on the sternum and the 4th rib (fig. 1). Each of these two parts has its own nerve - one of the two anterior pectoral nerves (see explanation to fig. 1). During further development, the part on the clavicle differentiates into two primordia which have their own neurovascular hilum each. The medial portion shifts its origin from the clavicle to the sternal manubrium (the manubrial primordium). The second part of the original mass gradually differentiates into 3 independent primordia - the sternocostal, the costal (from the 4th, later from the 5th rib) and the abdominal primordium - each of them having again their own neurovascular hilum with the branch of the 2nd (medial) anterior pectoral nerve. Thus five independent primordia in place of the pectoralis major muscle are developed and the sixth one is the pectoralis minor primordium (fig. 1). Only later the six independent individual muscles - according to individual primordia found during ontogenesis (Čihák, 1959) - were homologized in the comparative anatomy of lower mammals (Štěrba 1968). During the subsequent development the primordia fuse: the clavicular primordium remains relatively independent, the manubrial primordium fuses with the upper margin of the further sternocostal primordium; the other primordia fuse, partly overlapping one another. The form of the definitive pectoralis major is thus accomplished. The borderlines of portions can, however, be estimated in new-borns and their surgical separation is possible even in adults. This was employed

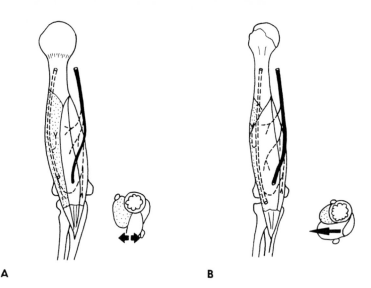

Fig. 2. Musculus brachialis and its primordia (and parts) in embryonal and adult muscle - anterior view and transverse section of the right side muscle; A - embryonal muscle: two main primordia, the medial one (white field), supplied by the musculocutaneous nerve, corresponds to the brachialis muscle of lower mammals, the lateral one (dotted), supplied by the radial nerve, corresponds to the musculus spiralis of the comparative anatomy; B - the brachialis muscle in adults; it is visible how the lateral (spiralis) part is reduced and overgrown by the rapidly increased medial part so that only several muscle bundles supplied by the radial nerve remain.

for surgical interventions reconstructing the lost functions of the deltoid muscle (Čihák and Eiselt, 1962). Also the partial congenital defects of the pectoralis major correspond individually with the above described primordia (Čihák and Popelka 1961). We can, therefore, say that the human pectoralis major phylogenetically developed by a secondary fusion of originally independent muscles and that in its ontogenesis this process is reflected in the original individualization and the subsequent joining of its primordia.

The brachialis muscle

is a seemingly very simple muscle in the adult, innervated by the musculocutaneous nerve. In its normal anatomy it is known that on its lateral side there are always several muscle bundles supplied by the radial nerve. We, therefore, observed the development of this muscle in embryonic serial sections to explain the component supplied by the radial nerve. It can be demonstrated that in young embryos up to 23 mm C-R-Length two main muscle primordia exist: one on the medial side, supplied by the musculocutaneous nerve, the second one on the lateral side, supplied by the radial nerve (fig. 2 A). According to its form, position and nerve supply this radial primordium corresponds to the musculus spiralis of lower mammals (the existence of which inside the human brachialis muscle was on the basis of the comparative anatomy supposed already by Ribbing (1938). During following development this radial primordium is suppressed and overgrown by rapidly enlarged medial primordium and finally only several muscle bundles supplied by the radial nerve remain on the lateral side of the musculus brachialis.

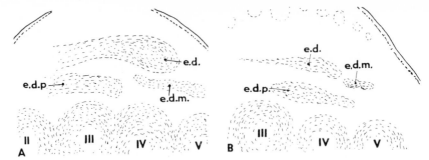

Fig. 3. The musculus extensor digiti minimi on transverse sections of human embryonic hands (redrawn from microphotographs).
A - embryo 14 mm C-R-Length; the extensor digiti minimi on the musculotendinous transition is larger - corresponding to the insertion on at least two fingers; B - embryo 17 mm C-R-Length; the extensor digiti minimi at the same level is already reduced to one (fifth) finger; II - V - the metacarpal primordia; e.d.m. - m. extensor digiti minimi primordium; e.d.p. - m. extensor digitorum profundus primordium (for radial fingers); e.d. - m. extensor digitorum primordium.

The musculus extensor digiti minimi

develops in human embryonal hand as the primordium of the belly as well as the primordium of the tendon in the form of a large flat muscle primordium on the dorsal sides of the 5th and the 4th metacarpal primordia. This corresponds to the form of the homologous muscle in the forearm of lower mammals - named in the comparative anatomy the musculus extensor digitorum lateralis (due to the fixed pronation of their forearm). In mammals (inclusive Prosimians) it is a deep long extensor of the 4th and 5th digit of the hand, or for more digits. In higher Primates and in Man this muscle is reduced to the 5th finger only. This reduction also occurs during ontogenesis of the human hand where the muscle belly and tendon are much larger in younger stages (fig. 3 A) and are then reduced very early (fig. 3 B) - retaining a suggestion of the split tendon - up to embryos 17 mm C-R-Length. In older embryos it is the slender primordium with a single tendon - only for the 5th finger. Also the varieties in adults are represented by the longitudinally split tendon for the 5th finger, or - rarely - by two tendons for the 4th and the 5th finger - i.e. by the phylogenetically ancient form (Kaneff and Čihák 1970).

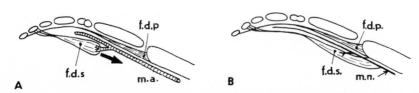

Fig. 4. Scheme of primordia of the flexor digitorum superficialis and profundus; A - embryo 15 mm C-R-Length - the flexor digitorum superficialis is represented by a palmar primordium growing proximalwards (arrow); B - embryo 21 mm C-R-Length; the primordium of the flexor digitorum superficialis is in its definitive position; f.d.s. - the flexor digitorum superficialis primordium; f.d.p. - the flexor digitorum profundus primordium; m.a. - the median artery (in the same position is the median nerve); m.n. - the median nerve (after the regression of the median artery).

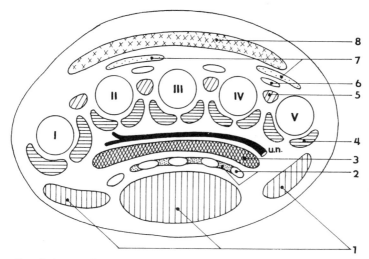

Fig. 5. Scheme of muscle and tendon layers of the primitive mammalian hand, shown in transverse section; this scheme can also be applied for the embryonal human hand; I - V - the metacarpals; 1 - layer of the flexor digitorum superficialis (flexores superficiales) together with the abductor pollicis and the abductor brevis digiti minimi; 2 -the tendons of the flexor digitorum profundus together with the lumbrical muscles; 3 - the layer of the contrahent muscles; 4 - the layer of the flexores breves profundi situated in twins on the palmar side of metacarpals; 5 - the intermetacarpal muscles; 6 - the deep short extensors of fingers; 7 - the layer of deep long extensors of the fingers (the extensor digiti minimi, the extensors of radial fingers); 8 - the extensor digitorum; u.n. - the ulnar nerve separating the layer of the contrahentes muscles from the flexores breves profundi layer.

The musculus flexor digitorum superficialis

originates as a bulky primordium only inside the palm of young embryos (13,5 - 16 mm C-R-Length), situated superficially to the tendons of the flexor digitorum profundus (fig. 4 A). In its external shape it is similar to the m. flexor digitorum brevis of the foot, not reaching proximally beyond the radiocarpal region. It is supplied by the median nerve and the median artery. Later on, this primordium grows, shifts and differentiates proximalwards (fig. 4 B), superficially to the median nerve and median artery (Dylevský 1967, 1968). The reduction and extinction of the median artery is joined with this process of the primordium shift (Mrázková 1986/87). The original palmar situation of the primordium and its proximal shift correspond with the situation of this muscle in lower vertebrates (mm.flexores breves superficiales) and with their following phylogenetic fate - up to the forms in the forearm of higher vertebrates.

The musculi contrahentes

form in the comparative anatomy of the mammalian hands a special layer consisting of more muscles (four in Marsupials /fig. 6 A /, Prosimians and

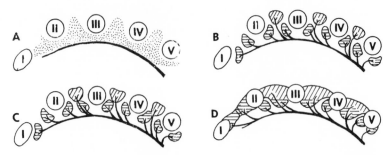

Fig. 6. The scheme of ontogenetic changes of the contrahentes layer, illustrated dark; palmar view, drawn according to plastic or graphic reconstructions; A´ - the shape of four contrahent muscles in the hand of Didelphis; A - embryo 20 mm C-R-Length - the contrahentes layer extends from the first to the fifth metacarpal anlagen; B - embryo 25 mm C-R-Length - the strips projecting from the contrahentes layer resemble the four contrahentes of mammals; C - embryo 26 mm C-R-Length - the ulnar half of the contrahentes layer disintegrates; D - embryo 30 mm C-R-Length - only the adductor pollicis primordium remains in place of the originally large contrahentes layer; u.n. - ulnar nerve.

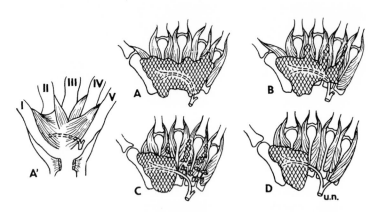

Fig. 7. Scheme of the development of human interossei in cross sections of embryonal hands; A - embryo 15 mm C-R-Length - the premuscular mass on the dorsal side of the ulnar nerve corresponds to the flexores breves profundi layer and to the intermetacarpal muscle layer of Marsupials; B - embryo 20 mm C-R-Length - individualization of muscle primordia with their nerve supply; C - embryo 25 mm C-R-Length - beginning fusion of intermetacarpal primordia, each with one member of the twins of deep short flexors primordia to form the dorsal interosseus of the 2nd - 4th intermetacarpal spaces; in the 1st space the dorsal interosseus originates by fusion of two neighbouring flexores breves profundi primordia; D - embryo 30 mm C-R-Length - finished fusion of two components for each dorsal interosseus; the remaining primordia of flexores breves profundi form the palmar interossei. Drawn according to the scheme of layers in fig. 5.

platyrrhine and catarrhine monkeys) except the hands of anthropoid Apes and
the human hand where the layer is reduced to the adductor pollicis muscle.
In human embryonal hand this layer is represented by a condensed premuscular mass forming in young embryos a large continuous sheet between the tendons of the long flexor of fingers (with lumbrical muscles) and the deep
palmar branch of the ulnar nerve (figs. 5, 6 A). In Primates the radial
part becomes gradually more bulky, while the ulnar part has two slender muscles which do not exist et all in the hand of Anthropoids and in the human
hand. The same rearrangement is seen in the course of human ontogenesis
(fig. 6). Whilst the radial half of the primordial layer becomes more bulky
and gradually obtains the form of the adductor pollicis (where probably two
radial contrahentes of the comparative anatomy are contained) the ulnar
part of the layer forms a transverse plate with two slender primordia projecting to the 4th and 5th fingers (fig. 6 B). It, in fact, at this moment
recapitulates the form of the layer in lower mammals or in Prosimians. The
ulnar part of the layer, together with the two strips,   then subsequently
disintegrates (fig. 6 C, D) and disappears (Čihák 1968, 1972). The manner
of differentiation and, especially, of the extinction of the ulnar part of
the layer is very interesting. There is no typical cell death in these parts
of the layer, but a special reduced form of myogenesis of short myotubes,
joined with the break-down of myofibrils, increase of glycogen areas and of
fat drops - signs of moderate degenerative changes. The number of cells decreases but not all cells die and they are subsequently lost in the surrounding mesenchyme. The process was electronmicroscopically studied by Grim
(1972 a, b) and was denoted the abortive myogenesis.

The musculi interossei

are a very controversial muscle group in the comparative anatomy (Ribbing
1938, Jouffroy 1962, 1971). In place of the interossei there is a group of
the flexores breves profundi (fig. 5), arranged in twins on the palmar sides of the metacarpal, partly projecting into intermetacarpal spaces. There
is also a group of intermetacarpal muscles, named by many authors the dorsal
interossei (fig. 5). Thus not only a nomenclatory confusion appears but there is also twice so much flexores breves profundi which by many authors are
named the palmar interossei. The comparative anatomy was unable to give a
satisfactory explanation. Therefore, embryological studies in human embryonic hands were performed in serial sections to elucidate the problem (Čihák
1963, 1972).

It can be shown that in the hands of human embryos and fetuses the material for the interossei is situated on the dorsal side of the ulnar nerve, separated by this nerve from the contrahentes layer (fig. 5, 7 A). During further development the primordia corresponding to the flexores breves
profundi are formed (fig. 7 B) and more dorsally, inside the 2nd - 4th intermetacarpal spaces the primordia corresponding to the intermetacarpal
muscles (with their independent nerve supply from the ulnar nerve) are formed. These primordia, formed in embryos between 16 and 22 mm of C-R-Length,
then begin to fuse: in the 2nd - 4th intermetacarpal spaces the intermetacarpal primordium fuses with one individual of the flexores breves profundi twins, adjacent to the axis of the 3rd finger. The results are the human
dorsal interossei (each of them having two nerve branches). In the 1st space two flexores breves profundi primordia fuse to form the 1st dorsal interosseus. In this mode the primordia are rearranged (fig. 7 C, D) and achieve the definitive form and composition of dorsal interossei. The remaining
primordia of the flexores breves profundi twins develop to form the palmar

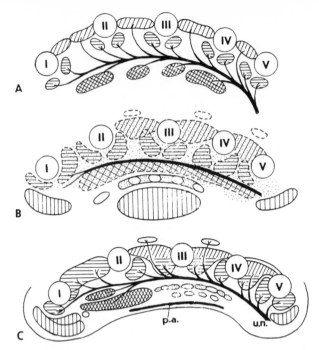

Fig. 8. concluding scheme with homologous muscles and muscle primordia, shown in transverse sections of the primitive mammalian hand, of the human embryonal hand and of the human hand in the adult; A - the hand of Didelphis; B - the embryonal human hand with already individualized primordia and their layers; C - adult human hand; all layers are dotted or hatched according to the common scheme of layers in fig. 5; u.n. - ulnar nerve; p.a. - palmar aponeurosis.

interossei and the deep marginal musculature of the 1st and 5th finger. The process thus recapitulates the form sequence known in the comparative anatomy and we are able to explain the complicated situation of the short muscles of the mammalian and of the human hand.

The musculi interossei dorsales accessorii

are represented in the human embryonal hand by relatively bulky primordia situated on the dorsal sides of the 2nd, 3rd and 4th intermetacarpal spaces. They are supplied from the intermetacarpal space by the branch going to and through the intermetacarpal primordium. This nerve then enters into the anastomosis with the branch of the ramus profundus nervi radialis. The further fate of these primordia is their reduction; they join from dorsal the dorsal interosseus, sometimes they remain separated, sometimes they are aponeurotically transformed (especially in the 4th space). They can be found during normal dissections. We, therefore, due to the fact of the nerve supply from the ulnar nerve, named these strips the interossei dorsales accessorii. Their origin, most probably, belongs to deep short extensors of the fingers (in lower vertebrates there are at least two layers of deep short extensors - according to Ribbing 1938). This primordium is a trace of them and was developed, subsequently reduced and joined the intermetacarpal space (Čihák, 1972).

Concluding remarks

We believe that the fact of observations of a "non direct" development of some muscles and their embryonal changes (even in man) is a simple fact. Since these changes resemble the phylogenetic fate of the muscles in question, they permit us to conclude that both, the phylogenesis and the onto- genesis procede on the basis of same or very similar formative mechanisms. The genetic and epigenetic mechanisms include the factors realizing the de- velopment. There came nothing new into the phylogenesis that did not enter before into the inherited chain of morphogenetic events. Thus for some chan- ges it is surely more simple to let the inherited sequence of morphogenetic mechanisms act and to add to them further mechanisms that change or correct the previous results (growth, cell death, fusion of primordia, etc.). The- refore, the ontogenetic changes resemble so much the phylogenesis, since older mechanisms are realized sooner. Also many adaptations and specializa- tions for various species can be explained on this basis.

References

Bardeen, Ch. R.: Development and variations of the nerves and musculature, etc. Amer. J. Anat. 6, 259 - 390, 1906/07.

Bardeen, Ch. R. and Lewis, W.H.: Development of the limbs, body wall and back in man. Amer. J. Anat. 1, 1 - 35, 1901.

Čihák, R.: Musculus pectoralis major und seine Komponenten in der Ontogene- se des Menschen. Čs. Morfologie 7, 174 - 191, 1959.

Čihák, R.: The development of the dorsal interossei in the human hand. Čs. Morfologie 8, 183 - 194, 1963.

Čihák, R.: Contribution à l´ontogénèse des muscles "contrahentes" de la main humaine. C.R.Ass. Anat. 1968, 141, 704 - 712, 1968.

Čihák, R.: Ontogenesis of the Skeleton and Intrinsic Muscles of the Human hand and foot. Ergebn. d. Anat. u. Entw. Gesch. Bd. 46, Heft 1, 1 - 194, 1972.

Čihák, R., Eiselt, B.: Proposition for replacement of the paralyzed deltoid by parts of the m. pectoralis major. Acta Univ. Carol. Med. 8, 367 - 381, 1962.

Čihák, R., Popelka, S.: Partial defects of the pectoralis major muscle. Acta Chir.orthop. traumat Čs. 28, 185 - 194, 1961.

Dylevský, I.: Contribution to the ontogenesis of the m.flexor digit. super- ficialis et profundus. Folia Morphol (Praha) 15, 330 - 335, 1967.

Dylevský, I.: Tendons of the m.flexor digitorum superf. et profundus in the ontogenesis of the human hand. Folia Morphol (Praha) 16, 124 - 130,1968.

Grim, M.: Ultrastructure of the Ulnar Portion of the Contrahent muscle lay- er in the Embryonic Human Hand. Folia Morphol.(Praha) 20,113 - 115,1972.

Grim, M.: Ergebn. Anat. Etw.-Gesch. 46, 1 - 194, 1972.

Kaneff, A, Čihák, R.: Die Umbildung des M.ext. digit. later. in der Phylo- genese und in der menschlichen Ontogenese. Acta Anat. (Basel) 77, 583 - 604, 1970.

Lewis, W. H.: The development of the arm in man. Amer. J. Anat. 1, 145 - 184, 1901.

Lewis, W. H.: Die Entwicklung des Muskelsystems; in Keibel - Mall, Hdb. d. Entw.-Gesch. des Menschen. Leipzig, S. Hirzel, 1910.

Mrázková, O.: Blood Vessel Ontogeny in Upper Extremity of Man, etc. Acta Univ. Carol. Medica, Monographia CXV, 1986/87.

O´Rahilly, R. and Gardner, E.: The Timing and Sequence of Events in the De- velopment of the Limbs in the Human Embryo. Anat. Embryol. 148, 1 - 23, 1975.

Ribbing, L.: Die Muskeln und Nerven der Extremitäten, in Bolk, Göppert, Kallius, Lubosch: Hdb. d. vergl. Anat. d. Wirbeltiere, Berlin, Wien, Urban u. Schwarzenberg, 1938.

Štěrba, O.: Pectoral muscles of some Insectivores. Zool listy 17, 149 - 156, 1968.

MUTATION AND LIMB EVOLUTION

Donald Ede·

Department of Zoology
University of Glasgow
Glasgow
Scotland, U.K.

INTRODUCTION

Evolution of the limb must have depended on the existence of mutations affecting its development. Here, a number of different "limb" mutants are considered, with a view to seeing to what extent such mutations might provide material for natural selection to act upon, or otherwise throw light on the evolutionary process. Goldschmidt (1940) coined the name "hopeful monsters" for major mutants which might produce sudden evolutionary advances. How hopeful are our hopeful monsters?

The answer depends on an analysis of the relationship between genotype and phenotype in each case, and we may be misled by the nicknames which have been given to these major mutants - wingless, ametapodia, amputated, etc., names which hark back to the outmoded idea of a one-to-one relation between genes and unit characters.

There is no harm in this when we are simply using these mutants as tools with which to analyse particular developmental processes, but we know that a gene affects a morphological unit only by its effect on the processes by which that unit arises in development. A complicated network of reactions connects genotype to phenotype, involving (Sewall Wright, 1980) universal multifactorial inheritance and universal pleiotropy. Almost always, many genes will affect a particular system, and any one gene will affect several different systems. We know that any gene maintained over a long period is liable to change considerably in its penetrance and expressivity because of changes in the genetic background, and that pleiotropy is the general rule.

This will have implications for the evolutionary potential of any mutant. Goldschmidt's idea would apply best if the organism were indeed made up of a mosaic of unit characters, and if a mutant affecting one of those characters in a favourable way were to arise occasionally, be viable and spread through the population under selection pressure. But major mutations hardly ever are as viable as the wildtype, are often in fact lethal, and natural selection is thought to be effective only where the alternative allele is present with an equilibrium frequency of at least 5% (Sewall Wright, 1980). However, lethal mutants can accumulate in a population where there is heterozygote advantage, and this appears to have applied to the talpid³ mutant. Multifactorial inheritance means that

Developmental Patterning of the Vertebrate Limb
Edited by J.R. Hinchliffe *et al.*, Plenum Press, New York, 1991

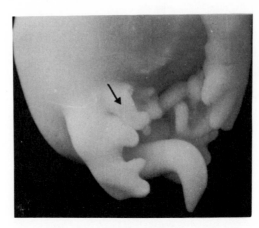

Fig. 1.   Left hind leg of <u>Disorganization</u> mouse embryo,
showing randomly arranged digits, one of which
(arrow) has developed at the base of anterior
to the main limb.

expression of major genes will vary according to whatever modifying genes
are present, so that what might seem a totally disruptive lethal gene may
be held in reserve in the population as its heteroallele and become in
time amenable to selection.

Pleiotropy may have its basis at the level of the chromosomal DNA,
but generally occurs through a single gene activity interfering in a
variety of ways with the developmental process. This may occur because
disruption of one event produces a cascade of consequence, e.g. in a
sequence of inductions, or because its effect is on some pervasive
fundamental property, e.g. of the cell surface, which affects developmental
processes throughout the embryo. Its significance for evolution is that,
where major genes are concerned, selection must operate on a wide range of
effects, which may be disadvantageous or advantageous. Thus a gene which
reduced the size of the wings in a bird for which this would be an
advantageous characteristic would be of no use if it also prevented the
development of a metanephros. On the other hand, a change in the shape of
a particular organ might be only advantageous if proportional changes
occurred in other parts of the body. Thus pleiotropy may have two effects:
it may make for harmonious change over the whole body, but it may mean that
one particular system cannot be isolated for evolutionary change.

MUTANTS AFFECTING DEVELOPMENT OF THE LIMB IN MOUSE AND FOWL EMBRYOS

Disorganization (Ds): Mouse (semidominant lethal)

Studied by Hummel (1958, 1959) and Ede (1980), this mutant has so far
baffled causal analysis. It has very low penetrance and extreme variability
of expression. Heterozygotes are characterized by duplications and ectopic
development of internal and external structures. The limbs are often
polydactylous, but extra digits are formed in no particular pattern and
are sometimes developed at the base of the limb (Fig. 1) or from the flank.
A similar condition is found rarely in man (Winter & Donnai, 1989). This
is a truly hopeless monster, since the pleiotropic manifestations of the
gene appear so capriciously.

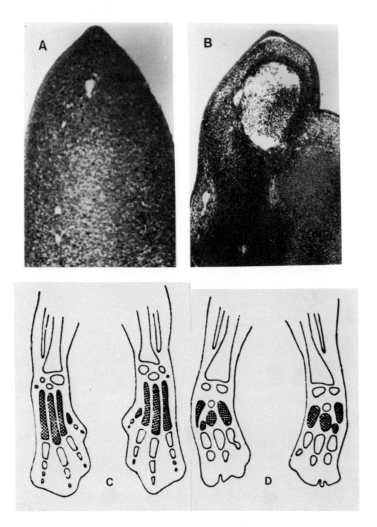

Fig. 2. A & B. L.S. hind-limb bud of normal (A) and <u>Ametapodia</u> (B)
6-day fowl embryos showing enlargement and necrosis of the
marginal blood vessel in the mutant.
C & D. Distal skeletal rudiments in legs of normal (C)
and <u>Ametapodia</u> (D) 10-day embryo. Metatarsal rudiments are
shaded.

Wingless (ws): Fowl (recessive)

       Studied by Hinchliffe and Ede (1973), this mutant's effects are
confined to the limbs. The wings are reduced or eliminated, which might
be evolutionarily advantageous; indeed Raynaud (1977) has shown that the
development of vestigial limbs in some reptiles is similar, i.e. depends
on the elimination of the apical ectodermal ridge, preceded by cell death
in the underlying mesenchyme. The legs are also affected, but not to the
same degree; digit number is reduced and the tibia is reduced or buckled.
The effect is not symmetrical, the left wing bud often developing much
further than the right. Though this particular wingless mutant shows no
pleiotropic effects, others do. Zwilling's (1949) eliminated the metanephros as well as the wings, and the first wingless mutant eliminated
the lungs also.

367

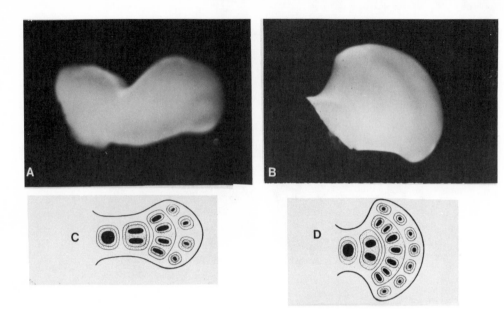

Fig. 3. A, B. Right limb bud of normal (A) and talpid³ (B) 6-day fowl
embryo. C, D. Diagram showing pattern of cartilage condensations
within limb bud of normal (C) and talpid³ (D) embryo. In normal
embryos cartilage development is confined to black areas; in
talpid³ matrix deposition also occurs in lightly-shaded areas.

The mechanism in wingless lies in a disturbance of the pattern of cell
death in the limb bud. Two areas of cell death, the anterior and posterior
necrotic zones are mysterious in having no known functional role, but in
wingless the ANZ appears precociously and comes to extend beneath the AER.
What governs the location and extent of these necrotic zones is not known,
but may be incidental to the pattern-determining gradients within the limb
bud. This mutant may be classified as a "pattern" gene, and this may be
the class that is most useful as the basis of evolutionary modification.

Ametapodia (am): Fowl (semidominant lethal)

This mutant, discovered by Cole (1967) and studied by Ede (1969),
similarly affects the appearance of the ANZ and the PNZ in the limb bud.
In this case, however, the ANZ is retarded in its appearance and the PNZ
precocious. The cell death of the PNZ extends as the limb develops into
the marginal blood vessel beneath the AER (Fig. 2A,B), and has the effect
of partially eliminating the potential metapodial (metacarpals in the wing
and metatarsals in the leg) region. The AER however remains intact, maintained by a band of healthy mesenchyme beneath it, from which reasonably
complete phalanges are formed.

An interesting feature of this mutant is that although the necrosis
appears to have no distinct boundaries, a well-defined set of abnormal
metapodial rudiments is formed (Fig. 2C,D). The pattern is not repeated
exactly in different individuals, but is very nearly so in contralateral
limbs on the same individual; there is in fact a marked bilateral symmetry
of development. Again, this may be defined as a pattern mutant and this,

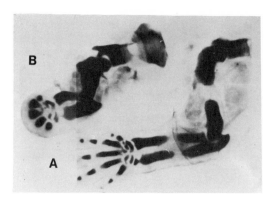

Fig. 4. Fore-limbs of normal (A) and <u>amputated</u> (B) 16-day mouse embryo showing cartilage rudiments of the limb skeleton.

together with the absence of pleiotropy and the symmetry of its modified development, would seem to make a mutant of this sort possible material for natural selection in particular circumstances.

talpid[3] (ta[3]): Fowl (recessive lethal)

To call this a limb mutant is somewhat misleading, since pleiotropy here is pervasive, i.e. due to a disturbance of a fundamental cellular property which affects developmental processes in all parts of the embryo. Postgastrulation homozygous talpid embryos can be distinguished by the form of the developing head and trunk. When the limb buds appear they are broader and shorter than normal, and soon develop a characteristic fan shape (Fig. 3 A,B).

Studies on the mutant limb cells <u>in vitro</u> (Ede & Flint, 1975) have shown that cell adhesion is increased and cell motility reduced; they are cells "which do not like to let go". This has effects on the form of the developing limb bud and on the pattern of precartilage condensations within it. It is most likely that it reflects some alteration of the cell membrane and it appears to have effects also on cell division, whose rate is increased (Mohammed, 1978).

Formation of the primitive streak and gastrulation is unaffected, which in view of the amount of cell movement involved is strange and suggests that this gene function is switched on late in development. Beyond gastrulation, all morphogenetic events involving cell movement are affected, e.g. the development of the somites (Ede & El-Gadi, 1986) the dispersal of neural crest cells and the formation of Rathke's Pouch in the head (Ede, unpublished). The mutant has no affect in heterozygote and is lethal in homozygotes, probably because the disturbance of cell adhesion produces a leaky vascular system. But in one respect it might be considered a hopeful monster, since the pervasiveness of its fundamental alteration of morphogenetic processes leads to an embryo whose shape is transformed, but in a self-consistent and symmetrical manner.

The polydactyly which characterizes the <u>talpid[3]</u> limb is interesting in that it suggests a dual mechanism of pattern control (Ede, 1982). The fan-shaped mutant limb bud provides an expanded region of mesoderm for digital development, and the response is that more digits are produced (Fig. 3C,D), indicating some sort of periodic determination of digit position rather than determination by a gradient of positional information where regulation of digit number would be predicted. Experiments by Ede and Shamslahidjani (Ede, 1982) show that dissociated normal limb mesenchyme in an ectodermal jacket will produce the periodic determination of digits without their individual characterization. Addition of normal ZPA will produce some degree of individual characterization of digits. Addition of talpid[3] ZPA-region mesoderm has only a very weak effect, and talpid[3] will not respond to normal ZPA. It appears that in the mutant the simple periodic mechanism is present, but not the positional information mechanism which leads to the characterization of specific digits.

It seems likely that the periodic mechanism is a more primitive one, likely to have determined the pattern of skeletal elements in the developing limbs of the crossopterygian fishes from which early land-living vertebrates are commonly thought to have evolved. One may suppose that in the evolution of the amphibians the pentadactyl-limb pattern was produced by the imposition of a positional informational signal on this primitive one. Recent discoveries (see Coates, 1990; Vorobyera, 1990, this volume) have shown that some early amphibians had limbs with more than five digits, and this variability would certainly be expected during the phase when a positional information mechanism of limb patterning was being evolved. Talpid[3] appears to be a mutant in which the secondarily-evolved positional information mechanism is lost, while the more primitive periodic mechanism persists.

This suggests a reason why, though mutant polydactly is common, evolution of digits additional to the pentadactyl pattern does not seem to have occurred except by way of modification of a carpal bone, e.g. in the 6th digit of the mole or in the Panda's thumb. Where a viable evolutionary form is to be produced, the ZPA positional mechanism must be functional, and this does not exist in at any rate some of the mutant polydactylys.

amputated (am): Mouse (recessive lethal)

This mutant is interesting in showing many of the features of the talpid[3] fowl embryo, notably that the basic defect is an increase in cell adhesiveness, demonstrated in vitro (Flint & Ede, 1982) and in studies on its somite development (Flint & Ede, 1978). The limb does not show such a dramatic fan-shaped development as in talpid[3], but its margin is extended. It is interesting that in this case (Fig. 4) the number of digits is not increased. The number of digits remains normal, but the rudiments are broader. This is what a positional information model would predict, and it would appear that in this mouse mutant the ZPA is functional and retains some control over digital development.

This mutant might also be considered a hopeful monster, in that while the form of the body is changed, some essential proportions are maintained. Thus, the body axis is much shortened, but the number of vertebrae remains constant (Flint, Ede, Wilby & Proctor, 1978), and a self consistent body shape is produced. Here again is an example of pervasive pleiotropy, where a basic cell character is changed, with consequences on development throughout the embryo.

REFERENCES

Coates, M.I., 1990, New palaeontological contributions to limb ontogeny and phylogeny, in: "Developmental Patterning of the Vertebrate Limb," J.R. Hinchliffe, J. Hurle and D. Summerbell, eds (in press), Plenum, New York.
Cole, R.K., 1967, Ametapodia, a dominant mutation in the fowl, J.Hered., 33: 82.

Ede, D.A., 1969, Abnormal development at the cellular level in talpid and other mutants, in: "The Fertility and Hatchability of the Hen's Egg," T.S. Carter and B.M. Freeman, eds, pp. 71-84, Oliver and Boyd, Edinburgh.

Ede, D.A., 1980, Role of the ectoderm in limb development of normal and mutant mouse (Disorganization, Pupoid foetus) and fowl (talpid[3]) embryos, in: "Teratology of the Limbs," H.-J. Merker, H. Nau and D. Neubert, eds, pp. 53-66, Walter de Gruyter, Berlin.

Ede, D.A., 1982, Levels of complexity in limb-mesoderm cell culture systems, in: "Differentiation In Vitro," M.M. Yeomann and D.E.S. Truman, eds, pp. 207-229, Cambridge University Press, London.

Ede, D.A. and El-Gadi, A.O.A., 1986, Genetic modifications of developmental acts in chick and mouse somite development, in: "Somites in Developing Embryos," R. Bellairs, D.A. Ede and J.W. Lash, eds, pp. 209-224, Plenum, New York.

Ede, D.A. and Flint, O.P., 1975, Cell movement and adhesion in the developing chick wing bud: Studies on cultured mesenchyme cells from normal and talpid[3] mutant embryos, J.Cell Sci., 18:301.

Flint, O.P. and Ede, D.A., 1978, Cell interactions in the developing somite: in vivo comparisons between amputated (am/am) and normal mouse embryos, J.Cell Sci., 31:275.

Flint, O.P. and Ede, D.A., 1982, Cell interactions in the developing somite: in vitro comparison between amputated (am/am) and normal mouse embryos, J.Embryol.Exp.Morphol., 67:113.

Flint, O.P., Ede, D.A., Wilby, O.K. and Proctor, J., 1978, Control of somite number in normal and amputated mutant mouse embryos: an experimental and a theoretical analysis, J.Embryol.Exp.Morphol., 45:189.

Goldschmidt, R., 1940, "The Material Basis of Evolution," Yale University Press.

Hinchliffe, J.R. and Ede, D.A., 1973, Cell death in the development of limb form and skeletal pattern in normal and wingless (ws) chick embryos, J.Embryol.Exp.Morphol., 30:753.

Hummel, K.P., 1958, The inheritance and expression of Disorganization, an unusual mutation in the mouse, J.Exp.Zool., 137:389.

Hummel, K.P., 1959, Developmental anomalies in mice resulting from action of the gene Disorganization, a semidominant lethal, Pediatrics, 23:212.

Mohammed, M.B.H., 1978,"Cell division in normal and mutant chick embryos," Ph.D. thesis, University of Glasgow.

Raynaud, A., 1977, Les differentes modalites de la rudimentation des membres chez les embryons de reptiles serpentiformes, in: "Mecanismes de al rudimentation des organes chez les embryons de vertebres," Colloques internationaux du CNRS No. 266, pp. 201-219, Editions du Centre National de la Recherche Scientifique, Paris.

Vorobyeva, E.I., 1990, Evolution of the paired fins in Saracopterygii and the origin of limbs in Tetrapoda, in: "Developmental Patterning of the Vertebrate Limb," J.R. Hinchliffe, J. Hurle and D. Summerbell, eds (in press), Plenum, New York.

Winter, R.M. and Donnai, D., 1989, A possible human homologue for the mouse mutant disorganisation, J.Med.Genet., 26:417.

Wright, S., 1980, Genic and organismic selection, Evolution, 34:825.

Zwilling, E., 1949, Role of the epithelial components in origin of wingless syndrome of chick embryos, J.Exp.Zool., 111:175.

INSIGHTS INTO LIMB DEVELOPMENT AND PATTERN FORMATION FROM STUDIES OF THE *LIMBLESS* AND *TALPID*[2] CHICK MUTANTS

John F. Fallon, Leah Dvorak, and B. Kay Simandl

Anatomy Department
University of Wisconsin-Madison
Madison, Wisconsin

INTRODUCTION

The amniote limb is a complex arrangement of cells and tissues derived from the embryonic somatopleure, somites, neural crest and ectoderm. A variety of tissue manipulation experiments have successfully elucidated some understanding of how the pattern of the adult limb structure is achieved. A hallmark of limb bud development that has emerged from this research is a series of interactions between the limb bud ectoderm and mesoderm and among mesoderm cells (Zwilling, 1961; Summerbell, 1979; Fallon et al., 1983). There is a sequence of these various interactions which permits the realization of the potential of the limb field cells. Obviously, this means a series of gene expressions must occur over developmental time and in a predictable sequence. Mutations of some of the genes involved in this cascade are available in mouse and chick. The use of these mutants, coupled with tissue manipulations and molecular techniques, should lead to fundamental insights into the control of pattern formation in the limb, and to a better understanding of the evolution of limb form and function.

THE *LIMBLESS* MUTANT

We have been working with several limb mutants found in the chicken and will describe experiments involving two of them here. The first is *limbless*, which is a simple Mendelian, autosomal, recessive mutation first described by Prahlad et al. (1979). Homozygous embryos have no limbs. However, these embryos live through the 21 days of incubation with no apparent ill effects. The mutation is effectively lethal, for without legs hatching is thwarted. When hatched by hand the chicks will survive for a limited time. Gross dissection and histological observations of the *limbless* embryo and hatchling indicate the single defect is the lack of limbs and their derivatives i.e. pectoral muscles (Lanser & Fallon, 1986). The interesting and useful aspect of *limbless* embryos is that limb buds actually form when and where they should. However, the buds lack an

Developmental Patterning of the Vertebrate Limb
Edited by J.R. Hinchliffe *et al.*, Plenum Press, New York, 1991

apical ectodermal ridge (Prahlad et al., 1979; Fallon et al.,
1983). The apical ridge is necessary for growth of the limb
mesoderm (Saunders, 1948; Summerbell, 1979; Rowe and Fallon,
1982) and the mutant limb mesoderm begins to die during
Hamburger & Hamilton (1951) stages 19 and 20; the entire limb
bud disappears by stage 23-24.

We have shown that the *limbless* limb bud mesoderm will make
a perfectly good limb if supplied with a normal apical ridge
through tissue recombinant procedures. However, the *limbless*
ectodermal jacket combined with normal mesoderm does not support
outgrowth, and the graft always fails to develop (Fallon et al.,
1983). While it was clear from this work that *limbless*
ectoderm had no ridge activity in the limb bud stages of
development, the possibility remained that, earlier in
development, the *limbless* mesoderm failed to give the inductive
signal(s) necessary for ridge formation by the ectoderm
(Carrington & Fallon, 1984). Alternatively, the *limbless*
mesoderm could have given the ridge an inductive signal, but the
limbless ectoderm was incompetent to respond.

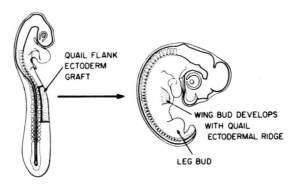

Fig. 1.   Drawing of early recombinant procedures. On the left
           is a drawing of a stage 14-15 chick host which has had
           ectoderm over the limb field mesoderm replaced with
           stage 14-15 quail flank ectoderm (stippled). The
           drawing on the right shows the resulting limb that
           develops covered with quail ectoderm. At the apex
           would be a quail apical ectodermal ridge.

To distinguish between these two alternatives we designed
the early limb recombinant technique shown in Figure 1
(Carrington & Fallon, 1984). At stage 14-15, ectoderm from a
prospective host was removed from the somatopleure mesoderm
opposite somites 14-21. In controls competent donor ectoderm
was grafted onto the denuded mesoderm; quail ectoderm was used
because the nuclei contain a heterochromatin marker, allowing
them to be distinguished from chick nuclei (LeDouarin & Barq,
1969). When normal (control) chick hosts were used, an apical
ridge was induced in the grafted quail ectoderm and the limb
that developed had quail ectoderm and ectodermal derivatives.

When the same experiment was done with a *limbless* host, an
apical ridge was induced in the quail ectoderm graft and a limb
formed at that site. Moreover, *limbless* ectoderm from stages
known to be competent to form an apical ridge in normal embryos
(cf. Carrington & Fallon, 1984) never formed a ridge when
grafted to normal embryos (Carrington & Fallon, 1986).

Conclusions about *Limbless*

        There are three conclusions from these data. First, the
limbless mesoderm gives the ridge-inductive signal at the right
developmental time. Second, the *limbless* ectoderm is incapable
of forming an apical ridge. Thus the simple mutation has
interfered with the ectoderm in a fundamental way, but the only
known defect is amelia. Therefore, there has likely been a
change in a limb development-specific gene. A third conclusion
is the strong inference that limb budding is autonomous to the
limb field mesoderm.

LIMB BUDDING IS INDEPENDENT OF THE APICAL RIDGE

    To test this hypothesis we carried out the following
experiments. Previously, we have demonstrated that stage 15-20
normal chick flank ectoderm can form a ridge when grafted to the
limb field. After stage 20, chick or quail flank becomes
incapable of making an apical ridge (Carrington & Fallon, 1984).
Therefore, we grafted stage 21 or 22 quail flank ectoderm onto
stage 15 normal chick wing field mesoderm. If ridge activity
were necessary for normal limb budding, no bud would form at
that site. However, a bud always formed, covered by the grafted

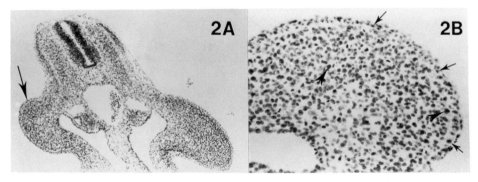

Fig. 2.   **A.**This is a cross section of a stage 20-21 embryo
           which had stage 22 quail flank grafted to the right
           wing field mesoderm at stage 15. The right wing bud
           (arrow) has formed, covered by quail ectoderm, but
           without an apical ridge. The left limb bud has formed
           normally.
           B.Higher magnification of right limb bud shown
           in Fig.2A. Quail nuclei with dense heterochromatin
           are shown by the small arrows. There are dividing
           cells in the mesoderm, but some cell death (arrow
           heads) has begun, presaging the elimination of the
           limb bud.

ectoderm (Figures 2A and 2B). The bud started to regress during stage 19 and disappeared by stage 24. No limb formed at that site. This grafting procedure, using normal ectoderm incompetent to form an apical ridge, results in a *limbless* phenocopy. This reinforces the conclusion that limb budding is autonomous to limb field mesoderm and apical ridge activity is not required for the budding process.

Experimental studies using *limbless* embryos have proven productive in analyzing lateral motor column–peripheral interactions during development (Lanser & Fallon, 1984, Lanser et al., 1986). We have also used the mutant to assess if and when the apical ridge is necessary for specific expression of a variety of genes (cf. Lyons et al., these Proceedings).

CHICK-REPTILE TISSUE INTERACTIONS

We have used *limbless* to demonstrate xenoplastic tissue interactions. Specifically, we have tested whether reptile flank ectoderm can respond to the ridge induction signal, and whether the reptile ridge will permit chick mesodermal elongation. Using the techniques given in Figure 1 we grafted flank ectoderm isolated from pre-limb bud and early limb bud stages of soft shell turtle or chameleon embryos, onto *limbless* wing field. We have been able to demonstrate that an apical ridge does form in the grafted reptile flank ectoderm (Figure 3A). Further elongation of the *limbless* mesoderm did occur and distally complete limbs formed (Figure 3B). These data demonstrate that turtle and lizard ectoderm do respond to the chick ridge-induction signal. Similarly, chick mesoderm will respond appropriately to the turtle and lizard elongation signal(s). We infer that these inductive signals for limb development have been evolutionarily conserved between reptiles and birds.

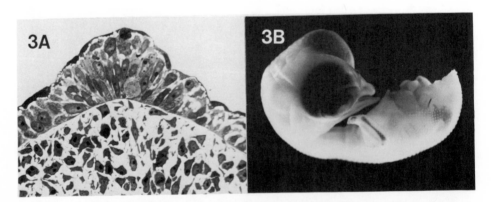

Fig. 3.   **A.**This photomicrograph shows a cross section of a
            turtle apical ridge induced by chick *limbless*
            mesoderm. Both ectoderm and mesoderm are healthy and
            the limb bud was growing normally.
            B.This photograph is of a whole 10 day *limbless*
            embryo. At stage 15 this embryo had turtle flank
            grafted to the limb field mesoderm. A distally
            complete limb formed demonstrating normal ridge
            activity.

Talpid² is a simple Mendelian, autosomal recessive, lethal mutation in the chick embryo. Tissue recombinant studies of *talpid²* limb buds (Fraser and Abbott, 1971) and early recombinants as in Figure 1 (Carrington and Fallon, unpublished) show that the limb mesoderm is affected in homozygous *talpid²* embryos. The ectoderm appears to be normal. The most distinguishing morphological feature of the *talpid²* embryo is their short, spade-like wings and legs which are polydactylous and syndactylous. Homozygous embryos begin to die at 5 days of incubation; most do not survive beyond 15 days. Unfortunately, more *talpid²* embryos die after microsurgery than normal embryos after similar procedures.

Developmental Anatomy of *Talpid²*

Because of the obvious changes in limb pattern formation we thought it useful to describe the developmental biology of the *talpid²* limb. We have measured the anteroposterior, proximodistal and dorsoventral dimensions of at least three wing and leg buds of *talpid²* and normal embryos at each stage between 18 and 25. The generalization that comes from these data is that from the onset of budding *talpid²* limb buds are wider along the anteroposterior axis than the normal bud.

We have observed that the usual zones of necrosis which sweep along the mesoderm of the normal anterior and posterior limb bud borders fail to occur in this mutant. Cell death between the digits was observed but was variable in extent and abnormal in pattern. For example, bands of necrosis were observed coursing from anterior to posterior across digital territories with the result that distal phalanges appeared without their proximal counterparts.

It is tempting to infer that the anteroposterior broadening of the *talpid²* limb bud is brought about by the failure of mesodermal cell death along the limb bud borders. We believe this is unlikely to be the sole reason. Cell death begins in the normal wing anterior border at stage 21 and along the posterior border at stage 24. However, the *talpid²* bud is significantly broader at stage 18. This inherent broadness may be enhanced by the failure of cells to die, but cell death cannot be thought of as causal.

We examined the skeleton of wings and legs from *talpid²* embryos at 10-21 days of incubation. An important observation was that *talpid²* limbs do have anteroposterior polarity. We give a description of the *talpid²* wing here (Figure 4) (a detailed description will appear elsewhere; Dvorak & Fallon, submitted). The humerus was short and wide with a slight anteriorward curvature. It articulated with a shoulder girdle and had two articulations with the forearm elements. The forearm elements were often separate proximally and fused distally. The extent of fusion varied among the embryos. The two forearm elements could be distinguished from one another based on morphology and staining characteristics (Figure 4). The anterior forearm bone (radius) was short and wide with

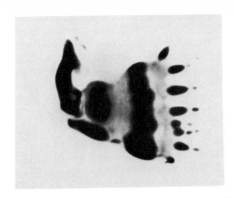

Fig. 4.　This is a photograph of a 10 day *talpid*[2] wing
stained with Victoria blue. Notice that the radius
and ulna are distinguishable and that the ulnar carpal
is easily distinguishable.

rounded edges. The posterior forearm bone (ulna) was longer in
the proximodistal direction and stained more intensely. In
addition, the ulna had a process corresponding to the olecranon
process of the normal ulna.

There were two carpal bones with the radial carpal
extending across the anteroposterior extent of the wing. The
smaller ulnar carpal was located at the posterior edge of the
forearm. The *talpid*[2] metacarpals were always fused; the
proximal edge was jagged, appearing to indicate the individual
elements. A similar pattern was evident distally in many
specimens. Often the most posterior portion of the metacarpal
mass stained more intensely than the anterior portion.

The wing digits were disorganized showing variable fusion
of the one or two phalanges present. None of the phalanges
resembled normal digits. However, in many cases the most
posterior digit was separate from the others and stained more
intensely. The number of digits that formed varied from five to
ten, with most wings having seven or eight digits. In the
majority of cases right and left wings had the same number of
digits, but when they varied it was by only one or two digits.

One important observation from these studies is that
talpid[2] has a definite anteroposterior polarity in the forearm
and hand. A second point is that the proximal elements are
closer to normal in *talpid*[2] than are the distal elements, with
distal elements becoming more and more disrupted and less like
the normal.

Experimental Studies

The observation of anterior to posterior polarity in
talpid[2] raises questions on the presence of polarizing activity
in the mutant. Summarizing a large body of data, we have
demonstrated the following. The mutant has a polarizing zone
which will cause respecification of normal wing bud anterior
mesoderm. However, the mutant polarizing activity was weaker

378

than that from the normal wing. Typically, the *talpid²* polarizing zone graft to a normal wing anterior border produced an extra digit 2, while normal polarizing zone grafted to a normal anterior border produced an extra digit 4.

We asked whether a normal polarizing zone (or retinoic acid) might change the *talpid²* phenotype, if grafted along the mutant <u>posterior</u> border thus augmenting the weak *talpid²* polarizing signal. This was not the case. The control left wing was always comparable to the right wing, which received the polarizing zone graft, with the single exception that retinoic acid invariably reduced the number of digits.

The *talpid²* anterior mesoderm does respond to normal polarizing zone (or retinoic acid) grafted to the anterior border. The result of such grafts is a respecification of the anterior forearm and hand into posterior elements within the *talpid²* phenotype. Extra digits did not develop. In fact, with retinoic acid there was always a reduction of the number of digits when the left (control) and right (experimental) sides were compared.

We conclude that polarizing activity is present but weak in *talpid²*. Transmission of the polarizing signal and retinoic acid does occur in the *talpid²* anterior mesoderm. Finally, simply boosting the putative morphogen from the posterior border, during limb bud stages, does not have an effect on the *talpid²* phenotype.

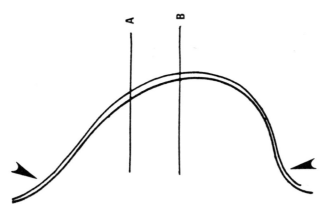

Fig. 5.   In this drawing of a normal stage 21 wing bud line A
           indicates a separation of the limb into an anterior
           third and posterior two-thirds; line B separated the
           limb into anterior and posterior halves. Arrowheads
           indicate anterior and posterior borders of the limb
           bud.

There is a body of evidence indicating that anterior limb bud mesoderm cells are dependent on the presence of posterior mesoderm for survival e.g. MacCabe & Parker, 1975; Summerbell, 1979; Rowe & Fallon, 1981; Hinchliffe & Gumpel-Pinot, 1981 and for an overview Todt & Fallon, 1987. The basic observation is that removal of the posterior third or more of the developing normal limb bud [with a cut made along line B of Figure 5] results in rapid onset of cell death (Figure 6A) in the remaining anterior mesoderm. In all cases a truncated limb develops. Interestingly enough, if a similar cut is made along line B in Figure 5 and the normal anterior limb bud is removed the posterior mesoderm continues to develop and distal elements are formed.

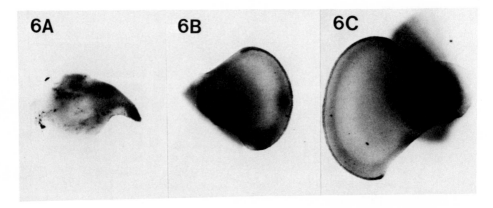

Fig. 6. **A.** This micrograph was taken 48 hours after the posterior half of a normal wing bud was removed at stage 19. The remaining anterior half has failed to grow. This specimen was stained supravitally with neutral red at 48 hours. The dark area at the tip indicates massive necrosis of the distal anterior mesoderm cells (cf. Todt and Fallon, 1987). Compare this with Fig. 6B.
B. This micrograph was taken 48 hours after the operation. The posterior half of a $talpid^2$ wing bud was removed at stage 19. The remaining anterior half has grown almost to the size of the contralateral left wing bud (see Fig. 6C). This specimen was stained supravitally with neutral red. The dark band at the top of the bud is the apical ridge; there is no cell death in the mesoderm. Compare this with Fig. 6A.
C. This micrograph is the contralateral left wing bud to the specimen shown in Fig. 6B.

We have carried out similar experiments on the $talpid^2$ mutant limb bud, where 1/2 to 2/3 of the posterior limb bud was removed [comparable to line B or line A in Figure 5]. To our surprise, no cell death was observed in $talpid^2$ anterior wing mesoderm even if 2/3 of the posterior bud was eliminated.

In fact after 48 hours the anterior *talpid²* bud had regenerated
to a mass that was similar to a comparably staged unoperated
talpid² bud (Figure 6B-C). When permitted to develop to day 10
these *talpid²* limbs were always proximodistally complete (Figure
7). Further, the number of digits between right and left wings
was comparable.

Fig. 7.   The posterior half-wing bud of this *talpid²* limb was
removed at stage 19. A proximodistally complete limb
formed from the remaining anterior limb bud.

It is clear from these experiments that *talpid²* anterior
mesoderm is independent of posterior mesoderm. With respect to
survival after separation of anterior and posterior limb bud
halves, the *talpid²* anterior mesoderm is similar to normal
posterior mesoderm. At this time it is not understood how
normal posterior mesoderm permits survival of the anterior
mesoderm. Todt and Fallon (1987) have proposed that
"maintenance" ability may be a property of posterior mesoderm
distinct from its polarizing ability (see also Frost and
Hinchliffe, these Proceedings). While the maintenance property
may overlap with polarizing ability, the former likely has a
much greater spatial distribution in posterior mesoderm. In
this context it is important to point out that the *talpid²*
anterior limb bud mesoderm does not have polarizing activity.
Our proposal for a specific interaction between posterior and
anterior limb bud mesoderm adds a dimension to our understanding
of growth control during limb development.

Conclusions about *Talpid²*

This mutant may be an important tool to gain insight into
how pattern is initiated and stabilized in the limb mesoderm.
The mutant limb mesoderm has properties which distinguish it
from the normal. Our present studies on *talpid²* include
retinoic acid receptor distribution, homeotic gene expression

and the capabilities of these cells in tissue and organ culture. The challenge is to try to relate what we learn about this mutant to normal limb development.

ACKNOWLEDGEMENTS

This work was supported in part by NIH Grant PO1-HD20743 and NIH Training Grant 5T32-HD07118. We thank Richard Hinchliffe for making unpublished data available to us. Special thanks to Karen Krabbenhoft and Mary Savage for reading the manuscript.

REFERENCES

Carrington, J.L. and Fallon, J.F., 1984, The stages of flank ectoderm capable of responding to ridge induction in the chick embryo. J. Embryol. exp. Morph. 84:19-34.

Carrington, J.L. and Fallon, J.F., 1986, Experimental manipulation leading to induction of dorsal ectodermal ridges on normal limb buds results in a phenocopy of the eudiplopodia chick mutant. Devel. Biol. 116:130-137.

Fallon, J.F., Rowe, D.A., Frederick, J.M. and Simandl, B.K., 1983, Studies on epithelial-mesenchymal interactions during limb development. In: "Epithelial-Mesenchymal Interactions in Development", R.H. Sawyer and J.F. Fallon, ed., Praeger Scientific Press, New York.

Fraser, R.A. and Abbott, U.K., 1971, Studies on limb morphogenesis V. The expression of eudiplopodia and its experimental modification. J.Exp. Zool. 176:219-236.

Frost, S. and Hinchliffe, J.R., 1991, Retinoic acid effects on experimental chick wing buds; patterns of cell death and skeletogenesis. These Proceedings.

Hamburger, V. and Hamilton, H.L., 1951, A series of normal stages in the development of the chick embryo. J.Morphol. 88:49-92.

Hinchliffe, J.R. and Gumpel-Pinot, M., 1981, Control of maintenance and anteroposterior skeletal differentiation of the anterior mesenchymne of the chick wing bud by its posterior margin (the ZPA). J.Embryol. exp. Morph. 62:53-82.

Lanser, M.E., Carrington, J.L. and Fallon, J.F., 1986, Survival of motoneurons in the brachial lateral motor column of limbless mutant chick embryos depends on the periphery. J. Neuroscience 6:2551-2557.

Lanser, M.E. and Fallon, J.F., 1984, Development of the lateral motor column in the limbless mutant chick embryo. J.Neuroscience 4:2043-2050.

LeDouarin, N. and Barq, G., 1969, Sur l'utilisation des cellules de la Caille japonaise comme'marqueurs biologiques' en embryologie experimentale. C.r.hebd.Seanc.Acad.Sci.,Paris D 269:1543-1546.

Lyons, G., Krabbenhoft, Simandl, B.K., Buckingham, M., Fallon, J.F. Robert. B., 1991, Temporal patterns of expression in normal and limbless chick embryos. These Proceedings.

MacCabe, J.A. and Parker, B.W., 1979, The target tissue of limb-bud polarizingactivity in the induction of supernumerary structures. J. Embryol. exp. Morph. 53:67-73.

Prahlad, R.B., Skala, G., Jones, D.B. and Briles, W., 1979,
    Limbless: A new genetic mutant in the chick.
    J.Exp.Zool. 209:427-434.
Rowe, D.A. and Fallon, J.F., 1981, The effect of removing
    posterior apical ectodermal ridge of the chick wing
    and leg bud on pattern formation. J.Embryol.exp.Morph.
    65 Supplement:309-325.
Rowe, D.A. and Fallon, J.F., 1982, The proximodistal
    determination of skeletal parts in the developing
    chick leg. J. Embryol. exp. Morph. 68:1-7.
Saunders, J.W., Jr., 1948, The proximo-distal sequence of origin
    of the parts of the chick wing and the role of the
    ectoderm. J. Exp. Zool. 108:363-404.
Summerbell, D., 1979, The zone of polarizing activity: evidence
    for a role in normal chick limb morphogenesis.
    J.Embryol.exp.Morph. 50:217-233.
Todt, W.L. and Fallon, J.F., 1987, Posterior apical ectodermal
    ridge removal in the chick wing bud triggers a series
    of events resulting in defective anterior pattern
    formation. Development 101:501-515.
Zwilling, E., 1961, Limb morphogenesis. Adv. Morph. I:301-330.

PROXIMAL ELEMENTS IN THE VERTEBRATE LIMB: EVOLUTIONARY

AND DEVELOPMENTAL ORIGIN OF THE PECTORAL GIRDLE

Ann Campbell Burke

Biology Department
Dalhousie University
Halifax, Nova Scotia
CANADA B3H 4J1

INTRODUCTION

The vertebrate skeletal *Bauplan* consists of two major systems – the axial skeleton (skull, vertebrae and ribs), and the appendicular skeleton(pectoral and pelvic girdles, and limb bones). There are distinctions made between these two skeletal systems on the grounds of history, function, and development.

The historical discrepancy arises from theories about the ancestral bodyplan of fishes. Phylogenetically, it is assumed that paired appendages were 'added' to an animal that was already a vertebrate, one that posessed the basis of an axial skeleton (Jarvik, 1965; and see Hall, this volume). Functionally, the post-cranial axial skeleton is morphologically more constrained than the appendicular skeleton. The limb elements demonstrate a wide range of morphological variation reflecting different locomotor adaptations. The axial skeleton serves more of a structural than locomotor function in most higher vertebrates and thus remains relatively conservative.

Embryonically there is an additional distinction between the appendicular and the post-cranial axial skeletons. Though all of the striated muscle of the body, including the limb muscles arises from the somitic myotomes, the vertebrae and ribs are formed from the somites, and the appendicular skeleton arises from the lateral plate mesoderm (somatopleure). Thus while the functioning limb is a product of two distinct embryonic cell populations, the skeletal portion was thought to belong solely to the lateral plate, in contrast to the axial skeleton (Balinsky, 1974, Gilbert, 1985). This assumed developmental segregation has been incorporated into the theories on the origin of the paired appendages. According to the lateral fin fold theory, the girdles develop as inward growths of the cartilagenous basiradials that are formed from the somatopleure (Goodrich, 1906, Zug, 1979).

The majority of studies on limb development and evolution ignore the girdles and concentrate on the more distal elements of the limb, the stylopod, zeugopod, and autopod. The girdles are,

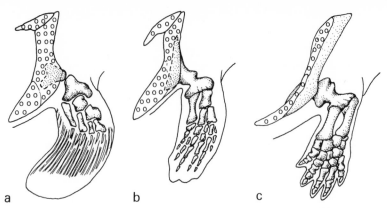

Fig. 1. Three conditions in the evolution of the pectoral
girdle. a) A rhipidistian fish. b) early tetrapod c)
Eryops Note the reduction in size of the dermal elements
and the growth of the scapulocoracoid from right to
left.. After Gregory and Raven, 1941.

however, integral parts of the appendicular skeleton serving to
anchor the limbs to the trunk. Morphological change of the
pectoral girdle has been considerable during the evolution of the
tetrapods. In Rhipidistian fishes and the earliest tetrapods, the
endochondral scapulocoracoid – the primary pectoral girdle, was
relatively small compared to the extensive dermal cleithrum,
clavicle and interclavical that comprise the secondary girdle. It
was the secondary girdle which was of primary functional
importance in these animals, serving to buttress the shoulder
against the skull. There has been a steady decrease in the dermal
elements and a complementary increase in the size of the
endochondral elements in all tetrapod clades (Figure 1).

There are experimental data on the development of the scapula
in two distantly related groups of amniotes. In 1977, Chevallier
demonstrated that the scapular blade of the chick is formed from
somitic mesoderm in a cranial to caudal mosaic. There is evidence
from work on turtle embryos that the chelonian scapula also has a
somitic component (Burke, 1989, 1990). Both of these organisms
contradict the assumed developmental segragation of axial and
appendicular systems. Given the assumptions about the evolution of
the paired appendages, and acknowledging the highly derived
locomotor systems in birds and turtles, it is important to know if
a somitic contribution to the scapula is primitive for tetrapods
or a unique adaptation of various clades within the amniotes.

The Lissamphibia represent the extant members of the amniote
outgroup. The generalized body plan of salamanders makes then good
candidates for investigation of the primitive developmental mode
of the tetrapod primary pectoral girdle. This study aims to
clarify the results of early studies on the shoulder girdle of
Ambystoma (Detwiler, 1918), and test a hypothesis based on the
development of the scapula in two distantly related amniotes
(Burke, 1990). I have predicted that the anamniote pectoral girdle
arises from lateral plate tissue only, that a somitic contribution
to the scapula is a derived character of amniotes, and that
urodeles retain the primitive condition. This approach
demonstrates how ontogeny can inform an interpretation of

evolutionary pattern and process, and point out errors in our assumptions about homology.

MATERIALS AND METHODS

Egg masses of *Ambystoma maculatum* were collected during March and April in the outskirts of Halifax, Nova Scotia. The eggs were kept at 17°C in dechlorinated tap water and allowed to develop to Harrison (1969) stages 19 through 27, at which time the embryos had between 3 and 13 somites. Embryos were dejellied manually and operations preformed in 100%, Calcium free, Steinberg's solution with 50 mg/liter gentamycin (Sigma) after the technique outlined by Asashima et al (1989) for axolotls. The ectoderm was peeled back and somites 1-2 (series "A"), 3-5 (series"B", Figure 2), or 6-8 (series "C") were unilaterally extirpated with tungsten needles and hair loops. The right side of the embryo was always the side operated upon. Some attempt was made to stretch the ectoderm over the wound, but shrinkage made this difficult. The embryos were moved after 30 minutes into sterile dechlorinated tap water with antibiotic. On the first day post op. the wounds were cleared of cellular debris with hair loops. The wound remained visible and free of ectoderm for up to several weeks. Larvae began to feed on brine shrimp at week three, and individuals were then separated to avoid cannibalism. They were sacrificed in MS 222 between 30 and 90 days post-op. After fixation in 10% buffered formalin, specimens were stained with Alcian blue and Alizarin red and cleared in KOH and glycerine, or embedded in paraffin, sectioned and stained with Mallory's Trichrome or hemotoxylin and eosin (Humason, 1972). Some specimens were stained in Alcian green and cleared with methyl salicylate. These specimens could be studied as whole mounts and then sectioned for additional information.

RESULTS

The normal development of the scapulocorocoid in *Ambystoma* begins with a pre-cartilaginous condensation in the region of the glenoid at Harrison's (1969) stage 42-43. At this stage the forelimb bud has elongated and bifurcated into two digits. The elbow is not yet visible and the only skeletal blastema are the girdle and humerus. The scapular condensation appears concurrently or slightly after the condensation of the humerus which is initially more robust. These two elements form a continuous condensation across the glenoid. The coracoid is very poorly developed.

In mature specimens (i.e. 4 full digits in the forelimb) the

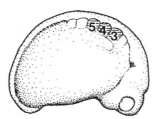

Fig. 2. A stage 22 *Ambystoma maculatum*. Drawing showing the tissue removed in series "B" embryos.

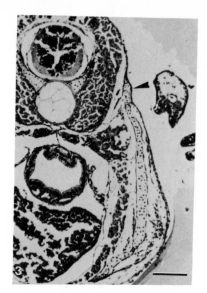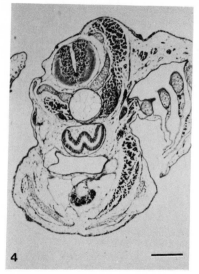

Fig.3. Cross-section through the shoulder region of a normal
   A. maculatum larvae. The arrow points to the dorsal
   extent of the scapular blade. The outlying structures
   are gills. Bar= 25 μ

Fig. 4. Cross-section through a specimen in series "B",
   anterior to the shoulder. Note the ill-formed neural
   arch and the absence of all epaxial and most hypaxial
   muscles on the animal's right side (left side of
   picture). Bar= 45 μ.

scapular blade lies in the region of the third myotome. It rises
vertically from the glenoid with a slight posterior curvature. It
is columnar in mid-shaft, and flares anterio-posteriorly at the
distal tip. In sectioned material, the distal scapula can be seen
extending dorsally beyond the nephric structures, and the distal
extreme lies adjacent to the base of the neural arch (Figure 3).
It lies just under the skin and is separated from the vertebrae by
a large mass of epaxial musculature, the *dorsalis trunci*.

   To test the assumption that somites 3-5 are the only brachial
somites, more anterior (1&2, series "A") and more posterior (6-8,
series "C") somites were removed and the effect on the forelimb
assessed. The specimens were scored for the condition of the
epaxial musculature, vertebral structure, and forelimb skeleton.
Operated side forelimbs were normal in all of series "A" and "C",
while 38% of those specimens in which somites 3-5 (series "B")
were removed had distal limb deformities on the operated side.
These deformities are most often varying degrees of axial
dupliction. Severely deformed right forelimb skeletons probably
result from unintentional damage to the lateral plate mesoderm
during surgery. As the experiment was designed to distinguish
between somitic and lateral plate derivatives, these specimens
were removed from further analysis.

   Virtually all of the experimental animals in all three series
show depletions to the dorsal epaxial musculature at a level
corresponding to the somites removed. The degree of muscular
depletion is quite variable in all three series. This is assumed
to result from two related factors: 1) variation in the amount of
myotomal material originally excised; and/or 2) variable amount of

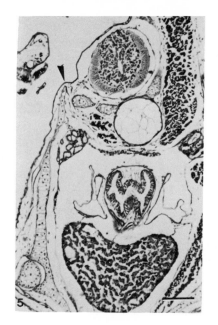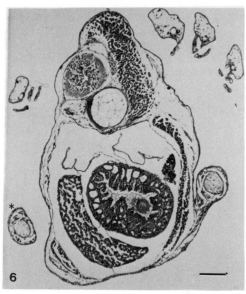

Fig. 5. Cross-section through the shoulder region in the
same specimen in Fig. 4. Arrow points to the dorsal
extent of the scapular blade on the operated side.
Bar =23 μ.

Fig. 6. The same specimen shown previously, at a level
posterior to the shoulder. Note that the animals right
limb (asterix) contains muscles. Bar= 25 μ.

regeneration, or spreading and compensation, by myotomal tissue in
and around the wound bed. Further studies will more accurately
assess the muscle deletions in relation to segmentation.

The effects on the skeleton can be seen in cleared and stained
animals. The vertebral arches and centra in the area of the muscle
depletion are reduced and abnormally formed, often showing
fusions. In series "A" the skeletal effects are minimal and
limited to the exoccipital region and the first trunk vertebrae
with a corresponding asymmetry in the spinal cord. In series "B"
the vertebral defects occur in some or all of trunk vertebrae 1
through 5, and in series "C", vertebrae 4-9 are effected.

Series "B" experiments are the main focus of this study. All of
these animals have depletions of the dorsal shoulder musculature.
Sixty two percent have normal distal forelimbs and were examined
further to determine the condition of the scapula. Cross-sections
of these specimens show the extent of the absent musculature
(Figure 4). All of the epaxial muscles in the shoulder region are
absent in all specimens examined. The hypaxial muscles are
diminished but a variably thin muscle sheet is present in most
specimens. Sections through the shoulder show the scapula on the
operated side rising to the same level attained by normal scapulae
– the level of the neural arch. The scapula approaches the axis
medially as there is no intervening muscle mass (Figure 4).

DISCUSSION

Ambystoma maculatum was studied in depth by numerous authors in the early and middle part of this century when the species was known as *Amblystoma punctatum*. In 1918, Detwiler published the results of a study on the development of the shoulder girdle in "*Amblystoma*" [sic], that attempted to isolate the rudiments of the girdle by means of surgical extirpations. He began with a notion of the position of the limb field based on the work of Harrison and others (see Harrison, 1918 for references). The limb bud in this salamander forms at the level of somites 3-5, immediately lateral or ventral to the pronephros. Detwiler's results and conclusions can be summarized as follows: The girdle behaves as a three part mosaic composed of; 1) the scapula, that portion dorsal to the glenoid; 2)the suprascapula, forming the main shaft and distal extreme, and; 3) the corocoid and procorocoid. Each 'unit' of the girdle can be perturbed independently of the limb proper by removal of 'rudimentary' tissues.

The timing of the surgery is critical in the interpretation of these results. Detwiler's experiments were performed on 'tail-bud' embryos, which have 12+ somites. Cells from somites 3-5 have already begun to disperse at these stages. Discriminating between the two mesodermal populations is thus impossible and Detwiler does not couch his conclusions in the specific terms of 'somitic' vs. 'lateral' mesoderm. He does conclude however, that ".... the suprascapula is derived from tissue other than that which gives rise to the limb and the remainder of the girdle." (Detwiler, 1918, p. 506). It could be interpreted therefore that as the remainder of the limb skeleton is lateral plate, the suprascapular portion arises from somitic mesoderm as it does in chicks and turtles. The experiments reported here test this possibility by removing only the somites at a stage before migration of somitic cells has commenced, being careful to leave the pronephros in place.

Development of the Scapula

Because so many studies focus on the distal aspects of the limb skeleton, very little is known about the development of the shoulder girdle. As part of the appendicular skeleton, it was thought to arise from lateral plate mesoderm, but the experimental data on the chicks and turtles disproves this assumption. The chick scapula takes contributions from somites 15-24, a series of nine somites that also give rise to cervical and thoracic vertebrae and ribs (Chevallier, 1977). The somitic contibution to the turtle scapula is limited to somites 8 through 12, a series of four somites that contribute to cervical *but not* thoracic vertebrae (Yntema, 1970; Burke, 1990).

Given that there is a minimum of 300 million years since the last common ancestor of turtles and birds stumbled about, it is not suprising that these two modern animals, with such different morphologies, should show incongruent development of homologous elements (i.e. non-equivalent sets of 'scapular' somites). The ancestral condition however, the state from which these two lineages diverged, is not discernable without a third reference point. The condition displayed by urodeles is the one most likely to approximate the ancestral condition for amniotes.

390

The analysis of the experimental data from *Ambystoma* presented here is in its early stages. Consistent qualitative results based on observations of whole mounted and sectioned specimen, however, compared with the results of Detwiler (1918) have led to the following conclusions. The condition of the scapula in series "B" animals is interpreted here to demonstrate that there is no significant contribution by the somites to the scapula in *Ambystoma*. Detwiler (1918) described the 'suprascapular' portion of the girdle as an independant entity, and illustrates experimental animals in which this portion of the scapula is absent after extirpation of somites. He stresses, furthermore, that the suprascapula is absent only when the pronephros is completely removed along with the somitic mesoderm. In the experiments sited here, only somitic tissue was taken, using the pronephros as the limiting boundary for the surgery (Fgure 2). When only somitic tissue is taken, at stages before migration of somitic cells, the scapula on the operated side is not reduced. This is true of all three series of animals. The scapular shaft on the operated side reaches the same dorsal level as the control scapula (Figure 5), even in the total absence of the epaxial and most of the hypaxial muscles. The scapula in *Ambystoma* thus appears to arise completely from the lateral plate along with the rest of the limb skeleton, in contrast to the situation seen in turtles and chicks.

Development of the Limb muscles

An interesting observation arises from these experiments concerning the origin of the limb muscles. There is an inconsistency in the literature concerning the origin of the appendicular musculature in vertebrates. In sharks and some teleosts, "muscle-buds" extending from the myotomes, can be seen penetrating the limb bud and have been interpreted as the *anlagen* of fin muscles. Many general references from the middle of this century list the limb muscles of all vertebrates as arising from the somites (e.g. Goodrich, 1930; Hyman 1942). Muscle buds have been reported in humans, lizards, anurans, and urodeles. Conversely, there also claims that muscle buds are absent in some teleosts, chicks, and amphibians (see Byrnes, 1898 for references). Several studies were interpreted to support the view that the limb muscles of urodeles arise *in situ* from the lateral plate mesoderm (Byrnes, 1898; Lewis, 1910; Harrison, 1915; Detwiler 1918). A lateral plate origin for the limb muscles was argued as late as 1955 (Detwiler).

It has now been conclusively shown in chicks that the appendicular muscles as well as the muscles of the ventro-lateral body wall arise from somitic mesoderm (see review by Gumpel-Pinot, 1984). The assumed necessity that fundamental developmental patterns are conserved in phylogeny has eclipsed the older opinions. Amphibian limb muscles, as homologues to amniote limb muscles, are commonly assumed to arise in a congruent fashion, i.e. from the somites (Balinsky, 1974; Duellman and Trueb, 1985; Gilbert, 1985). As far as I know, no experimental work has been done recently that specifically addresses this question in amphibians.

The results reported here are inconclusive regarding limb muscles. The degree and pattern of muscle depletion in the trunk is difficult to quantify without three dimensional reconstruction. The muscle-gap in the experimental animals often covers a more extensive region dorsally than ventrally, and the effects are concentrated at more anterior regions than would be predicted from the somite level excised. However, muscles are present in the forelimbs of all the experimental animals examined (e.g. Figure 6). This situation is not immediately compatible with a somitic ' origin of appendicular musculature and is very similar to the results reported in the older studies cited above. There are a number of possible explanations. There may be a significant degree of compensatory ability by the adjacent somites, but, in the absence of clear experimental data, some participation of the lateral plate in the formation of the limb muscles of *Ambystoma* should not be ruled out.

CONCLUSIONS

The variation in pattern of scapula formation, and the uncertainties concerning limb muscles, point to misconceptions that arise as the result of assuming strict conservation of developmental patterns in tetrapods. Such an outlook creates blindspots in our understanding of variation. A comparative study of development addresses how it can change during phylogeny. The developmental differences between taxa can be used to make inferences about the changes that took place in ancestral ontogenies to bring about novel morphologies.

The number and position of contributing segments (somites) in the appendages of vertebrates varies widely and is described by the theory of "transposition", the shifting participation of serially homologous segments (Goodrich, 1930). The participation of varying numbers of somites in the scapula of tetrapods is simply another example of transposition within the limb field. An 'embryonic field', strictly defined, is the set of tissues that can interact to form a structure, a necessarily larger set than the tissues in a fate map, which comprises those tissues that do interact to form a structure (Harrison, 1918; Huxley and De Beer, 1934). Embryonic fields are areas of developmental potential that represent paths of least resistance for morphological evolution.

The tetrapod limb has had over 400 million years to experiment with the limb field. I hypothesize that the original skeletal elements in the limb arose from lateral plate mesoderm, consistent with the theories about the origin of the paired appendages. One of the evolutionary transitions that took place during the move to a terrestrial existence was the novel participation of the somitic mesoderm - already included in the limb field - in the formation of the appendicular skeleton. This innovation allowed for 'transposition' to increase morphological variation and hence locomotor adaptations among amniote taxa.

ACKNOWLEDGEMENTS

I thank S. Friet who helped collect animals, T. Miyake who drew the illustrations and M. Primrose who printed the photographs. The manuscript was improved by comments from B.K.Hall and S. Smith. This work was supported by a I.W. Killam Postdoctoral Fellowship and a Research Development Grant from Dalhousie University.

REFERENCES

Asashima,M., G.M. Malacinski, and S.C, Smith, 1989, Surgical manipulation of embryos, in "Developmental biology of the Axolotl," J.B. Armstrong and G.M.Malacinski, eds. Oxford University Press, New York.

Balinsky, B.I. 1975, "An introduction to embryology," 4th edition W.B.Saunders Co., Philadelphia

Burke, A.C. 1990, The development and evolution of the turtle body plan: Inferring intrinsic aspects of the evolutionary process from experimental embryology. Am. Zool. In press.

Burke, A.C. 1989, "Critical features in chelonian development: Ontogeny and phylogeny of a unique tetrapod Bauplan," Ph.D. dissertation, Harvard University, Cambridge, Mass.

Byrnes, E.F. 1898, Experimental studies on the development of the limb-muscles in Amphibia. J. Morphol. 14:105-141.

Chevallier, A. 1977, Origine des ceintures scapulaires et pelviennes chez l'embryon d'oiseau. J. Embryol. Exp. Morphol. 42:275-292.

Detwiler, S. 1918, Experiments on the development of the shoulder girdle and the anterior limb of Amblystoma punctatum, J. Exp. Zoolol. 25:499-537.

Duellman,W.E. and L.Trueb, 1986, "Biology of the Amphibians," McGraw-Hill Book Co., New York.

Gilbert, S.F. 1985, "Developmental Biology," Sinauer Ass, Inc., Sunderland, Mass.

Goodrich,E.S. 1906, Notes on the development, structure and origin of the median and paired fins of fish. Quart. Jour. Micro. Sci. 50:24-376.

Goodrich, E.S. 1930, "Studies on the structure and development of the vertebrates," Macmillan, London.

Gregory, W.K. and Raven,H.R. 1941, Studies on the origin and early evolution of the paired fins and limbs. Ann. N.Y. Acad. Sci. 42:273

Gumpel-Pinot,M. 1984, Muscle and skeleton of the limbs and bodywall, in "Chimeras in Developmental Biology," N.Le Dourian and McLaren, eds. Academic Press. pp 281-310.

Harrison, R.G. 1915, Experiments on the development of the limbs in Amphibia. Pro. Nat. Ac. Sci. 1:539-544.

Harrison, R.G. 1918, Experiments on the development of the fore limb of Amblystoma punctatum, a self-differentiating equipotential systems. J. Exp. Zool. 25:413-461.

Harrison, R.G. 1969, Harrison stages and description of the normal development of the spotted salamander, Amblystoma punctatum, in "Organization and development of the embryo," R.G. Harrison ed., Yale University Press, New Haven.pp 44-66.

Humason,G. 1979, "Animal Tissue Techniques," W.H.Freeman & Co., San Fransisco.

Huxley, J.S. and De Beer, 1934, "The elements of experimental embryology," Cambridge University Press, Cambridge.

Hyman, L.H. 1942, "Comparative Vertebrate Anatomy," University of Chicago Press, Chicago.

Jarvik,E. (1965) On the origin of girdles and paired fins. Israel Jour. Zool. 14:141-172.

Lewis, W.H. 1910, The relation of the myotomes to the ventro-lateral musculature and to the anterior limbs in Amblystoma. Anat. Rec. 4:183-190.

Yntema,C.L. 1970, Extirpation experiments on the embryonic rudiments of the carapace of Chelydra serpentina. J. Morphol. 132:235-244.

Zug, G. 1979, The endoskeleton: the comparative anatomy of the
    girdles, the sternum, and the paired appendages, in "Hyman's
    Comparative Vertebrate Anatomy," M.H.Wake, ed., U.Chicago
    Press, Chicago.

EVOLUTIONARY TRANSFORMATION OF LIMB PATTERN: HETEROCHRONY AND SECONDARY FUSION

Gerd B. Müller

Department of Anatomy
University of Vienna
A-1090 Wien, Austria

The development of the vertebrate limb is a hierarchical process of temporal and spatial tissue organization. We begin to understand its cellular and molecular properties, its links to the gene level, and, through the processes of pattern formation, its relationship to phenotypic patterns. In addition, however, a comprehensive theory of limb development must provide a mechanistic basis for evolutionary transformations of limb patterns and for the origin of individualized parts of the limb during vertebrate phylogeny. This paper examines the present concepts about limb development with regard to their capacity to accommodate phylogenetic transformations. Examples from our work on archosaur limb development and evolution are used to illustrate the patterns of transformation and the problems arising from an evolutionary perspective.

THE TRANSFORMATION PROBLEM

The majority of evolutionary modifications of the tetrapod limb are changes of relative proportions of parts which, at a developmental level, can be understood as effects of altered growth rates. More difficult to explain are qualitative transformations of limb anatomy, such as changes in the number and composition of autopod elements. Evolution along many different lineages involves progressive numerical reduction of carpal and tarsal elements, and of the digits. For example, in the archosaur lineage – which includes the thecodontia, dinosaurs, pterosaurs, crocodylia, and the birds – the number of forelimb digits is reduced from five to three, and the number of ossified carpal elements is reduced from eleven to two. Similar reductions took place in the hindlimbs. This is different from the distal deletions seen in limb reduced species which have been shown to correspond with the cessation of AER activity (Raynaud, 1985). The problem of transformation is more complicated since not only distal elements are reduced but a number of elements at intermediate locations are missing or have been altered. Such modifications of the basic limb pattern require a developmental explanation. Several fundamentally different concepts have a bearing on the possible mechanisms of pattern transformation. Three must be taken into account.

Developmental Patterning of the Vertebrate Limb
Edited by J.R. Hinchliffe *et al.*, Plenum Press, New York, 1991

Early approaches to the problem were derived from the concept of recapitulation. Applied to vertebrate limb development, it was proposed that all tetrapod limbs pass through a stage at which an archetypic ground plan is layed down, and only during subsequent development of the derived limbs the species specific patterns are realized through complicated sets of fusions and deletions of elements (Holmgren, 1933; Steiner, 1934; and others). This concept of archetype recapitulation was rightfully rejected on the basis of the observation that already the initial chondrogenic patterns are class specific (Hinchliffe, 1977; and this volume). The mechanistic principle of evolutionary transformation embedded in the recapitulatory concept, however, is that secondary modifications during individual development generate some of the phylogenetically important alterations of pattern. This cannot be fully rejected since in many cases the primary chondrogenic patterns consist of more elements than the adult limb.

A second mode for pattern transformation emerges as a corollary of the chemical models of pattern formation. Gradients of diffusable morphogens are thought to set up positional information values or prepattern grids across the early limb fields and thus could specify the position and number of chondrogenic condensations (Summerbell, Lewis and Wolpert, 1973; Wilby and Ede, 1975). The initiation of cell condensations would occur at specific concentration levels or peaks of concentrations of one or more of such morphogens. The number of peaks or of concentration specific thresholds would directly translate into the number of carpal elements or digits. It follows that an evolutionary modification of the number of autopod elements would be based on a modification of the mechanism generating the positional values. Essentially, the whole limb field would have to be specified in a modified way, a process that we may term phylogenetic repatterning. The components of the patterning system are usually thought to include ectodermal-mesenchymal interactions and ZPA activity. These components are present in all tetrapod classes (Hinchliffe, this volume) and trans-class recombination experiments (Jorquera, 1971; Müller, 1991a; Fallon, this volume) show them to be very conservative. Thus, the mechanistic basis of pattern transformation would have to lie in very subtle changes of morphogen concentrations. Since the morphogens are unknown and models of chemical pattern formation are largely formal, it is difficult to test whether phylogenetic repatterning took place.

A third conceivable mode of pattern transformation can be derived from a different model of pattern formation. Oster et al. (1985) propose that the early chondrogenic condensations in the limb bud grow through cell aposition in a proximo-distal direction and produce the pattern of elements through branching and segmentation processes, resulting from the mechano-chemical properties of the condensations themselves. Branching and segmentation events of chondrogenic condensations have been demonstrated in all classes of tetrapods (Shubin and Alberch, 1986). One important component of the model is the asymmetry in branching capacity between the preaxial and the postaxial rays of chondrogenesis. The majority of the autopod elements results from branching and segmentation events of the postaxial ray. This asymmetry in branching behavior may be related to chemical differences between preaxial and postaxial regions of the limb bud often shown by molecular analyses (these proceedings), but branching itself is a physical property of growing condensations. It follows from this model that evolutionary transformations of limb patterns would be based on changes in the mode and number of branching and segmentation events.

The three hypotheses briefly introduced differ in several important aspects. They not only evoke different mechanisms as being instrumental in phylogenetic transformations of limb pattern, but they also lead to different predictions about the possible forms of such transformations, which can be tested against natural forms of limb patterns. Finally, the hypotheses also differ profoundly with respect to their capacity to explain homology of individual skeletal elements.

THE HOMOLOGY PROBLEM

Homology is a concept about the transspecific identity of individual anatomical parts between organisms of the same body plan. The correspondence of skeletal elements of the vertebrate limb represents a classic example of homology and has been used successfully for the identification of limb structures. Nevertheless, interpretations are often contradictory, especially so in cases of digital reduction or in major modifications of the carpus and the tarsus. For example, paleontologists and embryologists disagree on the identity of the three digits in the bird wing (Hinchliffe and Hecht, 1984) and a variety of conflicting interpretations were proposed for the carpus and tarsus of crocodilians (Müller and Alberch, 1990). On the basis of their lack of individuality it has recently been seriously questioned whether the skeletal elements of the carpus and the tarsus can be homologized at all (Wagner, 1989). Strictly chemical models of pattern formation do lend support to the notion that there is a principal fallacy in attempting to homologize elements of the autopod section of the limb (Goodwin and Trainor, 1983). If the positional values or prepatterns of a limb field are entirely rearranged for the specification of a new pattern, then no individual correspondence of elements may actually exist. Three digits would form instead of five, but no specific ones would have been lost, and the same would be true for modifications of carpal and tarsal regions. Resemblances would be a consequence of the homodynamics of the system but would not be based on a developmental individualization of specific elements.

If these assumptions were correct, then the idea of individual correspondences in the autopod, that is homology of its skeletal elements, would have to be profoundly challenged. Therefore, it is of major importance to determine whether reductive limb modifications involve the loss of specific elements and whether the remaining ones correspond to specific elements (or a composition thereof) of the primitive condition. If it can be shown that homologies exist, then it must be concluded that developmental models of pattern formation that do not incorporate a mechanistic basis for the origin of homologous structures are insufficient to explain limb evolution.

PRIMARY MODIFICATIONS OF PATTERN

Comparative studies of the developmental sequences of chondrogenic condensations can reveal where changes in primary skeletal patterning took place. In the case of archosaurs we can compare the chondrogenic sequences of bird limbs with those of their closest living reptilian relatives, the crocodiles. As a primitive outgroup to the archosaurs chelonians can be used, which are usually regarded as direct descendants of anapsids, the most primitive reptilian subclass. Although the limb patterns of turtles show some derived features, their general condition is conservative. Since the recent representatives do not markedly differ from the oldest known chelonians they can be regarded as having retained, in general, a primitive primary pattern of chondrogenic elements. Detailed descriptions of the development of the skeletal patterns are available for chelonians (Burke and Alberch, 1985), crocodilians (Müller and Alberch, 1990), and birds (Hinchliffe, 1977). These and personal observations of embryonic and adult limb patterns in all three taxa provide the basis for the following analysis which is schematically summarized in figure 1.

The autopod elements of forelimbs and hindlimbs can be divided into four groups: proximal elements, central elements, distal elements, and digits. The proximal group consists of the ulnare (fibulare), intermedium (intermedium), and radiale (tibiale). The central group consists of the centralia, and the distal group of the distal carpals and tarsals. Within each of these groups the appearance of the postaxial elements precedes that of the preaxial ones and the more proximal groups precede the distal ones, with the exception of the centralia which often form last. The comparison reveals changes that are related to the sequences of appearance of elements.

The distal group : In the distal groups the primitive sequence of element appearance is 4 - 3 - 2 - 1 (excluding distal tarsal 5 which is not part of the sequence but forms as a separate condensation). Both in fore- and hindlimbs, the last condensation of the distal sequence to appear fails to form in crocodiles. In birds the trend is taken further. The second last condensation of the forelimb is also lost and in the hindlimb the three distal tarsals fail to separate (Fig. 1).

The central group : A similar pattern of reduction can be seen. Three centralia form in a strict postaxial to preaxial sequence in the primitive chelonian forelimb. The last one to appear, which is the most preaxial one, always fails to undergo ossification in chelonians (personal observation) and has fully disappeared in crocodilians, which only possess one centrale condensation. This condensation is possibly made up of two unseparated foci and appears very late in limb development, about one week after all other carpal condensations. This element, appearing markedly late in crocodilians, is completely lost in birds. In the hindlimb the centrale sequence is also abbreviated, but less typically because no full series forms in turtles. In crocodiles the centrale fails to segment from the intermedium, and, again, no centrale forms in birds (Fig. 1).

The proximal group : is the most constant group, both in fore- and hindlimbs. The ulnare (fibulare) always precedes the formation of the intermedium and of the radiale (tibiale). While the ulnare (fibulare) forms in all cases, the minor modifications seen occur again preaxially. The preaxial condensations, that often appear as two separate foci in reptiles, are singular in birds (Fig. 1).

The digits : The sequence of digit formation in crocodilians is IV-III-II-V-I. The two digits that form last in the forlimbs of crocodilians are the ones that fail to appear in the skeletal patterning of the bird wing (Fig.1).

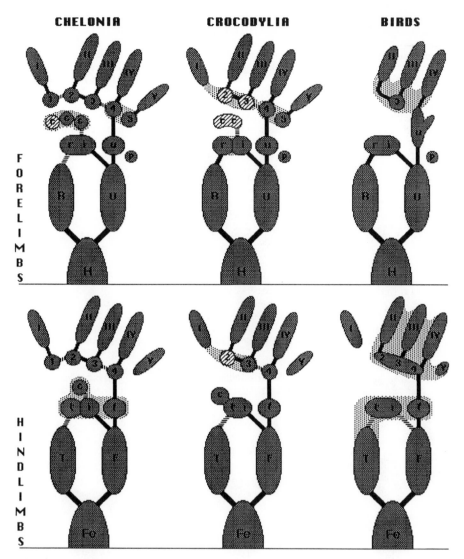

Fig.1. Primary chondrogenic patterns and their secondary modifications
in turtles, crocodiles, and birds. The initial connectivities between
chondrogenic elements are depicted according to the formalism proposed
by Shubin and Alberch (1986). Following the lines of connectivities
the temporal sequences of condensation appearances can be read from
the bottom (proximal) to the top (distal) of each diagram. Weak or
uncertain connectivities are depicted by interrupted lines. Stippled
areas indicate secondary fusions. Crosshatched elements do not ossify.
Humerus (H), radius (R), ulna (U), ulnare (u), pisiform (p), femur
(Fe), tibia (T), fibula (F), tibiale (t), fibulare (f), centralia (c),
distal carpals and tarsals (1-5), metacarpals and metatarsals (I-V).
Phalangeal elements are omitted. In the diagram of the bird wing the
disappearance of the ulnare and its replacement by the element "X" is
not shown (see Hinchliffe and Hecht, 1984). Further explanations in
the text.

The majority of the changes in the number of primary chondrogenic condensation of the carpus and the tarsus takes place in the central and distal groups. Invariably, elements that form at the end of the primitive sequences are the ones that fail to appear in the more derived limbs. These terminal deletions in chondrogenic sequences correspond to the end points of the branching and segmentation sequences described by Shubin and Alberch (1986). They showed that chondrogenic foci do not form as isolated condensations but are in most cases initially connected arrays. These initial connectivities are clearly visible in early limb buds (Fig. 2) and allow a precise identification of elements. The comparison of the patterns of connectivities shows that specific elements disappear in phylogeny and that the remaining ones can be identified as particular elements of the primitive pattern, i.e. they can be homologized. Some of the differences in primary chondrogenic patterns can be interpreted as the failure of chondrogenic arrays to segment, such as in the distal tarsal of birds or the intermedium-centrale in crocodiles (Fig. 1).

SECONDARY MODIFICATIONS OF PATTERN

The primary chondrogenic patterns of each of the three limb types undergo a number of secondary modifications. One is non-ossification. Through the failure of some of the cartilage rudiments to ossify, the adult osseous patterns appear more different than the embryonic patterns. Again, those elements that fail to ossify are terminal elements of a sequence. This can be seen in the carpus and tarsus of crocodiles (Fig. 1). Often non-ossification foreshadows the full elimination of elements, such as in the loss of distal carpal 2 and of the centralia in birds, which is preceded by the cartilaginous condition of these elements in crocodilians.

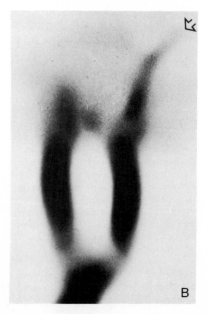

Fig.2. Connectedness of early chondrogenic condensations in the forelimb buds of *Alligator mississippiensis* . A) 18 days. Continuity of the humerus conden-sation with radius (left) and ulna (right). B) 20 days. Continuity of the postaxial ray (ulna) with postaxial elements of the carpus (ulnare, distal carpal 4) and digit IV (arrow).

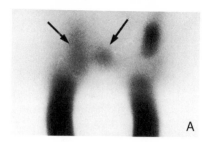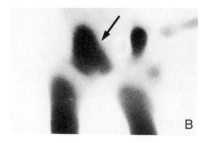

Fig.3. Fusion of two primary chondrogenic foci in the forelimb development of *Alligator mississippiensis* . A) 21 days, radiale (left arrow), intermedium (right arrow). B) 24 days, fused radiale-intermedium (arrow).

Another type of modification is realized through secondary fusions of primary elements. Fusions occur at different levels of skeletogenesis. Separate chondrogenic foci can fuse before demarcated cartilage elements have formed. This, for example, is the case in the fusion of radiale and intermedium in crocodiles (Fig. 3). Further, fusions can take place much later in ontogeny, after chondrogenic condensations have become fully demarcated cartilage rudiments. This is the case in the hindlimb of birds, where the proximal tarsals fuse to the distal tip of the tibia (Fig. 4), or in the chelonia and crocodylia with the fusion of distal carpals 4 and 5. Finally, fusions can transform the cartilaginous pattern even later, when ossification replaces the cartilage elements. In birds, for example, metatarsals two, three, and four only unite after they have existed as individual cartilage elements for most of the developmental period. The ossification process, starting at the central portion of each of the three metatarsals, bridges the space between them and results in the formation of a single tarso-metatarsus bone (Fig. 4). Secondary deletions of primary elements, that is the transitory appearance of a skeletal element and its subsequent disappearance, which play an important role in recapitulatory scenarios, never occur, with the exception of the ulnare in birds (Hinchliffe, 1977; Hinchliffe and Hecht, 1984).

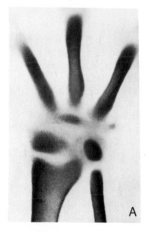

Fig.4. Fusions in the hindlimb of birds. Note the stepwise transformation of the primary pattern (A, 6 days) through cartilaginous fusion of the proximal tarsalia to the tibia and the distal tarsal to the metatarsals (B, 11 days), and through common ossification of the metatarsals (C, adult).

These secondary modifications add as much to the class specific differences between skeletal patterns as do the primary modifications. Fusions often occur along the lines of initial connectivity and can be interpreted as supporting the branching and segmentation hypothesis. They often confirm connectivities that are weak or that cannot be visualized through cartilage matrix staining. This can be seen in the connection of the terminal digital arch elements to the base of digit 1, in both the fore- and hindlimbs of crocodilians (Fig. 1). Similarly, the fusions between distal carpals 4 and 5 in chelonians and crocodilians suggest a developmental connectivity there, which is otherwise not seen.

HETEROCHRONY AND SECONDARY FUSION

The patterns of terminal deletions in chondrogenic sequences and the non-ossification of terminal condensations demonstrate that a truncation of skeletogenic processes, both at the level of chondrogenesis and of osteogenesis, underlies the phylogenetic modification of skeletal patterning in archosaur limbs. This mechanism can account for the basic reductions of carpal and tarsal elements and of the digits. Elements that form late in primitive limbs are the ones most easily lost in the derived forms. These truncations and the processes of non-segmentation of chondrogenic arrays suggest that a primary mechanism of limb transformation is paedomorphic heterochrony. Heterochrony, the evolutionary modification of rates and timing of developmental processes, thus may represent a major factor in vertebrate limb evolution.

The heterochrony hypothesis receives empirical support both from comparative studies of vertebrate limb patterns and from experimental perturbations of limb development. The studies of Greer (1987) on limb reduction in scincid lizards show a sequence of phylogenetic loss of digits (I-V-II-III-IV), which is an exact reversal of their sequence of ontogenetic appearance in crocodilians. This reversal of developmental sequences in phylogenetic reduction series is most clearly seen when limb reduction affects generalized limbs. When reductive trends set in after specializations have occurred the patterns of digital loss can be different. In reductions of generalized limbs, however, in most cases digit IV remains last. This is the digit that appears first in all tetrapod classes as a part of the primary chondrogenic axis (Shubin and Alberch, 1986) (see Fig. 2).

The heterochrony hypothesis also finds support in experimental analyses of limb development. Treatments with mitotic inhibitors of amphibians (Alberch and Gale, 1983; Bretscher and Tschumi, 1951), of lizards (Raynaud, 1985), of crocodilians (Müller, unpublished), and of mammals (Rooze, 1980) demonstrate that with increasing dosage invariably the first digits to disappear are those that form last in the developmental sequences, while the last remaining digit, if identifiable, was in most cases digit IV, the first one to form in development. Thus, phylogenetic sequences of digital loss can be mimiced to a good extent by artificial arrests of the chondrogenic process.

It is tempting to speculate that heterochrony in archosaur limb evolution is related to the general shortening of the developmental period which took place in sauropsid evolution (Fig. 5). The average chelonian incubation time of 99 days (maxima around 180 days) is about 4 times longer than in birds (average 25 days, mini-

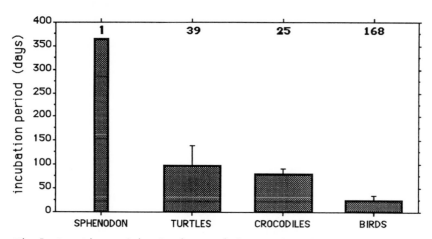

Fig.5. Durations of incubation period. Data based on Needham (1931)
(sphenodon), Ewert (1985) (turtles), Ferguson (1985) (crocodiles),
and Heinroth (1922) (birds). Numbers of species included are shown
above the columns.

mum 10 days). The time taken for the full sequence of chondrogenic
elements to appear in crocodilian tarsus is roughly three weeks
while it is completed within five days in birds. This speeding up
of development, however, cannot represent the only factor for
truncations in limb development, since many reptiles show limb
reduction while their sister taxa with similar breeding times have
none. But the shortening of breeding time as an overall trend may
facilitate shortcuts in skeletogenesis that underlie the radical
reduction of primary elements in birds.

The second major factor in phylogenetic limb transformation
must be seen in the secondary modifications of the primary chondrogenic patterns. As demonstrated above, a substantial number of
fusions take place at different times and levels of the skeletogenic process. Early chondrogenic foci, cartilage rudiments, and
ossifying elements become combined through different developmental
mechanisms. It is conceivable that precisely this switching of
mechanisms in the hierarchy of developmental processes represents a
general principle for the phylogenetic transformation of organismic
structures and for the generation of morphological novelty in evolution (Müller, 1990). With respect to the present theme, however,
it must be emphasized that not all class specific features of the
skeletal limb patterns are already expressed in the primary chondrogenic pattern. Substantial secondary modifications do take place
during individual development.

In addition, it is noteworthy that matching patterns between
forelimbs and hindlimbs are seen in primary and secondary modifications, suggesting a common internal constraint. In crocodilians
this is the case with the loss of distal carpal 1 and distal tarsal
1, as well as with the cartilaginous condition of distal carpal 2
and distal tarsal 2. In birds the absence of centralia in the wing
and leg, the fusion tendency in the metacarpus and metatarsus, and
the reduction of digits I and V illustrate these biases of transformation shared by fore- and hindlimbs.

CONCLUSIONS

The comparative analysis of skeletogenic sequences shows that evolutionary transformations of limb patterns are realized developmentally at two distinct levels, involving fundamentally different mechanisms. At a first level primary pattern formation is altered through heterochrony in terms of paedomorphic truncations of skeletogenic processes. At a second level several kinds of fusion processes modify the primary patterns. It has been proposed that the phylogenetic transformations resulting from first level changes are largely determined by internal constraints of the skeletogenic system itself, while second level transformations are more under the influence of adaptive factors (Müller, 1991b).

The hierarchical character of skeletal reduction in limb phylogeny reflects the hierarchy of processes effective in development. It is possible to identify the elements that are lost at each step of reduction and the remaining ones can be individually identified in most cases. It must be concluded that specific elements are lost in phylogenetic transformations, indicating that a trans-specific identity (homology) can be assigned to the majority of skeletal elements of the tetrapod limb.

With regard to the concepts of pattern transformation discussed in the introduction it seems as if elements of each concept are pertinent and can eventually become assembled to form an adequate developmental concept of phylogenetic limb transformation. While it is clear that no full recapitulations of archetype patterns take place, it must be emphasized that secondary modifications of the primary patterns, which represent the mechanistic essence of the recapitulation concept, do play an important role in limb evolution. Diffusable morphogens have an influence on the symmetries of limb patterning and could thus underlie some of the changes of primary patterns, but the chemical model alone cannot account for the patterns of secondary change, nor can it explain homology of limb elements. The patterns of heterochrony and secondary fusion observed above conform best with the physical model of branching and segmentation in limb skeletogenesis. A heterochrony and fusion mechanism based on branching and segmentation processes is the only model that allows to be predictive about evolutionary transformations of limb patterns and these predictions can be empirically tested. This model also provides a mechanistic basis for the homology of limb elements.

REFERENCES

Alberch, P., and Gale, E. A., 1983, Size dependence during the development of the amphibian foot. Colchicine-induced digital loss and reduction, J. Embryol. exp. Morph., 76:177.
Bretscher, A., and Tschumi, P., 1951, Gestufte Reduktion von chemisch behandelten *Xenopus* -Beinen, Revue Suisse de Zoologie, 58:391.
Burke, A. C., and Alberch, P., 1985, The development and homology of the chelonian carpus and tarsus, J. Morphol., 186:119.
Ewert, M. A., 1985, Embryology of turtles, in: "Biology of the Reptilia, Vol. 14," C. Gans, F. Billett, and P. F. A. Maderson, eds., John Wiley & Sons, New York.
Ferguson, M. W. J., 1985, Reproductive biology and embryology of the crocodilians, in: "Biology of the Reptilia, Vol. 14," C. Gans, F. Billett, and P. F. A. Maderson, eds., John Wiley & Sons, New York.
Goodwin, B. C., and Trainor, L., 1983, The ontogeny and phylogeny of the pentadactyl limb, in: "Development and Evolution", B. C. Goodwin,

N. Holder, and C. C. Wylie, eds., Cambridge University Press, Cambridge.

Greer, A. E., 1987, Limb reduction in the lizard genus *Lerista*. 1. Variation in the number of phalanges and presacral vertebrae, J. Herpetol., 21:267.

Heinroth, O., 1922, Die Beziehung zwischen Vogelgewicht, Eigewicht, Gelegegewicht und Brutdauer, Journal für Ornithologie, 70:172.

Hinchliffe, J. R., 1977, The chondrogenic pattern in chick limb morphogenesis: a problem of development and evolution, in: "Vertebrate Limb and Somite Morphogenesis," D. A. Ede, J. R. Hinchliffe, and M. Balls, eds., Cambridge University Press, Cambridge.

Hinchliffe, J. R., and Hecht, M. K., 1984, Homology of the bird wing skeleton, in: "Evolutionary Biology," M. K. Hecht, B. Wallace, and G. T. Prance, eds., Plenum Publishing Corporation, New York.

Holmgren, N., 1933, On the origin of the tetrapod limb, Acta Zoologica, 14:185.

Jorquera, B., and Pugin, E., 1971, Sur le comportement du mésoderme et de l'ectoderme du bourgeon de membre dans les échanges entre le poulet et le rat, C.R. Acad. Sc. Paris, 272:1522.

Müller, G. B., 1990, Developmental mechanisms at the origin of morphological novelty: a side-effect hypothesis, in: "Evolu-tionary Innovations," M. H. Nitecki, ed., The University of Chicago Press, Chicago.

Müller, G. B., 1991a, Experimental strategies in evolutionary embryology, Amer. Zool., in press.

Müller, G. B., 1991b, Developmental processes and phenotypic change in archosaur limb evolution, in: "The Unity of Evolutionary Biology," E. C. Dudley, ed., Dioscorides Press, Washington, in press.

Müller, G. B., and Alberch, P., 1990, Ontogeny of the limb skeleton in *Alligator mississippiensis* : developmental invariance and change in the evolution of archosaur limbs, J. Morphol., 203:151.

Needham, J., 1931, "Chemical Embryology," Cambridge University Press, Cambridge.

Oster, G. F., Murray, J. D., and Maini, P. K., 1985, A model for chondrogenic condensations in the developing limb: The role of extracellular matrix and cell tractions, J. Embryol. exp. Morph., 89:93.

Raynaud, A., 1985, Development of limbs and embryonic limb reduction, in: "Biology of the Reptilia, Vol. 15," C. Gans, and F. Billett, eds., John Wiley & Sons, New York.

Rooze, M. A., 1980, The effect of cytosine-arabinoside on limb morphogenesis in the mouse, in: "Teratology of the Limbs," H. J. Merker, H. Nau, and D. Neubert, eds., Walter de Gruyter & Co., Berlin.

Shubin, N. H., and Alberch, P., 1986, A morphogenetic approach to the origin and basic organization of the tetrapod limb, in: "Evolutionary Biology," M. K. Hecht, B. Wallace, and G. T. Prance, eds., Plenum Publishing Corporation, New York.

Steiner, H., 1934, Über die embryonale Hand- und Fuss-Skelett-Anlage bei den Crocodiliern, sowie über ihre Beziehungen zur Vogel-Flügelanlage und zur ursprünglichen Tetrapoden-Extremität, Rev. Suisse de Zool., 41:383.

Summerbell, D., Lewis, J. H., and Wolpert, L., 1973, Positional information in chick limb morphogenesis, Nature, 244:492.

Wagner, G. P., 1989, The origin of morphological characters and the biological basis of homology, Evolution, 43:1157-1171.

Wilby, O. K., and Ede, D. A., 1975, A model generating the pattern of cartilage skeletal elements in the embryonic chick limb, J. theoret. Biol., 52:47.

PLASTICITY IN SKELETAL DEVELOPMENT:

KNEE-JOINT MORPHOLOGY IN FIBULA-DEFICIENT CHICK EMBRYOS

Johannes Streicher

Department of Anatomy
University of Vienna
Vienna, Austria

INTRODUCTION

Most of the arguments for an intrinsic determination of skeletal morphology are inferred from results of chorio-allantoic grafts of limb segments (Fell and Canti 1934), from in vitro cultures of isolated skeletal elements (Carpenter 1950, Hicks 1982), and from amputations of whole distal segments at the joint level (Holder 1977).These approaches, however, do not unequivocally demonstrate an entirely intrinsic determination of the shape of skeletal elements, since most of the results reported were derived from developmental stages prior to the attainment of the final morphology.

This study explores the role of local interactions for the morphogenesis of individual skeletal elements. We experimentally eliminated one of the participating elements of the chick knee-joint by suppressing the development of the fibula through an early blastema reduction. Subsequently we studied the morphogenesis of the remaining elements in their in vivo position from their first appearance up to hatching. If the loss of the fibula does affect the individual shape of the related elements, developmental plasticity of skeletal morphogenesis due to the influence of local interactions can be inferred.

From an evolutionary point of view, the complete disappearance of the bird fibula represents a plausible continuation of the reductive trend from slim fibulae in dinosaurs to distally reduced ones in birds (Müller and Streicher 1989). Thus, this study also has a bearing on developmental plasticity as a potential mechanism for integrating evolutionary changes of single characters into an organ system.

METHODS AND MATERIALS

All experimental investigations were carried out on embryos of the domestic fowl (*gallus domesticus*). On the fourth day of incubation, at stages 21-23 of Hamburger & Hamilton (1951), eggs were windowed above the located embryo, and the right hindlimb bud was exposed. Subsequently, a central segment of the limb anlage was excised, thereby reducing the limb blastema to about one third of its original mass. The remaining distal part was fixed to the stump by a staple made of platinum wire. Following operation the eggs were resealed and further incubated up to stages between the sixth day of development and hatching. From a total of 96 operations 81 specimens were obtained for further investigation, the contralateral limbs serving as controls. Twentyeight sham operated limbs, where no blastema was removed, showed no skeletal alterations.

Hindlimb pairs of 62 specimens were cleared in glycerol after staining in toto with Alcian blue 8GX and Alizarin red S for cartilage and bone respectively. Nineteen other specimens were embedded in paraplast and histologically crossectioned at 5 µm thickness in 0.1 mm intervals. Mallory triple stain was used for differential staining of cartilaginous and osseous structures.

RESULTS

In 74.1% of the experimental hindlimbs the blastema reduction resulted in complete deficiency of the fibula. As this effect was observed both in early and in late stages of development, a transitional or delayed appearance of the fibula can be ruled out. Other elements of the limb skeleton were never found deficient, but the morphogenesis of several features related to the presence of the fibula was impaired. Consequently the morphology of the entire knee-joint was altered.

In normal limbs each of the femoral condyles is exclusively opposed to one of the zeugopodal elements until the seventh day of development. Subsequently, the lateral femoral condyle progressively shifts above the lateral part of the head of the tibia, while its outer portion remains in contact with the fibula and becomes transformed into a trochlea (Fig. 1a). The contact area between the medial femoral condyle and the tibio-tarsus becomes restricted to the medial two thirds of the tibial head.

In experimental limbs the distal extremity of the femur, with respect to size and shape, appears identical to the normal conditions up to the eighth day of development. The lateral femoral condyle, however, does not form in its normal position laterally to the tibial head, but both condyles are already initially located opposite to the tibia. In

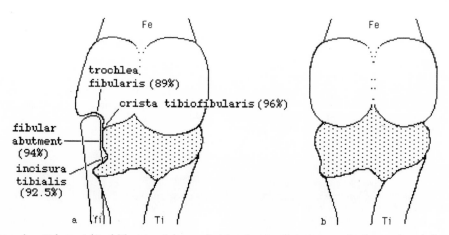

Fig. 1. Schematic illustration of the knee-joint morphology in (a) normal and (b) fibula-deficient right hindlimbs. In brackets: frequencies of nonappearance in experimental limbs. Fe=femur, Ti=tibia, fi=fibula.

contrast to normal conditions the lateral condyle retains its homogeneously curved contour. The partition into a medial (tibial) and a lateral (fibular) facet by the crista tibiofibularis as well as the subsequent formation of the trochlea does not occur. Until the twelfth day initial differences in size of the femoral condyles have disappeared and they both reside symmetrically above the tibial head (Fig.1b). The symmetry of the femoral condyles is matched by the articular surfaces of the tibia. On the tibia the incisura tibialis is much shallower than in normal tibiae and the postero-lateral part of the tibial head, normally serving as an abutment for the fibular head, does not flatten in experimental limbs (Fig.1b). In summary the final morphology of the knee-joint in fibula deficient limbs is characterized by the nonappearance of fibula related structures and the attainment of a symmetric shape.

DISCUSSION

This study demonstrates that the experimental elimination of the fibula profoundly affects the morphogenesis of the knee-joint. The remaining elements are not unspecifically impaired but exhibit a matching series of distinct modifications. This set of alterations comprises the nonappearance of fibula related structures as well as regulative modifications, such as the symmetric design of the femoro-tibial junction. Therefore the features of the fibula-deficient knee-joint do not result from an abortion of developmental programs but reflect the modification of morphogenetic pathways by altered local interactions. It is concluded, that in skeletal morphogenesis only the basic traits of individual elements, such as the bicondylar configuration of the distal end of the femur, are intrinsically controlled. Many morphological details, however, depend on local interactions during later stages of ontogeny. This developmental plasticity guarantees the fine tuning of functionally related characters.

In the evolutionary context outlined in the introduction, our results suggest that a phylogenetic loss of the fibula could initiate regulatory morphogenetic interactions that compensate the absence of an element and lead to the formation of a knee-joint composed of two bones only. Thus it is emphasized that the plasticity provided by the developmental coupling of related structures also represents an essential integrative factor in morphological evolution.

LITERATURE

Carpenter, E., 1950, Growth and form in vitro of avian femur rudiments on clots in which dried products are substituted for fresh fowl plasma, J exp Zool 113:301-317.
Fell, H. B. and Canti, R. G. 1934, Experiments on the development in vitro of the avian knee-joint. Proc R Soc B 116, 316-351.
Hamburger, V. and Hamilton, H. L., 1951, A series of normal stages in the development of the chick embryo, J Morphol 88:49-92.
Hicks, R, 1983; reported in: J. R. Hinchliffe and Johnson D. R., Growth of Cartilage", in: "Cartilage", Vol. 2, B. K. Hall, ed. Academic Press.
Holder, N., 1977, An experimental investigation into the early development of the chick elbow-joint, J Embryol exp Morph 39:115-27.
Müller, G. and Streicher, J. 1989, Ontogeny of the syndesmosis tibiofibularis and the evolution of the bird hindlimb: a caenogenetic feature triggers phenotypic novelty, Anat Embryol 179:327-339.

The Implications of "The Bauplan" for Development and Evolution of the

Tetrapod Limb

Neil H. Shubin

Department of Biology, Goddard Laboratories of Biology
University of Pennsylvania, Philadelphia, PA 19104

INTRODUCTION

We are witness to a 400 million-year-natural experiment that reveals the consequences of genetic and environmental changes on the process of limb development. In spite of genetic and environmental evolution, the pattern of the tetrapod limb has remained effectively similar throughout its history. This conserved pattern of morphological organization, or "bauplan", underlies the evolutionary diversity of tetrapods.

Conceptions of archetypes and "bauplane" have a long history in comparative biology. The observation of a "unity of type" was used to support both pre-evolutionary and evolutionary approaches to biological diversity (see Russell, 1916). For pre-Darwinian biologists, "bauplane" were evidence of design in nature, reflecting the plan of the creator (e.g., Owen, 1849). Evolutionists used the observed pattern of a "bauplan" as evidence of common descent of life. Similarly, Haeckel and others used this concept as the basis for a theory of recapitulation. For all these anatomists the morphological observations were similar-- conserved anatomical features in the face of organismic diversity.

Here, I evaluate: 1) the evidence for a "bauplan" of the tetrapod limb, 2) the developmental perturbations of this "bauplan" that are seen during the evolutionary history of tetrapods, and 3) the implications of these patterns for understanding the processes of both development and morphological evolution. I argue that when the concept of the "bauplan" is freed from its recapitulationist underpinnings it can link paleontology, phylogenetics, and developmental biology.

EVIDENCE FOR A BAUPLAN OF THE TETRAPOD LIMB

Comparative anatomists have long noted that several features of tetrapod limbs are seen in the fins of lobed-finned fish. The similarity of the adult structure of the tetrapod limb to the fins of osteolepiform fish (such as Sauripterus and Eusthenopteron) led to hypotheses of homology between fin and limb elements (Fig. 1). These hypotheses generally compared the adult pattern of fish fins to the embryonic pattern of tetrapod limbs (e.g. Holmgren, 1933,1939,1942,1949; Jarvik, 1965,1980). The two figured schemes reveal the common observations and inferences that characterize most attempts to compare the skeletal elements of fins and limbs: 1) particular elements of the tetrapod limb were homologized with either the "axis" or the radials of sarcopterygian fins; 2) this "axis" was hypothesized to include proximal limb elements and to extend through or between the digits; 3) the adult pattern of sarcopterygian fins was hypothesized to have made a transient appearance during the normal development of tetrapod limbs.

Recent morphological studies do not support the conclusions of these early attempts to compare fins and limbs. Developing limbs observed with S04 autoradiography and Alcian Blue (Hinchliffe, 1977; Hinchliffe and Griffiths, 1983; Shubin and Alberch, 1986) do not show the adult ancestral pattern at any stage of limb development. Furthermore, many species-specific patterns of variability are expressed early in limb development (Hinchliffe, 1977; Hinchliffe and Griffiths, 1983; Shubin and Alberch, 1986; Hinchliffe, 1989a,b; Muller and Alberch, 1989).

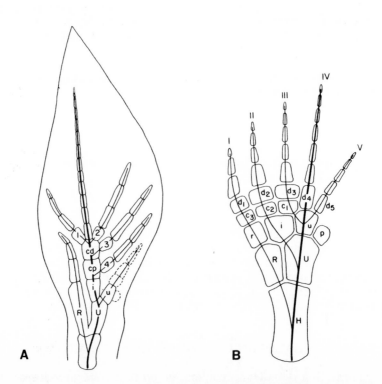

Figure 1. Two hypotheses of the homology of the axis and radials of the sarcopterygian fin with tetrapod limb elements. A) Holmgren's (1933) hypothesis of the homology of the urodele limb with lungfish fins, B) Watson's (1914) comparison of tetrapod limbs with the fins of osteolepiform fish. The bold line represents the homologue of the axis, lighter lines represent homologues of the radials of fins.

Whereas ancestral stages do not make a transient appearance during tetrapod limb development, the processes by which pre-cartilagenous cells condense and become skeletal elements is highly conserved. The initial condensation of pre-cartilagenous cells occurs before cartilage matrix products are secreted.   Limb pre-cartilage mesenchyme is characterized by its ability to branch (Hinchliffe, 1977) or segment new cartilage rudiments from existing ones (Oster et al., 1983,1985; Shubin and Alberch, 1986).   As they develop, pre-cartilage condensations generally are connected to one another and later cleave during the processes of joint formation (Sledge, 1981).

The patterns of embryonic connectivity reveal both invariant and labile patterns of limb chondrogenesis. Patterns of branching and segmentation are iterative events and delineate morphogenetically similar regions of developing limbs (Shubin and Alberch, 1986). The main invariant patterns of connectivity consist primarily of (Fig. 2):

Initial Branching Event

Development is initiated by a single proximal condensation, the humerus/femur. The humerus/ femur branches to produce the radius/ulna and the tibia/fibula.

Post-axial Dominance

Gregory (1915) observed a convergence of all the digits and associated distal carpals/tarsals on the ulna/fibula. Gregory's proposal of a "dominance" of the postaxial region of adult tetrapod limbs has a developmental basis (Shubin and Alberch,1986; Gardiner and Bryant, 1989). After the initial branching event, subsequent chondrogenesis is asymmetric. The pre-axial (radial/tibial) side generally does not branch; the postaxial (ulnar/fibular) side may branch to produce the ulnare/ fibulare and the intermedium. In amniotes, the ulnar and fibular sides are connected to the condensations of the digital arch (see below). In these taxa, the majority of carpal and tarsal elements arise from condensations whose pre-cartilage connections can be traced to postaxial region of the limb, hence the term, "post-axial dominance".

Primary axis

In amniotes, the ulnare, distal carpal/tarsal 4 and the metacarpal 4 develop more rapidly than elements at corresponding proximo-distal positions of the limb. This "primary axis" (Burke and Alberch, 1985; Shubin and Alberch, 1986) is present in amniotes but has not been observed in the urodele species examined to date.

Digital arch

Most of the distal branching events occur within a broad arch of branching mesenchyme, the digital arch.   The number of elements that arise within the digital arch varies widely but includes a variable number of distal carpals/tarsals and metacarpals/metatarsals. The amniote digital arch arises from condensations that originally branch from the postaxial proximal carpals/tarsals (the ulnare/fibulare). The urodele digital arch is apparently a separate field of condensed cells in most species.

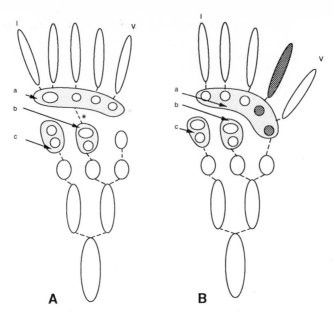

Figure 2. The major features of branching and segmentation of pre-cartilage condensations are depicted for A) urodeles and B) amniotes. The dotted lined depict embryonic connections between condensations. The number of elements within each series of condensation (stipple) may vary. Abbreviations: a= the digital arch, b= the intermedium/centrale series , and c= the pre-axial series of elements. The deeply stippled region in (B) depicts the primary axis. The asterisk represents a connection between the intermedium/centrale series and the digital arch that is seen in some urodeles.

<u>Digital development</u>

In spite of significant taxonomic variation of the number of phalanges, there is a profound uniformity in their morphogenetic origins. Most metacarpals/metatarsals arise from the digital arch whereas the phalanges arise by segmenting from more proximal digital elements.

This developmental "bauplan" suggests neither von Baerian nor Haekelian modes of recapitulation (for the different implications of the two types of recapitulation, see Ospovat, 1976 and Gould, 1977). Tetrapod limbs share common processes of chondrogenesis, not common "stages" of limb formation. Comparative studies of limb development suggest that even the earliest stages of limb development are labile; earlier branching and segmentation events are no more conservative than later ones. For example, the intermedium is an element that forms relatively early in development. This element is wholly absent in many frogs because the ulna segments a single element rather than bifurcating two elements. Both recapitulationist schemes would necessitate that the intermedium make a transient appearance during frog development only to be resorbed or lost in later stages.

These developmental similarities have provided a general framework for the analysis of the developmental transformation of sarcopterygian fins into tetrapod limbs (Shubin and Alberch, 1986). Developmental data have enabled the comparison of regions of the limb that develop by similar chondrogenetic processes. The processes of branching that characterize the axis of fish fins are homologous with those of the tetrapod digital arch. Recent discoveries of early tetrapods have revealed an early evolutionary diversity of the limb pattern and the number of digits. The pattern of the limb and the location of the digits is readily understood by the analysis tetrapod limb development: no new chondrogenetic processes need be proposed (Coates and Clack, 1990).

EVOLUTIONARY AND DEVELOPMENTAL VARIABILITY OF THE BAUPLAN

Variable features of the developmental "bauplan" characterize major evolutionary divisions of tetrapods. The evolutionary variability of the tetrapod limb results from a catalogue of processes that alter the "bauplan". These changes occur at any stage of development. The "constant" in tetrapod limb evolution is the process by which the pattern is built-- assymetric patterns of branching and segmented mesenchyme. These variable features of limb development make rigid hypotheses of evolutionary homology difficult (Goodwin and Trainor, 1983; Shubin and Alberch, 1986). In the cases where these patterns are conserved, however, they enable hypotheses of homology of specific limb elements. The existence of a "bauplan", and a small number of process that may perturb it, makes the existence of evolutionary parallelism and convergence highly likely. This, in fact, is the case and the parallelism that is observed is generally predictable from the "bauplan" itself (Shubin and Alberch, 1986; Oster et al, 1989; Muller and Alberch, 1989). The variable features of limb chondrogenesis include:

Sequence

The sequence of limb chondrogenesis is phylogenetically variable. Almost all tetrapods show a general proximal to distal sequence of limb chondrogenesis. Some urodele amphibians (Ambystoma, Triturus) violate this regularity in that several distal elements (Digits I,II, and the basale commune) appear precociously. Likewise, the primary axis includes several distal carpal/tarsal elements that develop more rapidly than other elements at a similar distal position. There is significant variability in the antero-posterior sequence of limb formation (Shubin and Alberch, 1986). In urodeles the digital arch develops in a general anterior to posterior sequence whereas in anurans and amniotes the reverse is true. The sequence of digital development is highly variable in urodeles. Some taxa exhibit a general antero-posterior sequence of digital formation; in others, the digits appear more simultaneously (Shubin, unpublished data).

To the extent that there has been a change in the developmental sequence there is little or no consequent alteration of the adult pattern. Hence, the actual sequence of development may not be an important factor in the specification of the pattern of the tetrapod limb.

Position of the Joints

Continuous mesenchymal condensations cleave early in limb development. These patterns of cleavage are variable in many taxa (Hinchliffe, 1989b) even though the earlier patterns of connectivity are invariant. The digital arch is a common feature of tetrapod limb development; yet, the patterns of its cleavage are highly taxon-specific. The chelonian digital arch divides into 4 distal tarsal elements whereas that of the chick only yields a single element. Similarly, the number of joints that appear in the intermedium-centrale and the pre-axial axes varies phylogenetically. There is evidence that the positions of these joints are determined prior to their actual morphological differentiation (Holder, 1977).

Number of Limb Elements

The number of skeletal elements in the limb is highly variable both within and among species. It is striking that 400 million years of limb evolution has only rarely seen an increase in the number of elements in the limb. Reduction in the number of elements has occurred in many different lineages and at many different times of tetrapod history (Alberch and Gale, 1985).

Reduction of the number of limb elements results from (Muller, 1990): 1) failure of continuous condensations to cleave during embryogenesis (as described above); 2) a switch from a branching event to a segmentation event; 3) failure of elements to ossify; 4) fusion of separate elements during either chondrogenesis or osteogenesis. The sequence of limb chondrogenesis is an excellent predictor of the skeletal elements that are lost during evolutionary reduction of the limb (Muller, 1990). The last elements to be formed are usually the first to be lost both in experimental and evolutionary perturbation.

Size and Shape of Elements

The size and shape of elements may vary widely in development. Frequently the differences in size and shape are expressed extremely early; during the initial phases of condensation. The anuran fibulare is a particular example of this early specification. Even in the earliest stages of condensation, the anuran fibulare is elongate and grows rapidly relative to the fibulare of most other tetrapods.

THE PARADOX: DEGENERACY AND CONSTRAINT

There are many instances where shared morphogenetic mechanisms have served to constrain morphological evolution (Maynard-Smith et al., 1985). It is paradoxical, however, that there are also a significant number of cases where similar adult patterns are produced by different morphogenetic processes. These cases appear to suggest a constraint on final form but not one on the developmental mechanisms themselves. This "degeneracy" of the developmental process (sensu, Edelman, 1988) requires a lability in the developmental process and a conservation of adult form. For example, the pronephric duct and the neural tube are conserved features of tetrapod evolution. Despite the evolutionary stability of these structures, the patterns of pronephric duct and neural tube formation are known to vary between species

(DeBeer, 1949; Poole and Steinberg, 1984). This apparent paradox between developmental degeneracy and constraint can be re-addressed by both experimental and comparative study of limb development.

One putative example of degeneracy is the amphibian tibiale and its putative homologue in turtles. In all tetrapods the element that lies distal to the tibia is generally hypothesized to be homologous to the amphibian tibiale. In amphibians the tibiale develops from mesenchyme that segments from the tibia. In adult turtles, however, the element that lies distal to the tibia has a different developmental history. This element arises as a branch of the intermedium-- the tibia does not segment any pre-cartilage. In turtles, the element that lies distal to the tibia is developmentally part of the intermedium/centrale axis-- not part of the tibia/tibiale axis. In a developmental sense, this element would be a homologue of the amphibian centrale-- not the tibiale.

This apparent case of "degeneracy" of the developmental process is illustrative of the resiliency of developmental mechanisms in the face of evolutionary change. A similar adult pattern is seen in turtles and amphibians: a single, enlarged element lies distal to the tibia. Developmentally, both amphibians and turtles possess an intermedium that branches or segments central elements. In some primitive taxa the intermedium segments a single central element, in others, extra centralia branch from the intermedium or from other centralia. The pattern of connections of the developing limbs of amphibians and turtles has remained constant but the <u>number</u> (the tibia of turtles no longer segments a single element) and the <u>location</u> (the centrale of turtles sits distal to the tibia, not the intermedium as in amphibians) of the branching and segmentation events has changed. This apparent degeneracy does not result from new mechanisms but from the elaboration of existing patterns of development-- it occurs within the framework of the "bauplan".

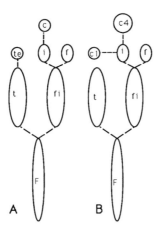

Figure 3. The proximal tarsus is depicted for A) urodeles and B) turtles. The element that sits distal to the tibia is developmentally part of the tibiale series in urodeles and is part of the intermedium/centrale series in turtles. The dotted lined connect embryonic connections. Abbreviations: c, centralia; F, femur; f, fibulare; fi, fibula; i. intermedium; t, tibia; te, tibiale.

IMPLICATIONS OF THE BAUPLAN FOR UNDERSTANDING MECHANISMS OF LIMB DEVELOPMENT

What does the evolutionary "bauplan" suggest about developmental mechanisms? The following evolutionary observations are germane:

1) Mesenchyme shows the ability to segment and branch in both tetrapod limbs and sarcopterygian fins. The patterns of branching are asymmetric, being localized on the posterior/post-axial side of the appendage.

2) The patterns of branching and segmentation between species are highly conserved and allow the recognition of a "bauplan" of the limb.

3) Variation between species may be encountered at all stages of limb development.

4) Condensations acquire a taxon-specific identity (characteristic sizes, shapes and growth rates) very early in limb development.

5) The breakup of the continuous condensations (such as the digital arch) is variable between species and the joints that demarcate individual elements may be specified very early in development.

The ability of mesenchymal cells to form spaced condensations (Ede, 1982,), to branch, and to segment may be based on mechanochemical interactions (Oster et al. 1985). These mechanisms are sensitive to a small set of developmental parameters such as size and shape of the developing limb and the mechanical properties of the extracellular matrix. There appear to be another set of cues, those that specify the identity and position of specific condensations and joints (Hinchliffe, 1989). Comparative and experimental evidence suggests that these cues act very early in development (Wolpert and Hornbruch, 1990).

The interaction of mechanochemical and "identity" cues may account for the patterns of stability and diversity seen in tetrapod limb evolution (Hinchliffe, 1989b). The mechanochemical constraints on mesenchymal condensation are properties of mesenchymal cells regardless of the specific taxon in which they are found. Positional cues may specify the identity and position of skeletal elements and joints (Wolpert, 1978; Bryant et al., 1987; Hinchliffe, 1989b). These identity cues are, in part, taxon-specific. Developmental similarities between taxa can arise from the common properties of mesenchymal cell condensations; many of the differences between taxa may relate alteration of the identity cues and from the perturbation of the physical parameters of cellular condensation.

CONCLUSIONS

Diversity and the <u>lack of it</u> are data for both developmental and evolutionary biology. It is striking that over 400 million years of limb evolution have resulted in only a small catalogue of limb types. The basic processes of cellular condensation have remained intact-- no new morphogenetic processes have arisen since at least the Devonian period. The existence of parallelism, convergence, and developmental degeneracy in limb evolution would be expected given the stability of process seen in the evolution of tetrapod limb development.

ACKNOWLEDGMENTS

I would like to thank R. Hinchliffe for giving me the opportunity to present this paper.   K. Padian and D. Hauser made helpful suggestions on earlier drafts of the manuscript.   This work was supported in part by a grant from the Research Foundation of the University of Pennsylvania.

REFERENCES

Alberch, P., and E. Gale, 1985. A developmental analysis of an evolutionary trend.   Digital reduction in amphibians. Evolution, 39: 8-23.

Bryant, S. V., D. M. Gardiner and K. Muneoka. 1987 Limb Development and Regeneration. American Zoologist 27:675-696.

Burke, A.C., and P. Alberch, 1985. The development and homologies of the chelonian carpus and tarsus. Journal of Morphology, 186: 119-131 .

Coates, M.I. and J.A. Clack, 1990. Polydactyly in the earliest known tetrapod limbs. Nature 347:66-68.

DeBeer, G. 1958. Embryos and Ancestors. Oxford University Press: Oxford.

Ede, D.A., 1982. Levels of complexity in limb-mesoderm cell culture systems, In M.M. Yeoman and D. Truman (Eds.) Differentiation in vitro, 207-229. Cambridge University Press: Cambridge.

Edelman, G. M. 1988. Topobiology: An Introduction to Molecular Embryology. Basic Books: New York.

Gardiner, D.M. and S. V. Bryant, 1989. Organization of positional information in the axolotl limb, J. Exp.Zool. 251:47-55.

Goodwin, B.C., and C. Trainor, 1983. The ontogeny and phylogeny of the pentadactyl limb, In B. Goodwin, N. Holder, and Wylie, C. (Eds.) Development and Evolution, 75-98. Cambridge University Press: Cambridge.

Gould, S.J., 1977. Ontogeny and Phylogeny. Harvard University Press: Cambridge, USA.

Gregory, W.K. 1915. Present status of the problem of the origin of the Tetrapoda with special reference to the skull and paired limbs. Annals of the New York Academy of Sciences, 26:317-383.

Hinchliffe, J.R., 1977. The chondrogenic pattern in chick limb morphogenesis: a problem of development and evolution. In D. Ede, J. R. Hinchliffe and M. Balls (Eds.) Vertebrate Limb and Somite Morphogenesis , 293-309. Cambridge: Cambridge University Press.

Hinchliffe, J.R., 1989a. An evolutionary perspective of the developmental mechanisms underlying the patterning of the limb skeleton in birds and other tetrapods. In J. Chaline (Ed.) Ontogenese et Evolution. 119-131. Geobios: Dijon.

Hinchliffe, J.R., 1989b. Reconstructing the archetype: Innovation and conservatism in the evolution and development of the pentadactyl limb. In D. Wake and G. Roth (Eds.) Complex Organismal Functions: Integration and Evolution in Vertebrates, 171-188. John Wiley and Sons Ltd: London.

Hinchliffe, J.R., and P. Griffiths., 1983. The pre-chondrogenic patterns in tetrapod limb develoment and their phylogenetic significance. In B. Goodwin, N. Holder and C. Wylie (Eds.) Development and Evolution, 99-121. Cambridge: Cambridge University Press.

Holder, N., 1977. An experimental investigation into the early development of the chick elbow joint. JEEM, 39:115-127.

Holmgren, N., 1933. On the origin of the tetrapod limb. Acta Zoologica, 14: 185-295.

Holmgren, N., 1939. Contribution to the question of the origin of the tetrapod limb. Acta Zoologica, 20:89-124.

Holmgren, N., 1942. On the tetrapod limb-- again. Acta Zoologica, 30:485-508.

Holmgren, N., 1949. Contribution to the question of the origin of the tetrapods. Acta Zoologica, 30:486-508.

Jarvik, E., 1965. On the origin of girdles and paired fins. Israel Journal of Zoology, 14:141-172.

Jarvik, E., 1980. Basic Structure and Evolution of Vertebrates, Academic Press: London.

Maynard Smith, J., R. Burian, S. Kauffman, P., Alberch, J. Cambell, B. Goodwin, R. Lande, D. Raup, and L. Wolpert, 1985. Developmental constraints and evolution. Quar. Rev. Biol, 60:265-287.

Muller, G.B., 1990. Developmental processes and phenotypic change in Archosaur limb evolution. abstracts, International Congress of Systematic and Evolutionary Biology.

Muller, G.B. and P. Alberch, 1990. Ontogeny of the limb skeleton in Alligator mississippiensis: Developmental invariance and change in the evolution of Archosaur limbs. Journal of Morphology, 203:151-164.

Ospovat, D. 1976. The influence of Karl Ernst von Baer's embryology, 1828-1859: a reappraisal in light of Richard Owen's and William B. Carpenter's paleontological application of von Baer's Law. J. Hist. Biol. 9:1-28.

Oster, G., J. Murray, and A. Harris, 1983. Mechanical aspects of mesenchymal morphogenesis. JEEM, 78:83-125.

Oster, G., J. Murray, and M. Miani, 1985. A model for chondrogenetic condensations in the developing limb. JEEM, 89:93-112.

Oster, G., N.H. Shubin, J. Murray, and P. Alberch, 1988. Morphogenetic rules and evolution. Evolution, 42:862-884.

Owen, R., 1849. On the Nature of Limbs, Van der Hoorst, London.

Poole, T.J. and M.S. Steinberg, 1984. Different modes of pronephric duct formation among vertebrates. Scanning Electron Microscopy, 1984/1:475:482.

Russell, E., 1916. Form and Function, Oxford University Press: Oxford.

Shubin, N.H., and P. Alberch. 1986. A morphogenetic approach to the origin and basic organization of the tetrapod limb. In M. K. Hecht, B. Wallace and G. Prance (Eds.), Evolutionary Biology. 319-387. Plenum Press:New York.

Sledge, C.B., 1981. The developmental anatomy of joints . In N. Resnick and K. Nwayama (Eds.). Diagnosis of Bone and Joint Disorders . W.B. Saunders: Philadelphia.

Watson, D.M.S., 1913. On the primitive tetrapod limb. Anatom. Anz., 44:24-27.

Wolpert, L. 1978. Pattern formation in biological development. Scientific American 239:124-135.

Wolpert, L. amd A. Hornbruch, 1990. Double anterior limbs and models for cartilage rudiment specification. Development, 109:961-966.

MOLECULAR BIOLOGY AND EVOLUTION: TWO PERSPECTIVES

A Review of Concepts

T.J. Horder

Department of Human Anatomy
South Parks Road
Oxford, U.K. OX1 3QX

PART 1. TODAY'S BIOLOGICAL IMPERATIVES AND THEIR LIMITS

Today two concepts shape our understanding of biology, evolution and molecular biology. Increasingly, our general perspective on what is important is dominated by this seemingly simple conceptual structure, and all the rest of biology is regarded as falling logically somewhere in between.

Perhaps the most clearly definable point of departure for this conceptual division was Weismann's simultaneous dismissal of Lamarckian inheritance and establishment of the pre-eminently genetic route for the transmission of the evolutionary endowment to successive generations. It then became easy to see evolution as a sequence of increasingly complex genomes, and to view adult form – with its determinants both during embryogenesis and in later supposedly non-genetic interactions with the environment – as a mere epi-phenomenon. Both concepts might well be thought to be nearing some sort of definitiveness. All biological structures can be said ultimately to be programmed by DNA and in the human genome project it appears that reduction is reaching a satisfyingly natural end point. Regarding the methods of reconstructing phylogenies, phenetics and cladistics have recently offered a computational rigour to evolution theory.

It seems entirely reasonable to view molecules as in some sense the explanation for all other types of phenomena, including evolution, and the temptation to link up these two grand themes as directly as possible is therefore great.

EVOLUTION

The raw data which all conclusions about evolution must ultimately go back to are individual organisms and their groupings in populations into species. By definition these entities are discrete from other species; the data consist of a vast collection of discrete, more or less invariant particulars. The idea of evolution implies that organisms simultaneously combine elements of novelty and elements shared as a result of common ancestry. All deductions about evolutionary relationships revolve around the concept of homology[1], i.e. the identification of evolutionarily shared characters in different organisms. Here one is faced with questions of the choice of characters (including the relative significance to be attached to adult, morphological, developmental or molecular characters), of character scoring (e.g. in an all-or-none fashion – as present or absent – or more quantitatively), of the number of characters to be used and of how they are to be combined together. The basis for character choice, being the need to separate out features which allow the distinctions between species to be defined, is a matter of measurable overt similarity or dissimilarity. In practice any attempt to specify differences leads

Developmental Patterning of the Vertebrate Limb
Edited by J.R. Hinchliffe *et al.*, Plenum Press, New York, 1991

to an artificial and arbitrary partitioning of the organism into discrete all-or-none items. This is notably true of modern (phenetic and cladistic) approaches to taxonomy in so far as they aim to itemize independent characters without making any assumptions about their relative merits and to use these listings to derive objective measures of degrees of relatedness[2].

Our ability to relate comparative data and to deduce phylogenetic relationships is totally dependent on the nature of the available linking concept of homology. All taxonomic procedures, including cladistics and phenetics, share the conceptual limitation that the linking operations are primarily directed towards classification; whether or not any assumptions about the nature of phylogenesis are included, linking is not based on any established set of rules of transition. In the absence of these, the vacuum tends to be filled by a saltatory way of thinking.

Conclusion

No serious biologist is likely to doubt the idea of evolution, even though it cannot be directly observed. However, it is difficult to avoid a certain confusion regarding the separability of various aspects of the subject, e.g. the concept itself, the tracing of phylogenetic paths and the mechanism of transitions. Ideally the last two should have a well-defined relationship, but the few available generalisations about phylogenetic transitions[3,23] have not been used in taxonomy in any explicit or even explanatory sense. Moreover, something is clearly wrong if there is a choice of approaches for analysing phylogeny as different as phenetics, cladistics and the alternative traditional methods, and this is no doubt a significant contribution to unease with evolution theory in general. If we understood the actual "mechanisms" of transitions, it remains conceivable that we could link the particulars in a new, more rigorous way.

MOLECULAR BIOLOGY

Darwinian selection emphasises variables external to the organism[4]. But it cannot account for evolution without variation internal to the organism on which to operate[5]. Increasingly molecular biology is carrying the burden of explanation; since it offers the prospect of objective, "mechanistic", even definitive, data it is an attractive route to take to define "mechanisms of transitions".

The molecular "facts"

As is well known, it has been difficult to translate the classical concept of "a gene" into molecular terms, because of the existence of a variety of DNA components that are not expressed (as proteins) and because of the fluidity of many (e.g. transposable) elements. Attempts so far to explain, find "functions" for and classify non-coding DNA components betray assumptions (as indicated in terms such as pseudogene, supergene, junk, parasitic, rubbish sequences), particularly regarding DNA as primarily devoted to protein coding, while leaving open the possibility that these could be essential, though perhaps remote, determinants of control or variation of expression. Thus it is by no means inherent in the data as to how DNA can be satisfactorily subdivided or its "units" delimited.

Granted that specific proteins are recognisably discrete and the most basic executive agents within organisms, what limitations arise if we use this level to describe the molecular programme? Any protein's function is difficult enough to define fully, whether by direct investigation or by prediction from structure; it is always possible that unsuspected aspects of function have not emerged; besides, how can the effects of all possible contexts be covered? But if an agent is being sought to fit an anticipated role, such as "developmental pattern control", its identification depends on the correctness of the developmental control model assumed. Even if a molecular species is found in the expected pattern in time or space, one still has to prove that it is a decision making molecule (rather than a corollary or consequence of the action of another pattern controlling molecule) and establish whether it is the primary responsible agent in vivo (i.e. exclude the possibility of other contributory agents).

The evidence from classical genetics and mutations

One does not have to go back to DNA to recognise that metazoan morphological pattern, as measured by specific distributions of differentiated cell types, is solely a matter of selecting specific forms of gene expression in a spatial array of cells. Since the pattern itself conforms to the genotype – the identity of monovular twins suggests that the genome exerts total and extraordinarily precise control of pattern – it would seem important and necessary that genes, in some sense, control the spatial expression of other genes.

The methods of classical genetics might be thought to offer some advantages over molecular methods; they may approximate unbiased screening of the whole genome, define discrete (Mendelian) units and encompass the effects of gene interactions, therefore demonstrating expression control. For straightforward functions (e.g. synthesis of specifiable proteins) the method can translate directly back to molecular genetics, but how far back can the controls be traced? The morphological distributions of mutant effects provide potent evidence of the general fact that gene expression is controlled in space and time. But beyond that the problems of detecting genes that might control pattern in higher organisms are formidable[1,6]. In most cases there is no definable, systematic, let alone one-to-one, relation between a specific gene and its spatial effects. Furthermore, the method still leaves ambiguous the distinction between coding and control; even in Drosophila the interpretation of mutant effects is not without its difficulties[7]. The complexity of the evidence leaves no doubt about the interactive nature of gene action.

The massive, apparently randomly dispersed, linear arrangement of DNA makes it unlikely that genetic elements can coordinate all their interactions simply on the basis of their spatial contiguity[7]. Thus the actual ways in which genetic interaction and integration are achieved assume key importance[8].

What can we deduce about DNA control from developmental cell interactions?

Of all major areas of biology the control processes of embryogenesis could well be considered among the least completely understood. However, I have argued elsewhere[1] that we know more than is generally realised and that the data are quite sufficient to define the principal underlying mechanisms. Experimental embryology shows that all cells start out with a virtually complete range of potentialities for differentiation open to them; commitment to particular fates through regional cytoplasmic specialisations in the egg is limited[9,10]. The actual route to commitment can, furthermore, be defined. Cells acquire their specialisations on the basis of spatial cues created during embryogenesis, by means of local cell-to-cell interactions involving induction and cell movement[11,16]. This permits accuracy at the single-cell level in the pattern formed and is consistent with what would be expected from the facts of cell biology (i.e. the single cell as the only unit of selection and expression of differentiation).

Thus experimental embryology provides us with the means, at one remove, of examining control of gene expression. It leads to the deduction that patterned differentiation is achieved through the control of gene expression in one cell by distinctive inductive properties (inferentially the product of other genes) expressed in another, adjacent cell, i.e. gene to gene interaction is mediated by way of interactions between whole cells. What this means about the way adult morphology is "mapped" in the genome is clearly going to be difficult to work out; given the complexity of embryogenesis it will be far removed from any one-to-one relationship.

Molecular characters as phylogenetic data

For the purposes of reconstructing phylogenetic paths, do molecular characters provide us with any ideal form of evidence, in terms perhaps of objectivity or closeness to the evolutionary record?[1,12]

The same problems (such as choice of characters and of methods of combining them) face us here as they do with traditional morphological measures of homology. But detecting homology is complicated because molecular change is primarily by substitution and transposition;

therefore evolving molecular characters automatically lose features shared through common ancestry in two senses, content and position. It is then hard to exclude convergence (the possible secondary evolution of two initially different molecules towards the same form), since there are few supporting cues (for example, position in relation to more stable ancestral cues which is important in the use of morphological characters[22]). Although molecular data is readily amenable to quantitative measures of degrees of similarity, the computational methods are various and at odds.

Irrespective of these problems, it is clear that different DNA components or proteins show vastly different rates of evolutionary change. This presumably reflects differing selective pressures, as well as variable mutability and repairability. As independent markers of absolute evolutionary time, molecular change would therefore not be helpful; on what basis could one chose one molecule over another, unless by calibration against morphology and geology[20]? But even from the point of view of using molecular change to date species relative to one another, there is a serious hazard. If single molecular types differ in rates of change, how can we be sure that any molecule used for purposes of dating species does not also vary its rates through time?

A new approach has been to seek out sequences (e.g. homeobox; homeotic sequences) which occur in multiply repeated or consistently related locations and to infer homology of function of corresponding gene sequences when they are found in different species.[13] This could be misleading since small changes in structure or context of genetic material may result in major changes in function. The criterion of position of repeated units of structure (serial homology) has, in morphology, no meaning as a criterion of evolutionary homology, which depends on content. Thus, as compared to its original use in the context of morphology, the concept of homology may be harder to define and apply in molecular terms.

Conclusion

Although "molecular biology" embraces many approaches and concerns, it is primarily defined by its emphasis on structure. The core of the data is static, descriptive and fragmented; it is in effect an extension of comparative anatomy. Compared to morphological data, problems arise from its sheer complexity, particularly the difficulty of interpreting the data in functionally meaningful terms. Apart from the obvious issues potentially amenable to molecular solutions (such as control of cell function, heredity or even the origin of life), molecular biology is a form of data which must run up against real limitations when it comes to understanding entities like whole cells, embryos or adult morphology. Even when we have a comprehensive knowledge of DNA structure, molecular characters do not offer any instant and automatic direct measure of evolutionary relationships, least of all any rules of evolutionary transitions at the species level.

THE MOLECULAR PERSPECTIVE

Being discrete and structure–based entities, genetic or molecular factors cannot readily explain the functional integrity and adaptatedness of organisms, in many ways the most central issues needing explanation in evolution theory, since they are the meeting point of heredity and selection. Adaptation is often "externalised"[4] in that it is interpreted primarily in terms of external selection forces or even environmental stimuli. But external environmental stimuli have a relatively trivial role in determining phenotypes[26,27] and the results of selection have to be transmitted through the genome to future generations; therefore they must acquire a solely internal basis. The modern synthesis is a conceptual framework which seeks to bridge externalist and internalist considerations, e.g. by linking genes with Darwinism and speciation through such concepts as allometry[1] and population genetics. It is difficult to say whether the synthesis achieves much more than pointing to the formal possibility of relational and causal connections between a variety of separable, abstract parameters[1,4]. It is plausible enough as regards speciation and species level issues, but increasingly unsatisfactory at higher taxonomic levels because of an inevitable tendency towards increasing abstraction, reflecting inadequacies in available specific laws of evolutionary change[3] and a divorce from the raw taxonomic data. If nothing else, the synthesis makes it clear that any "mechanisms of transition" must involve a highly complex set of concepts covering all levels of biological organisation and that internal and external considerations must always be operating together.

Conceptually molecular evidence is no different from morphological as regards the reconstruction of phylogenetic paths, because both are essentially descriptions of structure. Both forms of evidence are doubly limited; the data they generate are discrete both because of the discreteness of known species and because structural entities are spatially separable constituents within organisms. Evolutionary transitions are, on the other hand, the result of combinations of structures and of their functions. Evolution theory as a whole has failed to penetrate beyond the raw data of the particular species available; by implication or by default, it provides no alternative to saltation as the mode of transition.

The molecular approach is an extension, towards greater detail, of the genetic foundations that characterise the synthesis. As such it is purely internalist. It can be argued that it adds a new level of complexity, without itself generating new evolutionary principles or laws. Few operational principles are to be expected at the DNA level[14] and these will generate few explanations, predictions or limitations applicable at the organismal or macroevolutionary levels. As for the modern synthesis as a whole, a serious conceptual hole remains if genes are not understood sufficiently to fully and rigorously explain adult morphology and its adaptedness[1]; without this it is impossible to relate molecules and genes to selection.

Conclusion

Like reductionism in general, the molecular approach may entail more than seeking an ultimate account of all biological entities and phenomena; it may also come to appear a complete and sufficient account. Such a restricted view would in itself limit the way further questions were asked. It would risk a failure to fit in all the known contributing variables relevant to evolution. It is clear that other sources of data (such as embryology) are equally objective and precise; they may be more conclusive on certain issues. Whatever the formal merits of the molecular perspective, one has to conclude that, in the face of the molecular complexity and in terms of practicalities, it would be merely an article of faith to regard the molecular approach alone as a satisfactory way of understanding evolutionary transitions.

PART 2. IS THERE ANY ALTERNATIVE PERSPECTIVE?

We have, then, considered two types of approach to conceptualising the "mechanisms of evolutionary transitions". The first is reduction to its material basis in DNA. Intrinsically this has major limitations. Indeed the central problem has now almost reversed itself; the issue of greatest interest has now become how evolution can explain the massively complex accumulation of specific and yet coordinated biological molecules, rather than the reverse. The second approach also fails, but for quite opposite reasons; the concepts of the modern synthesis are so disjointed, abstract and generalised that they have become detached from much of the data they refer to. We need to get back to the essence of the evolutionary process and to reconsider the relation between general features of that process and the totality of available specific data as provided in particular, known organisms.

AN EVOLUTIONARY PERSPECTIVE

If we return to the logic of the process, it is an obvious but fundamental feature of evolution that it is progressive; the only alternative is creationism. To be more precise, it must be cumulative, any new evolutionary acquisition being conditional on its compatibility with already acquired characters of the species. As complexity increases two tendencies will become evident.

Firstly, some characters (e.g. the DNA code) will become increasingly stabilised, and occasionally even invariant[14], as they become preconditions for the reproduction, development and future evolution of the species. It follows that any one organism must consist of a combination of characters including some which exactly repeat an ancestral condition, and others which range over all possible degrees of evolutionary novelty. This is immediately evident in the varying rates of evolution of molecules. Equally, on the cellular level, multicellular organisms evolve new cell specialisations at varying rates, but the basic structure of cells as such remains a fixed, obligatory prerequisite – the egg cell literally ensures that this prerequisite is fulfilled through the whole life cycle and through succeeding generations.

Secondly, the avenues for further evolutionary change will become increasingly circumscribed in relative terms, e.g. as variants, recombinations, losses or additions permissable within the restrictions represented by the stabilised characters. Evolution of metazoans required not just the initial attainment of the unicellular condition, but also the use of potentialities available to single cells; a number of these, which remain permanent restrictions through subsequent evolution, have been discussed[15]. Metazoan evolution could not have advanced far without mechanisms for structural integration and conformity to the same genotype; these are, of course, ensured by the precondition of mitotic division out of a single initial cell, i.e. the egg. The coordinated processes of embryogenesis must therefore have evolved as variations based on this starting point. Adult morphological complexity in turn could only evolve through variables made possible by embryogenesis.

It is of particular interest that these principles would be expected to appear in explicit form in embryogenesis. Morphogenesis – the process (defined in[1]) whereby cells come to occupy specific positions, through mitosis or movement – imposes a major evolutionary limitation. It is the process which makes possible, and coordinates, the inductive interactions leading to differentiation; together, morphogenesis and induction build up an integrated causal cascade[16]. Inherently morphogenesis is a spatially continuous causal chain of events in time; each step is conditional on prior steps and surrounding mechanical interdependencies. Evolutionary increases in complexity will only be repeatable in future embryos given a re-creation of comparable conditions in subsequent embryos. Thus, phylogenetically, as new complexity is added on primarily at the end of development, earlier morphogenetic conditions will become increasingly stabilised. Despite any superimposed modifications and even changes in cell mechanisms[17], the morphogenetic chain cannot be broken in subsequent embryos. As a result, evolutionary novelties can only occur in spatial continuity with ancestral anatomy, and will be increasingly small scale.

Given these considerations, one would expect to see a parallelism between phylogenesis on the one hand and embryogenesis (and consequential adult morphological organisation) in advanced species on the other[14,17]. More specifically, relatively novel types of organisation (i.e. as seen in early metazoans) would, in advanced species, be expected to be superimposed (relatively late in development) on early developmental organisation which reflects stabilised versions of preceding phylogeny. Early morphogenetic mechanisms will be highly conserved and shared in common widely among species.

We can perhaps recognise, in the most general of terms, three grades of organisation reflecting stages of increased morphogenetic elaboration, in order of increasing evolutionary antiquity;

(i)   Differentiated cell types can emerge by direct cell to cell (inductive) interactions within a static epithelium or mesenchyme; e.g. organisms differentiating when the embryo has few cells (as in "mosaic" invertebrates), mosaics of cell types within and across the epithelia composing Hydra, small scale transitions in connective tissue (in the types of cell differentiation and in variations in the numbers of cells composing each component, for example cartilage/perichondrium/synovial membrane). Adjacent cells may show totally distinct differentiations (the number of alternatives usually being restricted); some may subsequently mitose to form varying sized groups of identical cell type. This grade occurs terminally in any given developmental sequence, when the relevant cells have ceased interacting morphogenetically with cells other than those immediately adjacent.

(ii)   Differentiation of arrays of repeated, multicellular differentiative modules (involving several cell types and cell collaboration in attaining complex anatomical forms) e.g. feathers, hairs, scales, segments, digits, phalanges. Typically, differentiation is preceded by small scale morphogenesis, e.g. condensation, epithelialisation (as in somites or joint formation), epithelial folding (as in a feather bud). Modules are repeated in varying numbers (meristic variation) with characterising spacing rules (often morphogenetic "competition").

428

(iii)   Differentiation of singular, specifically located, specialised organs e.g. limbs, eye. All degrees of approximation of Grade (ii) towards this grade can be found; the transition is a result of gradual reduction in numbers (Williston's Law), and increasingly large scale variants of the original modules (e.g. midline fins become localised and specialised bilateral fins in fish). In more advanced organisms this grade is associated with increasingly large scale, integrated prior morphogenesis (accompanied by increasing delay of differentiation, larger cell numbers, more spatially and temporally extensive morphogenetic movements and spreading out of elements of the causal cascade), e.g. limb bud outgrowth mediated by the AER which itself is localised through dependence on epithelium traceable back in continuity to the gastrula (and ultimately the egg) and is therefore linked to the total "Bauplan"[18]; similarly mesenchyme depends on somites and their antecedents. In vertebrates early gastrulation initiates and coordinates the laying down of the Bauplan; all adult structural elements are dependent on this event. By comparison, gastrulation in invertebrates is often later and less important for integrating the whole embryo.

Thus, increases in complexity throughout all stages of metazoan evolution can be understood in uniform terms on the basis of morphogenetic mechanisms and their universality throughout all species. In any given species each morphogenetic grade merges with the next in continuity; during evolution, through a succession of embryos, increasing complexity must also be built up, by way of extension and variation in form of terminal patterns of morphogenesis and differentiation, again in continuity. There is no reason to exclude the possibility of variation at any stage of morphogenesis but the earliest stages will be maximally stabilised: hence the sharing between species of a Bauplan.

Differentiation is dependent on inductive interactions only made possible by morphogenesis[16]. Therefore no evolutionary change in the position or type of differentiation can occur that is not coordinated through morphogenetic controls; the continuity of morphogenesis will ensure that such changes occur gradually. Change brought about gradually is compatible with ongoing functional integration of the organism and therefore with a succession of viable organisms. Selection can then also operate gradually. For example, variation in the number of repeated modules of structure at specific sites (such as tail segments, or terminal phalanges) makes it possible to vary morphology to significant structural degrees, yet with minimal effects on function or adaptedness. The scheme proposed here fits and explains many of the puzzling phenomena of embryology. Morphogenesis can be seen to play a central role in the tight, integrative control of pattern. It, in effect, sums up the rules, built up through evolution, which mediate the readout of the unit events of differentiation. But perhaps most significantly of all, the early stages of morphogenesis explain the Bauplan: this evolutionarily stable coordination of major body structures directly reflects the nature of the evolution of embryogenesis.

Conclusion

Accepting that every biological component is potentially open to selection, we must resist teleology[19] and see all properties and components of any given species as the mechanism for continuing and possible future evolution, as well as the product of earlier evolution. The evolutionary perspective, applicable as it is to all spatial and temporal scales within organisms, views evolutionary change as a tightly integrated and, overall, effectively continuous process. Embryogenesis implies a continuity of evolutionary changes in morphology; a relative stability, and cross-species universality, of the underlying mechanisms can be inferred. The example of organisational grades shows how, in principle, a systematic mechanism-based specification of evolutionary transitions, and their manifestations in adult morphology, is attainable - at least for the metazoa.

AN EVOLUTIONARY BASIS FOR PHYLOGENETIC RECONSTRUCTIONS

When we examine the phylogenetic status of a species we face the problem that each organism is structurally integrated. Each species has the overall appearance of uniqueness and indeed, given the complexity of morphogenesis, it is improbable that any definable anatomical entity in higher vertebrates is composed of absolutely constant numbers or arrangements of cells even within a species. What is required is some rigorous procedure by which to identify and separate characters according to grade of evolutionary stability or novelty. Finally the data must be integrated to represent the totality.

What are the practical difficulties to be overcome if, for example, one sets out to account for the evolution of the pentadactyl limb from a fish fin?

(a)   The discreteness of adult anatomical structures, such as skeletal elements, imposes an inevitable limitation on one's ability to identify particular adult forms as ancestors. Morphologies with discretely different patterns, in which no possible intermediate forms can be envisaged (e.g. limbs differing in the numbers of digits), could in fact be closely related. Trivial quantitative differences in morphogenesis could account for the phenotypes, but there is no obvious way that one can quantify the relationship from the adult data alone or go backwards from the end results to the embryological process. Clearly there is something inherently unsatisfactory about trying to represent an evolutionary sequence by means of a series of static adult morphologies given that evolution operates by way of changes in the dynamic process of embryogenesis.

(b)   If we consider embryological characters (e.g. prechondrogenic condensation patterns) we appear to get notably nearer to an ancestral form. The close similarities of chondrogenic patterns in the embryos of species which become widely different as adults is a striking fact. These are, however, arbitrarily frozen stages in a whole series of dynamic developmental processes. It is important to recognise that a fundamental conceptual error arises whenever an actual developmental pattern is equated with an ancestral (adult or embryonic) form[17]. Condensation patterns necessarily differ (particularly in timing) between species at every stage. The interdependencies of morphogenesis have the effect that, in order to get a new final morphology, all prior stages of embryogenesis must to some extent reflect the species difference. Moreover, the rules under which limb buds are transforming into adult structures are themselves changing multidimensionally in the course of evolution. This means that, although transformation between limb forms can perhaps be easily understood in one species, the identical procedure cannot be used in comparing different species[24]. Thus what is relevant is not so much the invariant aspects of developmental patterns but the identification of modifiable features which would permit evolutionary change.

(c)   Is it realistic to consider the evolution of one specific organ in isolation? Like meristic characters in general, skeletal elements in the limb can increase or decrease in number. Limbs differentiate late and relatively autonomously, which correlates with a relative freedom to vary pattern independently of the Bauplan of the whole organism. This means that convergence is possible (as in the evolution of paddle features in secondarily aquatic species). The detection of convergence[22] depends on giving more weight to the sum of characters (e.g. the Bauplan) than to any individual character. It is not the limb that evolves but the whole species. The concept of limb evolution only has meaning if ancestral limb forms are detected on the basis of a knowledge of the sequence of ancestral species rather than of features of the limbs themselves. The actual sequence of species may only have manifested a few, discrete limb forms rather than a series that appears more continuous or more probable when limb forms are compared in isolation. Furthermore, not only do we have to rely on the evidence of the whole species, but we have no alternative but to rely on the classification of species into higher taxa to give us evidence on the form of ancestors in sequence through phylogeny.

It was argued in the previous section that developmental and morphological organisation are effectively interchangeable considerations and that organisational features as such (i.e. collections of individual characters) give direct access to evolutionary conclusions. The foregoing analysis of the evolutionary process makes it inevitable that no absolute features of any single component structure (such as a particular molecule or form of cell differentiation) can be relied upon to date or characterise stages of evolution[20,23]. Yet the analysis illustrates how metazoan species can be directly compared because of the universal character of embryological mechanisms. The gradings described above provide a basis for rational character choice indicative of relative evolutionary flexibility and applicable across species. However, given the possibility of change at the level of even the most stable ancestral characters[17], all gradings will

430

necessarily be relative and any actual procedures of phylogenetic analysis will have to reflect this feature.

The way in which a grading system can be used, given these limitations, can be illustrated, as it might apply to an analysis of the evolutionary sequence from fins to limbs – though the same would also apply to the analysis of the whole morphology of a species, comparisons of species and thence groupings of species into higher taxa – in the following way;

(i)   It is first necessary to isolate grossly invariant features, relative to the span of the evolutionary stages being considered, e.g. identification of the fin/limb character as a specifically positioned (i.e. bilateral, pectoral, pelvic), articulated appendage establishes a relatively stable character by reference to the broadest possible criterion of stability (using all integrated "Bauplan" characters of the vertebrate condition). Position and pattern in this case are more useful than cellular differentiation (e.g. cartilage or bone) or its timing. Differentiation or gene products, as such, may be useful characters, but they must be judged in relation to the context of the organisational grade involved.

(ii)   It is then possible to identify novel characters; e.g. pentadactyl limb (stylopod, branching metapod, segmented digits of relatively defined number) as distinct from alternative forms of appendage. The relative grades of the types of variation give a measure of relative evolutionary distance. Thus the fish appendage represents an ancestral condition because it shows wide variations in number and arrangement of skeletal elements (the only consistent restriction may be in the distinctions between girdle/basal/radial cartilages). Pectoral and pelvic appendages are often different in one species and their locations are only loosely defined. All patterns appear easily interconvertible. In general, fins contain large numbers of skeletal units, especially distally. Approximations to the pentadactyl plan occur but only in restricted components of the whole pattern (e.g. single proximal element, five radials; even the crossopterygian pattern deviates distally). These appendages are relatively unspecialised in that they share many features with other parts of the anatomy, especially midline fins and tail. By comparison, the differences between limbs in different tetrapods are a matter of small-scale changes in number and shape of a highly constrained array of skeletal modules. The pentadactyl format has achieved the status of a Bauplan for the whole limb[24] and therefore represents a higher grade of organisation.

(iii)   An important aspect of the procedure is that it must be one of successive approximation. Because no characters in themselves identify any particular grade, each step must be continually returned to in the light of conclusions reached at the others and individual characters must be evaluated in contextual terms. Judgements about grade are relative with respect to other grades. The end result of the procedure will only be probabilistic – homology will no longer be all-or-none[1]– but this does not mean arbitrary. Embryologically-based choice and use of characters provide the procedure with a rational foundation.

Developmental considerations and the evidence of comparative anatomy, with palaeontology, offer the possibility, when suitably combined, of a rigorous interpretation of evolutionary relationships. As understanding of embryonic development increases one can anticipate a closer definition of the evolutionary implications of specific aspects of limb form. Even though they are arbitrarily frozen stages in dynamic sequences of events, prechondrogenic condensation patterns come close to depicting a key process, since condensation is the morphogenetic event leading to chondrogenesis. These events are initially continuous (and depend on potentially definable rules based on the availability of mesoderm in the limb-bud; Hinchliffe, this volume) and the emergence of discrete structures from them occurs according to definite rules; e.g. branching patterns, secondary emergence of phalanges and joints. Therefore – given that limb pattern is essentially a matter of the distribution and sizes of repeated skeletal units   –   it might be possible to describe evolutionary change as a single coherent quantitative change in the initial morphogenetic conditions, such as the shape of the limb bud[1] (Fig. 10),[24,26]. It will be particularly important to compare the developmental rules in a variety of taxa; a few living representatives of a taxa will be sufficient. In this way the direct evidence of embryology and the indirect evidence of taxonomy can be seen as interdependent and mutually supportive.

The idea of depicting limb evolution is in itself to some degree misconceived when considered in isolation. In any event, in practical terms it is a matter of great complexity[24]; e.g. as regards the use of the actual data in the form of sample adult or embryological morphologies or abstract and graphic representations of multidimensional changes occurring through time. Nevertheless, one can perhaps be optimistic that comparative embryology will eventually leave little doubt about the nature and course of metazoan phylogenesis. Despite the many possible sources of error – and the inevitable discontinuities in the data due to the succession of generations, morphological particulars, variation within populations composing a species and gaps in the phylogenetic record – the likelihood remains that an effectively continuous series of adult morphologies occurred in evolution and that this series can eventually be closely predicted[24]. This follows from the interdependencies and trends towards stabilisation in morphogenesis which imply minimally graded changes in the evolutionary sequence of morphologies.

Conclusion

An attempt has been made here to show how it is possible to make use of rules about the nature and consequences of the evolutionary process to generate a method of establishing phylogenetic relationships among morphological patterns. It must never be forgotten that one cannot expect a definitive phylogenetic classification, universally valid evolutionary rules or rigidly applicable methods. Evolution is so complex that phylogenetic reconstruction can at best be a matter of probability and relativistic methods. However, the above considerations provide a rational foundation for what amounts, almost certainly, to the traditional taxonomic method – based up to now on a pragmatic best fitting (including weighting) of all the data.

THE PLACE OF MOLECULAR BIOLOGY IN EVOLUTION THEORY

What about molecular characters? Given the discrete, saltatory and independent changes possible in DNA, how can phenotypic integration, let alone evolutionary continuity, be possible? How can we avoid genetic preformationism[10,19]?

The logic of evolution requires that DNA control mechanisms needed in metazoa must – at least potentially – already have been present in unicellular ancestors. We know[25] that protozoa have a complexity quite sufficient to encompass most multicellular organising phenomena. The prediction must therefore be that all the developmental gene control events necessary to generate adult morphology must be mediated on the basis of decisions made by and within single cells.

Experimental embryology confirms this prediction. All differentiation is the product of individual cells triggered by induction, single cells being, of course, the only units of differentiative expression. The accumulated evidence shows that all such developmental decisions are based on signals present in the immediate environment of the cell[11,26], e.g. inductive cues from adjacent cells. The phenomenology is consistent with the idea that the cues are products of the inductor cells, related to their own prior commitment to specific courses of differentiation. Thus stimulus and response are both effected at the individual cell level. It is the all–important mediating process of morphogenesis which, in cascade fashion, orchestrates the location of specific inductive interactions and which therefore establishes the link up of the separate genetic events in the cells involved.

I have argued elsewhere[1] that a pattern as complex and specific as that of the vertebrate limb, for example, may be entirely explicable in terms of the way morphogenesis deploys a small number of cell types, which are all also found elsewhere throughout the embryo. Although the nature of developmental (such as inductive) stimuli remains undefined in detail, there is considerable evidence that cells are responsive to function-based contextual cues[21,26]. For example, bone cells appear to differentiate in accordance with mechanical stresses; bone structure throughout the body could therefore be traceable back to a single, intrinsic (gene-based) potentiality of bone cells to respond to their mechanical context. The extent to which considerations of this kind could account for the patterning of a structure like the limb[21,26] could turn out to be remarkable. I also argued[1] that no limb-specific genes of any kind may be required. This, together with the inevitable involvement of very large numbers of more generally employed genetic factors, implies that it may be almost meaningless to speak of a class of "genes controlling limb development" or "pattern genes".

This is not to suggest that function-based mechanisms, even though in a sense self-assembly, are in any way uncontrolled by the genome. What is important is that pattern formation and evolution depend crucially on the nature of the capacities of the genetic mechanisms mediating expression to respond to cues and stimuli outside the cell. Significant implications follow from the foregoing propositions; relatively few genes may be sufficient to programme the phenotype (perhaps just structural genes and their immediate enhancers switched by inductors or function); genes are free to evolve and function independently because of the way integration is mediated epigenetically; adaptedness of structure is largely "automatic" and internal to the organism. In this way embryology can explain how a linear array of dispersed, independent and relatively freely mutable genetic units, common to all cells in the organism, can programme a three dimensional, coordinated morphological pattern. As regards the central evolutionary issues of adaptation and selection, it is evident that whole organism morphology plays an integral part in the causal chain of the evolutionary process, as the interface with external selection forces which in turn determine the genome itself[27]. Through time these considerations can explain the cumulative acquisition of DNA in the complexity of organisation in which we find it.

Conclusion

Viewed within the broad evolutionary perspective, discrete structures, including varying forms of DNA, add up to become a continuous process. Structure, function and adaptation all become coextensive concepts. In protists, molecular change may come close to fully describing evolutionary change, but as regards metazoa, molecular mechanisms may be of relatively less interest, since the mechanisms as such are unlikely to be new; metazoan evolution will, rather, reflect increased combinatorial complexity and variation made possible by pre-existing mechanisms[14]. Instead, evolutionary change can be more directly understood in other, more functional and interactive terms, such as are particularly available in embryology.

As regards estimating the most probable phylogenies, similar methods and limitations apply to all characters. It seems very likely that morphological and increasingly detailed molecular-based estimates will - after "averaging" multiple characters to take into account the variable rates of evolution of individual characters, be they limbs or molecules - eventually converge on the same conclusions. They are, after all, only two aspects of the same thing. But morphology may remain for some time to come the most convenient way of summing up the genotype and of identifying multicellular species.

SUMMATION

What we take as our raw data is not just defined by observational methods but by the conceptual framework we start off with. It is a measure of the significance of our frames of reference that the two perspectives discussed here may seem so different, even opposed[28]. Yet they refer to the same problem and must in fact be simultaneously true. The arguments presented in this paper seek to go beyond the traditional terminology of reduction, organicism, holism or emergence and to define the broadest possible view of processes and data. Although, given the nature of evolution, almost every generalisation about it is likely to have some exceptions, we can at least be sure that the evolutionary perspective is as well-founded as the molecular; it is the most universally valid concept within biology, because it refers to its most basic precondition, subsuming even the molecular.

Concepts of the most generality and profundity often appear simple but are, in a sense, the hardest to handle. They are likely to be the most important in setting priorities, and yet, being "obvious", are likely to be taken for granted; biological science lacks mechanisms to review concepts[29]. Many of the greatest issues in biology today are as much conceptual as discovery problems, and it is often easier to invent new concepts than to modify old ones. In such a complex subject the concept of proof may have limited relevance. The reason why the fact of evolution and the broad outlines of phylogenesis are so generally accepted is simply that these are the only ways to fit the evidence, and all the evidence, together.

NOTES AND REFERENCES

1.   With particular reference to the limb, I have reviewed the conceptual structure relating evolution, genetics and embryology: T.J. Horder, Syllabus for an embryological synthesis. in: "Complex Organismal Functions; Integration and Evolution in Vertebrates", D.B. Wake and G. Roth, eds. Dahlem Conference, Wiley, Chichester (1989). Key references can be found therein.

2.   Although phenetics and cladistics have highlighted the need for taxonomic methods to define their procedures, neither avoids dependence on assumptions, e.g. cladism on speciation by bifurcation, parsimony, polarity[23], prior identification of outgroup, exclusive use of novel characters unique to a group. Both give equal weighting to each character selected. Neither explains the relation of novel and unchanged characters; phenetics does not even distinguish them.

3.   There has been a historical tendency (most recently in transformed cladistics) for systematics to become separated from considerations concerning the general features of evolution, e.g. "trends", rates, extinction, "tempo and mode", and certain "laws" (such as Williston's or Dollo's laws)[23]. Such laws are essentially descriptive rather than explanatory and far from universally valid. See: B. Rensch, "Evolution above the Species Level". Methuen, London (1959). The most widely applicable laws, i.e. heterochrony or Cope's law, are perhaps the least predictive or explanatory in practice. Concepts such as hopeful monster, correlated progression, macroevolution or key innovation depend on a dichotomous distinction (from short term, normal, species-level evolutionary change) yet are more likely to be matters of degree.

4.   The distinction between "internal" and "external" is one of the more pervasive yet unexpressed of conceptual dichotomies[26,27]. Neo-Darwinism, while recognising the role of the organism itself (i.e. in genes and mutations), emphasises the external in its concentration on fitness, adaptation, selection and population forces determining gene frequences, and in its neglect of the specifics of gene actions by characterising genes or mutations as "random", neutral, clocks or drift. Evolution theorists have conceptualised the internal contribution through terms such as constraint, canalisation or structuralism. Not only are these abstract and unmeasurable – often not distinguishing between embryonic or genetic bases – but, by implying that internal factors restrict evolution, beg the question of their origin or role in later evolution. For a critique of such concepts see: S.C. Stearns, Natural selection, fitness, adaptation and constraint, in: "Patterns and Processes in the History of Life", D.M. Raup and D. Jablonski, eds, Dahlem Conference, Springer, Berlin (1986).

5.   This is an example of an important and potentially confusing distinction between "elective" and "instructive" causes. Other situations in which the character of a biological change depends on options already fully elaborated in the responding cells or organism (the stimulus being merely a trigger/selector)[9,11,27] include embryonic induction, canalisation, mutations[6], hormone actions.

6.   Genetic pattern abnormalities[1] are few (they fall into a small number of classes), non-specific (the same patterns are caused by many distinct mutations), spatially variable and ill-defined (e.g. they are usually pleiotropic), conditional on other genes (they are often polygenic) and on developmental conditions (they are frequently mimicked and rescued by, or summate with, extraneous factors, as in phenocopies) and often atavistic (e.g. polydactyly). These complexities allow few conclusions about evolutionary advance or normal genetic control of pattern, other than its general temporal and spatial dependence. Mimicking of pattern mutation effects by teratogens suggests susceptibility of specifically unstable or finely balanced developmental decision points, an elective situation[5] and an epigenetic step in the mutant action.

7.   The relatively frequent evolutionary variations in DNA partitioning (e.g. into new chromosomes) and other transpositions, suggest that, to a degree, a gene's position relative to others is not important. However, it is hard to evaluate the role of smaller scale positioning; the possibility of its significance for control is raised by the colinearity of segments in Drosophila and corresponding homeotic gene loci (although this may only reflect relative recency of gene duplications). Even here differential expression of serially arranged sequences presumably requires signals from outside the cells involved; for patterning not directly mediated by the

maternal cytoplasmic gradient[13] there would be no other basis for the appropriate gene to be selected. Merely to describe temporo-spatial patterns of expression of gene products or patterns of interdependency between genes (the methods so far employed) does not alone make it possible to define the localising signal and how it itself achieves its position.

8. For a discussion of major issues concerning genetic integration, see: E. Mayr, "Animal Species and Evolution". Belknap Press, Cambridge, Chapter 10 (1963).

9. The importance of maternal egg cytoplasmic factors hinges on the amount of pre-formed[10] pattern organisation ascribed to them. All evidence suggests this is extremely restricted (i.e. only establishment of embryo "polarity") and that the nature of the factors varies considerably between species. Such factors may be elective[5]. Eggs (and other early specialisations such as trophoblast) can vary widely in character without necessarily affecting the evolutionary stability of later basic embryological organisation, provided the initial stages in forming the embryo body plan itself are diffuse and regulative.

10. "Preformationism" is an unilluminating form of argument which, in effect, "explains" a phenomenon in terms of conditions already in existence, yet themselves unexplained. Examples include mosaic development, prepattern and pre-adaptation. "Genetic preformation" explains phenomena by referring them directly back to the genome (e.g. "pattern genes"); i.e. to explain morphological pattern (in the absence of other considerations) this implies a number and specificity of genes equal to all elements of the pattern. In relying directly on the genome, one step (gradient-based or positional information) models of development fall into this category.

11. It is sometimes, though unreasonably, thought that embryonic induction is "unproven", particularly because its chemical basis remains so problematical. The fact of induction is massively documented and fully defined operationally, as an interaction between adjacent cells, brought together through morphogenesis, and leading to a specific form of differentiation in the induced cells. Requirements (e.g. specificity or complexity) of the inductor cell stimulus are minimal in relation to the contribution of the potentialities of the responding cells[5]. Further chemical definition is difficult because inductive interactions are reciprocal and multifactorial (e.g. double assurance, involvement of morphogenesis) and are probably based on different factors depending on which cells types are involved.

12. For a recent review of DNA organisation and evolution, see: B. John and G. Miklos, "The Eukaryote Genome in Development and Evolution". Allen and Unwin, London (1988).

13. Basic aspects of genomic organisation would be expected to be evolutionarily stable: the occurrence of homeobox domains in both vertebrates and insects is not itself surprising. However it is more risky to make a parallel judgement of homology regarding features like morphological segmentation; this is an abstract concept with potentially widely different cellular and developmental bases in phyla as different as insects and vertebrates, i.e. it is possibly a convergent feature. There are additional, more specific reasons to question the applicability of Drosophila segmentation genetics to vertebrates; homeotic pattern mutations are (virtually[1]) unknown in vertebrates though common in insects; the initial generation of body plan in insects is fundamentally different from vertebrates (i.e. establishment of main segments by subdivision in the blastoderm stage rather than by serial addition; segment specification occurs early due to a maternally-derived cytoplasmic gradient - related perhaps to the blastoderm being initially a syncytium - whereas in vertebrates cells retain totipotentiality until gastrulation).

14. During evolution, mechanisms or organisational principles (such as the cell, morphogenesis or induction) become obligatory and universal among species. It is important to recognise that, in contrast, the ways in which they are actually manifested may remain constantly flexible. Thus the DNA code (i.e. use of four nucleotides), as such, is an ancestral feature which has become totally invariant, but subsequent evolution involves constant change of any one specific DNA sequence. Given the essential simplicity of DNA only a limited number of general organisational principles is likely to apply to it; e.g. rules of replication, transposition and transcription, control by sequence proximity or by gene product feedback, single versus multiple sequence copying.

15.   L.W. Buss, "The Evolution of Individuality". Princeton University Press, Princeton (1987).

16.   Induction is inseparable from morphogenesis. As described elsewhere[1] inductive interactions occur at spatially and temporally discrete points in development as a direct consequence of cellular conjunctions brought about by morphogenesis. Early rounds of induction and resulting differentiation in turn modify morphogenesis as it continues into later stages. The resulting integrated causal cascade is built up gradually during evolution, based primarily on variation and elaboration made possible in morphogenesis. Because induction depends on conjunctions resulting from morphogenesis, new forms of differentiation or the inductive receptivities to spatial locations needed to evoke them, will only be able to evolve in the context of morphogenetic coordination and continuity. The nature of the inductive stimulus designating a new location is relatively unimportant; freedom of the inductive response to change depends on the responding cell's ability to discriminate a new location and to link a modified form of differentiation to this. Inductive mechanisms probably evolve slowly; few forms of it are to be expected given the limited number of types of cell differentiation in major metazoan groups.

17.   A degree of "recapitulation" of phylogeny during embryogenesis is inevitable; for the reasons presented in this paper; the concept reflects a key truth about the evolution of embryogenesis. This idea has often been misrepresented because the following points have been insufficiently clarified; an embryo stage and an adult ancestor can never exactly correspond because evolutionary advance involves losses and changes in differentiation; even the embryo of the ancestor cannot correspond because evolutionary advance into a new adult form requires a changed embryogenesis – all stages of embryogenesis are potentially open to change, even the earliest[9]. What, however, is important is that early developmental stages change only slowly. The concept of recapitulation depends on relative evolutionary stability of ancestral characters. Matters are further complicated by the fact that the retained characters are processes (the morphogenetic causal chain) rather than any specific structure or mechanism (e.g. the cellular mechanisms of blastula or gastrula formation may change; only the organisational processes remain invariant[14]). The closest we can get to identifying what is recapitulated is that it is a modified version of a series of ancestral embryonic stages[24]; the egg is an example[9]. The concept is clearly exemplified in the way in which embryogenesis proceeds sequentially from Bauplan organisation towards more species-specific organisation, and is consistent with many other concepts[23].

18.   The concept of Bauplan (or archetype) is an ill-defined but indispensable one. It refers particularly to the general morphological common denominators of a major taxonomic group. Clearly it can only be depicted as an "average" or in abstract form. It is difficult to distinguish the concept from that of the group ancestor or from representations of early stages of morphogenesis, which members of the group frequently share more closely in common than adult features.

19.   In this field one is constantly at risk of teleology on two fronts; i.e. a tendency to derive explanations for development, and for evolution, on the basis of their end results. The consequence is preformationism[10] and many arbitrary distinctions; e.g. the adult state being seen as an end point distinct from its development (rather than merely a relatively stationary phase of the life cycle retaining most of the potentialities of the embryo, e.g. plasticity of cells (as seen in regeneration), turnover, growth, metaplasia, budding and metamorphosis); the egg being seen as a unique cell (rather than just one form of cell specialisation linked to one phase of the life cycle but in other ways little different from other cell specialisations or a permanently preserved form of the unicellular ancestor).

20.   Ideal dating or classificatory criteria are rare, irreversible binary events, such as gene duplications or transpositions, karyotype transformations, changed relations to parasites and symbiosis, or continental drift. Such events give an absolute marker for one major point of division in the phylogenetic tree, but would not allow any further discriminations among species.

21.   The concept of growth is one of great difficulty. Growth is controlled multifactorially and it must be extremely remote from direct genomic programming. It may well be biologically unrealistic to distinguish it from a number of other concepts; size, shape, proportion, function (functional demand being a major determinant of growth), morphogenesis, cell death and

turnover, and even induction (inductive dependencies may evolve from functional dependencies). Morphogenetic continuity leads one to expect a direct relation between morphogenetic change and proportionate size changes in adult morphologies through evolution (allometry). However, any simple rules of proportion will only hold in the short term. Long term, morphogenetic complexity will break up initial allometric relationships; the concept of allometry has significant limitations[1].

22.   Convergence is, in a sense, the mirror image of homology. Some characters, while appearing to be shared between species and therefore ancestral, can be shown to be novelties (derived from a different ancestral condition) on the basis of collateral evidence; namely the test of the consistency of that character (and its development) with other characters judged (according to criteria such as those listed in[23]) to be of greater evolutionary stability. In practice this essentially amounts to consistency of position in relation to the majority of surrounding (Bauplan) structures.

23.   The evolutionary relations of characters are often (e.g. as in "polarity"[2]) judged on poorly defined, law-like concepts of the following kinds; relative complexity, specialisation, primitiveness, relative time of appearance in development and adaptive versus non-adaptive, phyletic or conservative characteristics. These are not independent factors and, as we have shown in this paper, the reason that they are indicators of the direction of evolution can be understood.   Grades of morphological organisation can be defined and they correspond necessarily to stages in the developmental events leading up to them. Thus, the Bauplan concept[18] identifies a relatively invariant, apparently non-adaptive[26], complex of structures; these will correlate closely with phyletic features (i.e. identifying a higher taxonomic group) and a potentially identifiable, relatively constant, early sequence of developmental events.

24.   The most frequently attempted representations of principles underlying limb evolution are schemes (e.g. canonical elements, archetype or Bauplan) defining the common denominators of the pentadactyl limb (see: Shubin, this volume). These schemes usually combine the pooled evidence of adult comparative anatomy and the evidence of early morphogenesis (e.g. condensation patterns).   They are often influenced by assumptions regarding the supposed ancestral (e.g. crossopterygian) form.   They address a limited aspect of limb pattern, e.g. identifying individual digits by reference to the carpus/tarsus, or defining organisation in the carpus/tarsus itself according to some organising principle (e.g. a branching pattern or primary axis) that can be applied to the limb as a whole. They therefore make simplifying assumptions, which may not be justified given the undoubtedly complex interactions actually governing development of the carpus/tarsus.

Despite the complexities of the detailed evidence, it might eventually be possible to summarise the fin/limb transition in terms of the following, necessarily speculative, kind. Early fin cartilages probably evolved as part of the substance of the body wall. Fin localisation probably depended on the definition of the girdle, which clearly evolved early (it is the most constant skeletal feature in fish fins and has reached an advanced grade of specialisation and singularity); the numbers and shapes of associated radial cartilages and fin rays (both typically reflecting simple body segmental organisation) match the extent of the more highly individuated basals and girdle. Branching and segmented patterns of cartilages, as seen in fins, appear to be easily interconvertible variants in the setting up of arrays of centres of chondrogenesis (as still seen in the carpus/tarsus). But achievement of the stylopod stage (and the corresponding pentadactyl array of distal elements) reflects a general trend towards reduced numbers of elements together with greater specialisation and interdependence among them. Long bone growth patterns and synovial joint specialisation were part of this trend. The transition could have occurred gradually by replacement of fin rays by cartilages, which came to occupy positions increasingly peripheral to, and eventually outside, the trunk. This could have been brought about by increasing control by the AER of the mesoderm: as the dermal skeleton and fin rays were lost, the AER may have achieved a more direct role in integrating chondrogenic mesoderm affecting earlier stages in the sequence of cartilage differentiation. The crossopterygian fin is widely accepted as ancestral for the tetrapod limb because it manifests more approximations (especially proximally) to the pentadactyl plan than other fins. Its transitional status may be indicated by the still variable pattern of the most distal elements.

If the developmental mechanisms are known, the combining of adult and developmental evidence makes it possible to assess the likelihood that a hypothesised sequence of adult forms could have been achieved through gradual developmental changes. In particular "recapitulatory" features seen in advanced embryologies should relate obviously to the proposed sequences of adult forms[17]. Limb development is dominated by a proximo–distal sequence of laying down structures in a way that would correlate well with the likely (above described) sequence of evolutionary acquisitions of skeletal specialisation and with an early dependence (in the fin) of more peripheral elements on basal cartilages; it may also reflect an evolutionary increase in the involvement of the AER in mediating and integrating the specific pattern of peripheral cartilages in tetrapods (it remains uninvolved in girdle formation). The trend towards reduction is seen in the numerous tetrapods showing reduced numbers of digits; the initial appearance of the pentadactyl condition at embryo stages in these forms is perhaps the best available illustration of the recapitulation of highly conserved, ancestral embryonic sequences. Examples of recapitulation of this kind must be interpreted in context. Different aspects apply at different stages of development and refer to different scales of evolutionary change. It would be an error[17] to translate them directly into inferences about ancestral adult morphologies; it does not necessarily follow that a recognisable humerus evolved before a recognisable carpus or even the five digit array.

A striking feature of these considerations is the fact that all the basic mechanisms pre-date the pentadactyl form, which must be seen as a minor variant of the fin, requiring few evolutionary changes in cell potentialities (such as joint specialisations and more direct ectodermal control of cartilage pattern), all of which may be anatomically generalised features rather than limb-specific.

25.   L. Wolpert, The evolution of development, Biol. J. Linn. Soc. 39: 109 (1990).

26.   Recognition   of the role of functional interactions as determinants of pattern has important implications; adaptedness does not have to depend on factors external to the organism; the dichotomy between structure and function can be bridged; internal and external stimuli can combine and interact on a similar basis (e.g. internal and external mechanical forces in bone modelling); although developmental decisions result from conditions impinging locally on individual cells, these stimulus conditions may themselves be influenced by functional interactions over long distances, e.g. hormones, mechanics; the distinction between adaptative and nonadaptive characters[23] breaks down if the latter are the result of developmental processes "adapted" to internal developmental conditions.

The extent to which limb patterning might be accountable in terms of functional (including morphogenetic) interactions is inherently difficult to specify and so may well have been underplayed.   The meaning of "function" is itself problematical[21].   Many relevant epigenetic interactions have been demonstrated; see: T.J. Horder, Embryological bases of evolution, in: "Development and Evolution", B.C. Goodwin, N. Holder and C.C. Wylie, eds. Cambridge University Press, Cambridge (1983); J. Bowen, J.R. Hinchliffe, T.J. Horder and A.M.F.Reeve, The fate map of the chick forelimb-bud and its bearing on hypothesized developmental control mechanisms, Anat. Embryol. 179: 269 (1989).

27.   "Genetic assimilation" illustrates the critical conceptual borderland between genetics, selection and "environment", liable to lead to suggestions of Lamarckism. It depends on finding rare, finely balanced situations in which a character, seen to depend directly on external environmental factors, can be selected for and become purely genetically determined.   The terminology is perhaps unfortunate because the character must (like all responses to external stimuli and phenocopies) have been a potentiality already available in the genotype; the external stimulus (and the genetic background selected for) merely affect the threshold for expression.

28.   In the course of this paper a number of types of difficulty arising in the use of concepts have been considered.   False dichotomies (e.g. saltation/gradualism,   genetic/epigenetic, genetic/environmental) seem to be fundamental to many of the greatest areas of confusion in biology. This is well illustrated in: J.H. Woodger, "Biological Principles". Kegan Paul, London (1929).

29.   On the importance and neglect of conceptual issues, see: J.H. Woodger, Observations on the present state of embryology, in "Growth", 2nd Symposium, Soc. Exp. Biol., Cambridge University Press, Cambridge (1948).

THE DEVELOPMENTAL BASIS OF LIMB EVOLUTION :   A REVIEW

Pere Alberch and
J Richard Hinchliffe

Museo Nacional de Ciencias Naturales, CSIC
MADRID, SPAIN
and
Department of Biological Sciences, University College of
Wales, ABERYSTWYTH, Wales, UK

Embryologists played a major role in the early days of evolutionary
theory.   Darwin repeatedly cites the work of Von Baer as the best evidence
in support of a "unity of structural plan" among organisms and as a source
of information about their genealogical relationships.   But it was Haeckel
with his law of recapitulation who made embryology a consubstantial part of
evolution.   The influence of Haeckel's views was so persuasive that for many
year the goal of most embryologists was to elucidate phylogenetic
relationships.

Paradoxically, it was Haeckel's former disciple Wilhelm Roux, with his
interest in the immediate causes of development (his Entwicklungsmechanik),
that marks the progressive schism between development and evolution.   Roux's
concentration on experimental approaches to examining the chemical and
physiological aspects of development led him to ignore Haeckel's biogenetic
law.   It is precisely this lack of communication among the two disciplines
that marks the next century of scientific progress in both development and
evolution (Horder, these proceedings).   Haeckel's legacy was heterochrony: a
field dominated by a comparative approach and with little interest in the
mechanistic basis of the phenomenon (Alberch, 1985).   Developmental biology,
on the other hand, adopted a reductionist approach and became closely
associated with physiology, biochemistry, and more recently, with cell and
molecular biology.   In spite of de Beer's emphasis on heterochrony, and
especially neoteny as a source of major structural change in evolution, and
Goldschmidt's unorthodox 'hopeful monsters', developmental biology was
largely absent from the so-called neo-Darwinian synthesis of the 1940's,
which essentially was the integration of population aspects of the Darwinian
theory of natural selection with Mendelian genetics.   This new evolutionary
synthesis was largely concerned in practice with identifying selection
pressures and their effect on gene frequencies within populations.   Recent
years, however, have witnessed a progressive rapprochement between evolution
and development.   This is due to the realisation by evolutionists that
regulation of developmental processes, rather than a simple accumulation of
mutations, is the motor that drives morphological evolution.

This volume can be placed in a series of similar meetings on limb
development (see preface).   Considerations about evolution were practically
absent in the previous volumes.   The fact that this time a third of the
conference was devoted to evolution and limb development is a reflection of
this rapprochement among the two disciplines.

Developmental Patterning of the Vertebrate Limb
Edited by J.R. Hinchliffe *et al.*, Plenum Press, New York, 1991

The articles published within this section are representative of the kinds of insights that an evolutionary approach can provide to our understanding of limb development – a system particularly suited for the integration of development and evolution, since it combines easy access to experimental manipulation with a high degree of morphological diversification. In this summary article, we highlight a few of the main issues raised in the conference that are likely to have an impact in the coming years.

In the past, palaeontological evidence had been of little use to the problem of the development and evolution of the vertebrate limb. Well preserved extremities of early tetrapods, such as Eryops, exhibited a pentadactyl limb not too distinct from generalized modern forms except for the fact that there were a few more tarsal and carpal elements. M.I. Coates reviews the extraordinary new palaentological findings that indicate that the earliest tetrapods from the Soviet Union, Tulerpeton, and from Greenland, Ichthyostega and Acanthostega, were not pentadactyl, but in fact had a number of digits ranging from 6 to 8.

This result suggests that the early evolution of the tetrapod limb was characterized by a period of phenotypic liability and experimentation. The pentadactyl condition is a secondary character resulting from a late stabilization. Also remarked on at the meeting, was the striking similarity of the polydactylous patterns described by Coates and Clack (1990) and the ones experimentally obtained by ZPA implantation in the anterior portion of the limb bud. It is tempting to speculate that modifications in the action of the ZPA region played a role in the process of evolutionary stabilization of pentadactyly.

Coates and Clack (1990) interpreted their observations on early polydactyly on the basis of the model by Shubin and Alberch (1986). This formal scheme describing and comparing patterns of limb development and evolution was referred to by several other contributors in different contexts. Of particular interest is the fact that chrondrogenic patterns of branching and segmentation – the basis of the Shubin and Alberch model – can be perturbed experimentally by altering the number of cells in the early limb bud (Raynaud, 1990; Alberch and Gale, 1985; Hinchliffe, Muller, these proceedings).

Two contributors to this volume, Ede and Fallon, discussed the information provided by the analysis of mutants. Mutants have sometimes been dismissed as throwing little light on normal developmental mechanisms, but they are now clearly important subjects for the study of homeobox gene expression in control of patterning (Blundell *et al.*, these proceedings, Fallon *et al.*, these proceedings). As Ede stresses in his contribution, mutants such as talpid make it possible to attempt to dissect out periodic patterning mechanisms and positional information signals (e.g. from the ZPA) which act to impose individuality on initially modular structures. Ede sees the basis of limb structural transformation in evolution as a primitive periodic spacing mechanism and a secondarily-evolved positional information mechanism (see also Wolpert, these proceedings). Fallon in his contribution describes cross-class interaction between chick mutant mesoderm and turtle ectoderm, demonstrating conservation of limb induction mechanisms between birds and reptiles and providing important evidence of the similarity of the mechanisms of limb development throughout amniotes (see also Hinchliffe, these proceedings). Particularly intriguing, however, was the observation made in his talk by T. Horder, which showed that there is a predominance of mutations that result in polydactyly. This pattern may be a reflection of a trend towards the generation of the atavistic condition as postulated by the palaeontological data.

More comparative data are required to study the evolution of tissue interactions involved in the genesis of the tetrapod limb. Hall's review (these proceedings) is a particularly appropriate introduction to the subject. It is evident that to study the origin of the tetrapod limb the logical outgroup to study is the fish fin. Unfortunately, there have been very few studies on this sytem. Thorogood's research (these proceedings) is an important contribution at this level. He makes the proposal that a critical change in teleost fin development is from the AER phase, inducing the mesoderm outgrowth necessary for endoskeleton formation, to the apical fold phase, concerned with dermal (fin-ray) skeletal development. Heterochrony, by delaying the AER to apical fold transition would be a means of transforming the ray fin into the lobe fin of the Sarcopterygians, which include the Dipnoans and the Rhipidistian Crossopterygians, with their limb-like paired fins (Coates, Vorobyeva, these proceedings).

Recent advances in molecular techniques permit us to carry out more detailed studies of the temporal and spatial patterns of gene expression that accompany tissue interactions (e.g. see reports on Hox gene expression in this volume). We believe that a comparative study among different organ systems of these patterns of morphogenetic gene expression may be a very productive field of inquiry in the next few years. In particular, it may throw light on the still unsolved problem of the developmental basis of limb homology, possibly by distinguishing 'pattern control genes' common to limb development throughout tetrapods, from those specific to particular variations on the limb theme.

As reviewed by Hinchliffe and by Hall in this volume the rules of interaction that control epithelial-mesenchymal interactions are few and general. The interaction of these rules in different contexts can generate an enormous diversity of patterns. This conceptual scheme was discussed by Oster and Alberch (1982) in an evolutionary context. As an illustrative example we suggested that a wide variety of epithelial derivatives, such as feathers, scales, glands and teeth could be viewed as variations within a single morphogenetic process. Our example was based on simple observation of morphogenetic events such as folding of epithelial layers of cells (Odell et al., 1981). Edelman et al. (1986) have recently proposed a specific model of epithelial morphogenesis based on their data on cell adhesion molecules (CAMs) that integrates gene action, mediated by morphogenetic molecules, with morphogenesis. Burke (1988 and in prep) using this conceptual perspective has speculated that a wide variety of structures, such as hairs, glands, feathers, scales and even limbs and the carapace of turtles, can be viewed as variations within a single system of epithelial-mesenchymal interactions. The key issue here is that morphological novelty is not the result of addition of new genes but of the interaction of the same rules in different contexts.

One feature of this section of the conference was the number of papers which emphasised the morphological results of variation in the interaction of these rules in specific examples, for example in teleost/tetrapod (Thorogood), in crocodile/bird (Müller), in pentadactyl/reduced digit lizards (Hinchliffe, data from Raynaud, 1990) and in normal/mutant (Ede) comparisons. In bits and pieces, both traditional descriptive and modern experimental embryology, together with the new molecular technology for investigating homeobox gene expression and retinoic acid, are gradually advancing our understanding of the relation of genes, molecules and tissue interactions to that ancient problem of biology: structural transformation in evolution.

REFERENCES

ALBERCH, P. (1985). Problems with the interpretation of developmental sequences. Syst. Zool. **34**, 46-58.

ALBERCH, P. and GALE, E. (1985). A developmental analysis of an evolutionary trend. Digital reduction in amphibians. Evolution **39**, 8-23.

BURKE, A.C. (1989). Epithelial-mesenchymal interactions in the development of the chelonian bauplan. Prog. in Zool. **35**, 206-209.

COATES, M.I. and CLACK, J.A. (1990). Polydactyly in the earliest known tetrapod limbs. Nature, **347**, 66-69.

EDELMAN, G.M. (1986). Cell adhesion molecules in the regulation of animal form and tissue pattern. Ann. Rev. Cell Biol. **2**, 81-116.

ODELL, G., OSTER, P., ALBERCH, P. and BURNSIDE, B. (1981). The mechanical basis of morphogenesis. I: a model for epithelial tissue folding. Devel. Biol. **85**, 446-462.

OSTER, G. and ALBERCH, P. (1982). Evolution and bifurcation of developmental programs. Evolution, **36**, 444-459.

RAYNAUD, A. (1990). Developmental mechanisms involved in the embryonic reduction of limbs in reptiles. Int. J. Dev. Biol. **34**, 233-243.

SHUBIN, A. and ALBERCH, P. (1986). A morphogenetic approach to the origin and basic organisation of the tetrapod limb. Evol. Biol. **2**, 319-387.

PARTICIPANTS

AKIMENKO, Marie-Andree
Institute of Neuroscience
University of Oregon
219 Huestis Hall, Eugene
OREGON 97403, USA

ALBERCH, Pere
CSIC
Museo Nacional de Ciencias Naturales
J Gutierrez Abascal 2
28006, MADRID, SPAIN

BRAND-SABERI, Beate
Abt f Anat u Embryol
Institut fur Anatomie
Geb MA 50G/148, Ruhr-Universitat
4630 BOCHUM, GERMANY

BRICKELL, Paul
Medical Molecular Biology Unit
Dept of Biochemistry
University College and Middlesex School
    of Medicine,
Windeyer Building, Cleveland Street
LONDON W1P 6DB, UK

BROCKES, Jeremy
Ludwig Institute for Cancer Research
Middlesex Hospital, University College
    Branch
Courtauld Building
91 Riding House Street
LONDON W1P 8DB, UK

BRYANT, Susan V
Developmental Biology Center
University of California
IRVINE, California 92717, USA

BURKE, Ann
Department of Biology
Dalhousie University, HALIFAX
Nova Scotia B3H 4II, CANADA

BURT, David
Institute of Animal Physiology
    and Genetics Research
Edinburgh Research Station
ROSLIN, Midlothian EH25 9PS
SCOTLAND, UK

CAMPA, Juan S
Biochemistry Unit, National Heart
    and Lung Institute
Emmanuel Kaye Building
Manresa Road
LONDON SW3 6LR, UK

CAPLAN, Arnold
Biology Department
Case Western Reserve University
Skeletal Research Centre
2080 Adelbert Road
CLEVELAND, Ohio 44

CHRIST, Bodo
Anatomisch Institut der
    Albert-Ludwigs-Universitat
Albertstrasse 17
D-7800 FREIBURG i.Br.
GERMANY

CIHAK, Radomir
Institute of Anatomy
First Faculty of Medicine
Charles University in Prague
U Nemocnice
PRAHA, CS12800 CZECHOSLOVAKIA

COATES, Mike
Zoology Department
Downing Street
CAMBRIDGE CB2 3EJ, UK

CRITCHLOW, Matthew
Anatomy Department
St George's Hospital Medical
    School
Cranmer Terrace, Tooting
LONDON SW17 0RE, UK

DUBOULE, Denis
European Molecular Biology
  Laboratory
Meyerhoffstrasse 1, Postfach 10
6900 HEIDELBERG, GERMANY

EDE, Donald
Developmental Biology Building
Zoology Department
Glasgow University
124 Observatory Road

FALLON, John
Anatomy Department
University of Wisconsin-Madison
  Medical School
325 Services Memorial Institute
1300 University Avenue
MADISON 53706, USA

FERNANDEZ-TERAN, Mariangeles
Departamento de Anatomia
Facultad de Medicina
SANTANDER, Cantabria, SPAIN

FROST, Sharon
Biological Sciences Department
University College of Wales
Penglais, ABERYSTWYTH
Dyfed SY23 3DA, Wales, UK

GANAN, Yolanda
Departamento de Ciencias
  Morfologicas
Universidad de Extremadura
BADAJOZ, SPAIN

GRIBBIN, Clare
IAPGR
Edinburgh Research Station
ROSLIN
Midlothian EH25 9PS
Scotland, UK

GRIM, Milos
Institute of Anatomy
Charles University, Medical Faculty
U Nemocnice
PRAHA, CS12800 CZECHOSLOVAKIA

HALL, Brian
Department of Biology
Dalhousie University
HALIFAX
Nova Scotia B3H 4J1, CANADA

HARRIS, Albert K
Department of Biology
Campus Box 3280, Coker Hall
University of North Carolina
Chapel Hill, North Carolina
27599-3280, USA

HILL, Robert
MRC Human Genetics Unit
Western General Hospital
Crewe Road
EDINBURGH EH4 2XU
Scotland, UK

HINCHLIFFE, J Richard
Biological Sciences Department
University College of Wales
Penglais, ABERYSTWYTH
Dyfed SY23 3DA, Wales, UK

HOLDER, Nigel
Anatomy and Human Biology Group
Biomedical Sciences Division
King's College, The Strand
LONDON WC2 LS, UK

HORDER, Tim
Department of Human Anatomy
University of Oxford
South Parks Road
OXFORD OX1 3QX, UK

HORNBRUCH. Amata
Department of Anatomy
University College and
  Middlesex School of Medicine
Windeyer Building
Cleveland Street
LONDON W1P 6DB, UK

HURLE, Juan
Departamento de Anatomia
Facultad de Medicina
Poligono de Cazona, SANTANDER
Cantabria, SPAIN

IDE, Hiroyuki
Biological Institute
Faculty of Science
Tohoku University
AOBA-YAMA
Sendai 980, JAPAN

JANOCHA, Reinhold
Department of Cellular and
  Molecular Physiology
Harvard Medical School
25 Shattuck Street
BOSTON MA 02115, USA

JOHNSON, David
Department of Anatomy
Medical School
University of Leeds
LEEDS LS2 9JT, UK

KASTNER, Philippe
INSERM-U 184
Institut de Chimie Biologique
Faculte de Medicine
11 Rue Humann 67085
STRASBOURG, FRANCE

KOSHER, Robert A
Department of Anatomy
University of Connecticut Health
  Center, FARMINGTON
Connecticut 06032, USA

LYONS, Gary E
Department of Molecular Biology
Pasteur Institute
28 Rue de Docteur Roux
F 75724 PARIS
Cedex 15, FRANCE

MADEN, Malcolm
Anatomy and Human Biology Group
Biomedical Sciences Division
King's College, Strand
LONDON WC2R 2LS, UK

MAINI, Philip K
Centre for Mathematical Biology
Mathematical Institute
24-29 St Giles'
OXFORD OX1 3LB, UK

McCLACHLAN, John C
Biology and Preclinical Medicine
  Department
University of St Andrews
ST ANDREWS, Fife KY16 9TS
Scotland, UK

MULLER, Gerd
Department of Anatomy
University of Vienna
Wahringerstrasse 13
A-1090 WIEN, AUSTRIA

PAPAGEORGIOU, Spyros
National Research Center for
  Physical Sciences
Demokritos, GR-153 10 Ag Paraskevi
PO Box 60228, ATHENS, GREECE

ROS, Mariangeles
Departamento de Anatomia
Facultad de Medicina
SANTANDER, Cantabria, SPAIN

ROWE, Ann
Medical Molecular Biology Unit
University College and Middlesex
  School of Medicine
Windeyer Building
Cleveland Street
LONDON W1P 6DB, UK

RUBERTE, Ester
INSERM-U 184
Institute de Chimie Biologique
Faculte de Medicine
11 Rue Humann
67085, STRASBOURG, FRANCE

SHUBIN, Neil
Department of Biology
University of Pennsylvania
Goddard Laboratories
PHILADELPHIA
PA 19104-6059, USA

SOLURSH, Michael
Biology Department
University of Iowa
IOWA CITY
Iowa 52242, USA

STREICHER, Johannes
Institute of Anatomy
Wahringerstrasse 13
A-1090 WIEN, AUSTRIA

SUMMERBELL, Dennis
National Institute for Medical
  Research
The Ridgeway, Mill Hill
LONDON NW7 1AA, UK

TABIN, Cliff
Department of Genetics
Harvard Medical School
25 Shattuck Street
BOSTON, MA 02115, USA

THOROGOOD, Peter
Department of Oral Biology
Institute of Dental Surgery
Eastman Dental Hospital
256 Gray's Inn Road
LONDON WC1X 8LD, UK

von THULEN, Berend
Institut fur Molekularbiologie
Karl von Frisch Strasse
D 3550 Marburg/Lahn, GERMANY

TICKLE, Cheryll
Anatomy and Developmental Biology
  Department
University College and Middlesex
  School of Medicine
Windeyer Building
Cleveland Street
LONDON W1P 6DB, UK

TOOLE, Bryan
Anatomy and Cellular Biology
  Department
Tufts University
136 Harrison Avenue
BOSTON MA 0211, USA

VOROBYEVA, Emilia
A N Severtzov Institute of
  Evolutionary Morphology
USSR Academy of Science
33 Leninsky Prospekt
MOSCOW 117071, USSR

WEDDEN, Sarah
Anatomy Department
University Medical School
Teviot Place
EDINBURGH EH8 9AG
Scotland, UK

WILSON, David
Anatomy Department
Queen's University of Belfast
97 Lisburn Road
BELFAST BT9 7BL
Northern Ireland, UK

WOLPERT, Lewis
Anatomy and Developmental
  Biology Department
University College and Middlesex
  School of Medicine
Windeyer Building
Cleveland Street
LONDON W1P 6DB, UK

ZELLER, Rolf
EMBL, Postfach 10
Meyerhofstrasse 1
HEIDELBERG 6900, GERMANY

ZNOIKO, Sergei L
Koltzov Institute of Developmental
  Biology
USSR Academy of Science
Vavilov Street 26
117808 MOSCOW, USSR

ADDITIONAL SPANISH POSTGRADUATE AND POSTDOCTORAL PARTICIPANTS

ARRECHEDERA, Hector
GARCIA, Isidro
GARCIA, Pilar
del PINO GIL, Maria
LAJO, Alberto
MACIAS, Domingo
MARTINEX, Virginio Garcia
TAVIRA, Juana
DIEZ ULLOA, Alberto